Hydrogen Technologies for Energy

Other related titles:

You may also like

- PBPO0890 | Angelo Basile, Francesco Dalena, Jianhua Tong and T. Nejat Veziroğlu | Hydrogen Production, Separation and Purification for Energy | 2017
- PBPO1010 | Rupp Carriveau and David S-K. Ting | Methane and Hydrogen for Energy Storage | 2016

We also publish a wide range of books on the following topics:
Computing and Networks
Control, Robotics and Sensors
Electrical Regulations
Electromagnetics and Radar
Energy Engineering
Healthcare Technologies
History and Management of Technology
IET Codes and Guidance
Materials, Circuits and Devices
Model Forms
Nanomaterials and Nanotechnologies
Optics, Photonics and Lasers
Production, Design and Manufacturing
Security
Telecommunications
Transportation

All books are available in print via https://shop.theiet.org or as eBooks via our Digital Library https://digital-library.theiet.org.

IET ENERGY ENGINEERING SERIES 205

Hydrogen Technologies for Energy

Daniele Candelaresi, Giulio Raimondi and Giuseppe Spazzafumo

The Institution of Engineering and Technology

About the IET

This book is published by the Institution of Engineering and Technology (The IET).

We inspire, inform and influence the global engineering community to engineer a better world. As a diverse home across engineering and technology, we share knowledge that helps make better sense of the world, to accelerate innovation and solve the global challenges that matter.

The IET is a not-for-profit organisation. The surplus we make from our books is used to support activities and products for the engineering community and promote the positive role of science, engineering and technology in the world. This includes education resources and outreach, scholarships and awards, events and courses, publications, professional development and mentoring, and advocacy to governments.

To discover more about the IET please visit https://www.theiet.org/.

About IET books

The IET publishes books across many engineering and technology disciplines. Our authors and editors offer fresh perspectives from universities and industry. Within our subject areas, we have several book series steered by editorial boards made up of leading subject experts.

We peer review each book at the proposal stage to ensure the quality and relevance of our publications.

Get involved

If you are interested in becoming an author, editor, series advisor, or peer reviewer please visit https://www.theiet.org/publishing/publishing-with-iet-books/ or contact author_support@theiet.org.

Discovering our electronic content

All of our books are available online via the IET's Digital Library. Our Digital Library is the home of technical documents, eBooks, conference publications, real-life case studies and journal articles. To find out more, please visit https://digital-library.theiet.org.

In collaboration with the United Nations and the International Publishers Association, the IET is a Signatory member of the Sustainable Development Goal (SDG) Publishers Compact. The Compact aims to accelerate progress to achieve the SDGs by 2030. Signatories aspire to develop sustainable practices and act as champions of the SDGs during the Decade of Action (2020–30), publishing books and journals that will help inform, develop, and inspire action in that direction.

In line with our sustainable goals, our UK printing partner has FSC accreditation, which is reducing our environmental impact to the planet. We use a print-on-demand model to further reduce our carbon footprint.

Published by The Institution of Engineering and Technology, London, United Kingdom

The Institution of Engineering and Technology (the "**Publisher**") is registered as a Charity in England & Wales (no. 211014) and Scotland (no. SC038698).

First published 2024

The Institution of Engineering and Technology
Futures Place,
Kings Way,
Stevenage,
Herts, SG1 2UA
United Kingdom

www.theiet.org

British Library Cataloguing in Publication Data
A catalogue record for this product is available from the British Library

ISBN 978-1-83953-445-4 (hardback)
ISBN 978-1-83953-446-1 (PDF)

Typeset in India by MPS Limited

Cover image credit: top left (hydrogen refuelling station, Hiroshima, Japan): Daniele Candelaresi

bottom right (fuel cell bus, Himeji, Japan): Daniele Candelaresi

top right (hydrogen powered car, Konin area, Poland): Giuseppe Spazzafumo

bottom (hydrogen tanks, Tulum, Mexico): Giuseppe Spazzafumo

Contents

Preface

'We have to leave the Earth better than we found it' is a famous sentence from an American Indian chief.

Certainly, thanks to technological progress, the level of well-being that generations have left to subsequent generations over the centuries has improved. But, at the same time, especially since the Industrial Revolution, the exploitation of resources and the impact of human activities do not allow us to say that we have been left a better planet and do not guarantee that we will leave it better to our descendants unless appropriate measures are implemented to reduce the exploitation of resources and the production of waste.

Among the most exploited and also most impactful resources are certainly fossil fuels. They are used above all in the energy sector and for the production of chemicals, metals and cement. Their main constituents are hydrogen and carbon. Therefore, to avoid using fossil fuels and to reduce the environmental impact, it is necessary to find sources of energy that do not increase carbon content in the atmosphere. Hydrogen, for the most part, is an energy vector, that is, it must be produced starting from energy and substances that contain a lot of it such as water. The exception is hydrogen that can be extracted directly from underground reserves, whose availability is still unknown, but very probably insufficient to cover global needs and, in any case, unevenly distributed across the planet. If hydrogen must be 'green', the energy source must be renewable and net carbon emissions must be zero. We could also accept hydrogen produced by electrolysis or thermolysis using fissile resources. However, moving from the current energy system to one based entirely on electricity and hydrogen presents more than a few problems. Sustainable and green electricity is not so abundant nowadays and is not easily stored, especially in the long term. Hydrogen in perspective is a solution, but even in this case its accumulation and distribution still present important challenges. An intermediate step, through synthetic fuels more similar to current ones, could be useful, although still expensive. This involves the use of 'renewable' or 'recycled' carbon that can derive from biomass, waste, air, or sequestration of dioxide that would otherwise be released into the atmosphere.

This book aims to present an overview of the entire hydrogen supply chain, from its production to its uses in the energy field, taking into account both the technologies that are already mature and those that, still in the development phase, could become so in the next 10–20 years. The first chapter is dedicated to sustainability. It is an introduction to the complex relationship that exists between the use of energy by mankind and its impact on the environment in terms of emissions and consumption of resources. The second chapter is dedicated to the physical-chemical characteristics of hydrogen and also presents a comparison with other fuels. The production of hydrogen is covered in the third chapter. From this point of view, hydrogen is extremely flexible because it can be produced using both fossil, nuclear (both fission and fusion) and renewable energy. And, using the appropriate technology, it can be extracted from any substance that contains it (including biomass and waste). The first section of the chapter deals with production from hydrocarbons, which is currently the dominant one, although it is gradually destined to disappear unless adequate carbon dioxide sequestration systems are found. The next section deals with production through gasification, both of coal and biomass. These two sections are followed by one that presents the existing technologies for the necessary purification of the hydrogen produced through the processes discussed in the two previous sections. The fourth section deals with the issue of cogeneration of electricity and hydrogen, a valid method for increasing efficiency and reducing environmental impact even when starting from fossil fuels. The following section abandons fossil fuels to deal with the various technologies that, by exploiting heat and/or electricity, allow water to be split into its constituent elements. Hydrogen is the main constituent of water and water is the most available substance on Earth. It follows that the technologies discussed in this section are the most important in perspective, especially if capable of extracting hydrogen from seawater or wastewater. The chapter ends with two sections dedicated, respectively, to the natural processes operated by microorganisms and to the chemical reactions of some hydrides with water. The fourth chapter deals with the most critical aspects of the hydrogen supply chain: storage and distribution. Hydrogen can be stored as a pure substance, in the form of compressed gas or cryogenic liquid, which are the subject of the first two sections. Or it can be chemically combined with other substances or physically trapped in suitable structures. Metal hydrides and complex hydrides are presented in the third section of the chapter, while the fourth is dedicated to organic liquid compounds and ammonia and the fifth to materials capable of adsorbing hydrogen. Once the overview of storage technologies has been completed, they are compared in the next section considering various aspects including energy density and safety. The chapter ends with a section dedicated to hydrogen distribution methods, by road, by sea and by pipeline. The fifth chapter deals with the technologies for using hydrogen for energy purposes, technologies that are grouped into three sections: combustion, electrochemical conversion and production of synthetic fuels. The first section deals with its use in reciprocating and rotary internal combustion engines, in gas turbines, and in steam cycles by oxy-combustion. The next section deals with the different types of fuel cells and their operation. Finally, the third section presents the different options for generating fuels that bypass the problems of accumulation and distribution. Hydrogen could also be used to produce many other substances of industrial interest, but this is beyond the scope of this book. The book ends with a chapter that reports both historical and current examples of applications in various sectors ranging from transport to portable systems, to residential systems and industrial plants. The chapter ends with a section dedicated to the potential application of hydrogen at a local level within renewable energy communities.

Finally, we want to thank Dr. Sarah D'Adamo, for reviewing the section on biological hydrogen, and Dr. Guido Galeno and Mr. Michele Corte for their assistance in generating some original pictures. We also thank all the companies and journals which allowed us to use their pictures to illustrate our book. And also many thanks to the IET staff for the great work in helping us during all the various steps for realising this book in the best possible way.

Daniele Candelaresi
Giulio Raimondi
Giuseppe Spazzafumo

Nomenclature

Please note: Symbols and acronyms mentioned in only one section of the book are not listed here and may have a different meaning even if they appear identical.

a	activity of a chemical species
AEMFC	Anionic Exchange Membrane Fuel Cell
AFC	Alkaline Fuel Cell
DMFC	Direct Methanol Fuel Cell
F	Faraday constant
G	free energy
H	enthalpy
HHV	higher heating value
I	electric current
IR-MCFC	Internal Reforming MCFC
k	ratio between specific heats at constant pressure and volume
K_{eq}	Equilibrium constant
LHV	lower heating value
m	mass
$\dot{m}$	mass flow
MCFC	Molten Carbonate Fuel Cell
p	pressure or partial pressure
PAFC	Phosphoric Acid Fuel Cell
PEMEC	Proton Exchange Membrane Electrolyser Cell
PEMFC	Proton Exchange Membrane Fuel Cell
R	gas constant
S	entropy
SOFC	Solid Oxide Fuel Cell
T	temperature
t	time
v	volume, capacity
V	voltage
W	power

Greek letters

η efficiency
ρ density

Superscripts

$^{\circ}$ standard

Subscripts

a activated
c concentrated
e electric
f fuel
I current
marg marginal
o ohmic
t thermal or thermodynamic
V voltage

About the authors

Daniele Candelaresi is the chief technology officer (CTO) at EnTraT (Energy Transition Team), an academic spin-off focused on hydrogen energy systems and life cycle sustainability assessments for energy projects. He obtained his PhD working on life cycle assessment of hydrogen energy systems, carbon neutral fuels and sustainable mobility.

Giulio Raimondi is a freelance engineer with Innoind, a consulting brand based in Italy. He provides technical and strategic support for product and process development, as well as for the creation of hydrogen supply chains and for energy efficiency. He obtained his PhD working on development of local hydrogen production and use chains.

Giuseppe Spazzafumo is an associate professor at the University of Cassino and Southern Lazio, Italy. His main research field is hydrogen energy systems. He serves as a member of the IAHE (International Association for Hydrogen Energy) Board of Directors since 1999 and as coordinator of HYPOTHESIS (HYdrogen POwer THeoretical and Engineering Solution International Symposium) series.

Chapter 1
Introduction: hydrogen, energy and environment

The title of this introductory chapter suggests that 'hydrogen is energy in respect of the environment'. Obviously, hydrogen can also be produced without respecting the environment, as it has been done in the past and as it is still predominantly done today, but there are many technologies that can make this statement real.

1.1 Sustainability and development

The conversion of energy has had, in the history of the interaction between man and the environment, an increasing role with the increase of the world population and with the development of technology.

Probably on many books the term 'machine' is defined as an object that necessarily has at least one moving organ. Actually, the term, which derives from the ancient Greek *μηχανή* through Latin *machina*, means 'tool to make'. But 'to make' something, it is necessary to use energy, because without using energy, nothing is made. So what really characterises a machine is the transformation of energy.

Some examples of machines that have no moving solid parts are the fuel cell, the magnetoplasmadynamic generator and the thermoelectric generator. The fuel cell will be discussed in detail in a later chapter. The magnetoplasmadynamic generator is an electrical energy generator that uses a velocity field and a magnetic field. In practice, it is a convergent-divergent conduit, to which a magnetic field is applied from the outside; an ionised gas that has a certain speed is sent into it and this generates an electric field. Although this machine has no moving solid parts, it can also be used as a motor, in which case it is supplied with energy and the magnetic field accelerates the gas. The thermoelectric generator, however, is a static device that exploits the Seebeck effect, i.e. the ability to generate electricity by metal conductors or semiconductors subjected to a temperature difference. And others could be mentioned, for example the much more common photovoltaic cells.

The term energy also derives from the ancient Greek *ἐνέργεια* through the Latin *energia* and means 'effective action', that is the ability to produce changes. One might think that machines only have to do with engineering, but that is not the case. In fact, we, like every other living being, are machines. We convert energy and our fuel is called food, and it is a fuel which, in turn, is produced by other machines. If one goes back along the food chain, it comes to plants, which are machines that use solar energy and convert it to grow and produce fruit. So everything we see around us that somehow exchanges energy is a machine. The energy balance of a human being is: 70% of basal metabolism, 15% of thermogenesis and 15% of physical activity. So even if we stand still, we still continue to exchange an 85% of energy.

In past times, the correlation between environment and energy was never considered too much. This has led to a whole series of imbalances which we have only recently begun to realise and which, perhaps, we have not yet framed too well. The growing energy consumption of the last centuries has led to increasing emissions of polluting substances and greenhouse gases which could be climate-altering substances.

The concept of sustainability derives from scientific and naturalistic literature: the management of a resource is defined as sustainable when, once its reproductive capacity is known, its exploitation is not exceeded beyond a certain threshold. It is therefore a concept referable to the management of renewable natural resources, that is, having the ability to reproduce. In this regard, it should be noted the particular meaning attributed to the word renewable. In fact, the primary sources of the available energy considered renewable are almost entirely of solar derivation. But the Sun is a star and lives thanks to a series of fusion reactions of hydrogen atoms. That hydrogen is not renewable and in fact even the stars die. But when the Sun will begin to lose power, the Earth will become not liveable for humanity, so the Sun will last longer than us and for this reason solar energy is considered renewable even if it is not in an absolute sense.

The other resources, which are defined as exhaustible, are characterised by the scarcity, that is the ratio between reserves and consumption. This is a variable ratio both to the numerator and to the denominator in relation to several parameters: the progressive depletion of reserves, the ability to identify new reserves, the increase in population, the changes in population behaviour, the new technologies developed and the replacement of the resource used with another (concept of substitutability).

The concept of development, however, is closely connected to the social sciences, in particular to economics. At the beginning, it was essentially linked to economic growth, that is, to the increase in gross domestic product (GDP) which measures the production of goods and services traded at market prices. This index is an objective parameter, but very unsuitable, even in its purchasing power parity (GDP-PPP) version, to represent the overall development of a nation and to make comparisons between different nations. It is also completely detached from the concept of sustainability.

The term sustainable development became popular in the late 1980s and its definition is as follows: ‘development that meets the needs of the present without compromising the ability of future generations to meet their own needs’. In other words, we speak of infragenerational equity and intergenerational equity: the first implies equal access to resources by the current citizens of the planet without distinction of place, while the second implies equal access to resources between successive generations. It is clear that our society is currently very far from this concept.

Until the end of the 18th century, the progress of humanity has been rather slow and the use of primary sources of energy has been modest and essentially based on renewable resources (sun, wind, water, wood, etc.). This happened despite the fact that coal was already known to the Romans and that the Chinese knew about natural gas well before the 18th century. Neither one nor the other had valid means of transportation and distribution. And the wood did its duty adequately without impacting the amount of carbon dioxide in the atmosphere. The quality of life, while progressing, was at a much lower level than the current one, essentially because the work was done by man with the help of animals only, while the environmental impact of anthropogenic activities was negligible. There were also strong social imbalances within each community, but those that were created between different peoples as a result of new discoveries were reabsorbed before other discoveries widened the gap between one people and another too much.

The Industrial Revolution has begun to shift work from humans, animals and simple machines (e.g. windmills) to more complex thermal machines, thus creating the need to use fuels with a higher energy density than wood and, consequently, a growing energy market. But this would not have been enough if it had not also created the means suitable for the distribution of fossil fuels. In fact, one of the first innovations of the time was the steam locomotive, mainly intended for the transport of coal. So at first coal began to replace wood and then was joined by oil and, finally, natural gas. The ability to distribute and use energy has transformed all manufacturing sectors and initiated a very rapid process of technological development, producing an impressive series of improvements in the condition of human life, thanks to the increase in available goods and services and the reduction of costs of production. The different forms of energy have caused important effects in various sectors: coal has allowed industrialisation, then oil has caused a huge increase in mobility, while electricity has allowed a more distributed and intensive energy use and more recently given rise to a computerised society.

Unfortunately, this did not happen in a uniform way and the speed of technological development has led to a progressive widening of the gap between the rich and industrialised countries and those countries that remained more backward and that indeed often saw their resources exploited by others. Ultimately, a small number of countries have enjoyed a lot of development by employing exhaustible resources (and produced waste), regardless of the lack of sustainability of such development which was to the detriment of all. This also happened because there was still no awareness of this problem, but, judging by the current trend, it does not seem that having become aware of the problem has had significant consequences and the use of exhaustible resources continues to grow.

Therefore, development has also begun to refer to a series of social variables by introducing, for example, the genuine progress indicator (GPI) which takes into account, on the one hand, positive components such as work that directly produces goods without producing income, and, on the other hand, social costs such as environmental costs or those related to crime. Few countries have calculated their own GPI and, therefore, there is no world ranking as for GDP. However, it is of considerable impact that global GDP has increased more than three-fold since 1950, whereas GPI decreased since 1978 [1]. This confirms the poor representativeness of GDP.

Among the various indices defined to take into account the various aspects of development understood as well-being, at least two have been evaluated for almost all nations: the human development index (HDI), which takes into account GDP, life expectancy and the level of education, and the happy planet index (HPI), which does not directly take into account economic aspects, but rather life expectancy, personal satisfaction and ecological footprint. These indices are able to represent development in a more complete way precisely because they take into consideration aspects of human life that are much more important than the economic condition; however, they are less objective than GDP, both because they introduce variables that are not quantifiable in an exact way and because they attribute to each variable a certain weight which is often subjective. The differences that are found therefore increase as the number of parameters increases and the importance assigned to the various parameters considered varies. This is evident if the various indices referring to entire geographical areas are compared: Figures 1.1, 1.2 and 1.3, respectively, shows the distribution of GDP per capita PPP, HDI and HPI. The first thing that stands out is how the GDP gap is much wide between Africa and a large part of Asia, well below the world average, and Europe, Oceania, Canada and the USA far above. The HDI has a similar distribution, whereas the HPI shows a situation completely reversed in some cases.

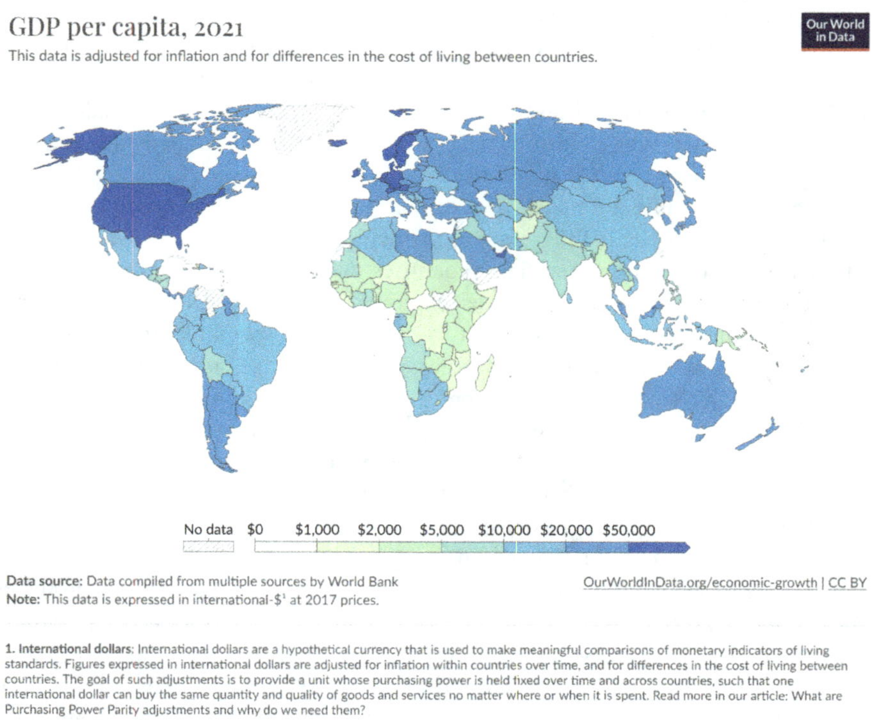

Figure 1.1 Distribution of gross domestic product, adjusted with purchasing power parity, per capita (GDP-PPP per capita) [2]

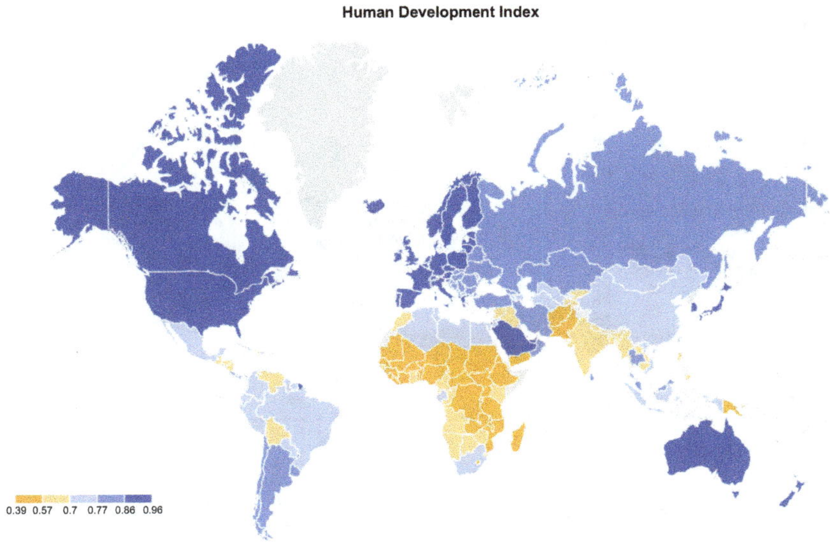

Figure 1.2 Distribution of human development index (HDI) [3]

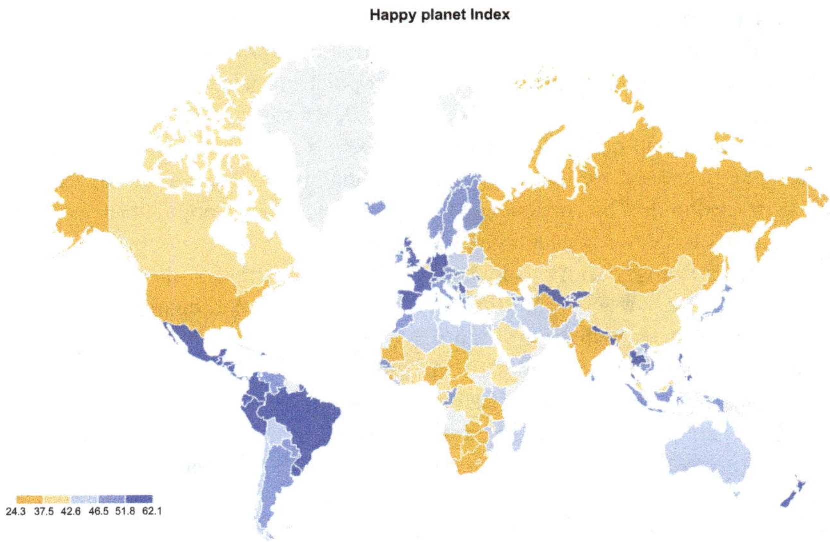

Figure 1.3 Distribution of happy planet index (HPI) [4]

Leaving aside any comment on the validity of these indices and on which one is the most representative, it is important to underline that these indices highlight that current development is not sustainable either at an infragenerational level (large differences on a geographical basis of the various indices) or at an intergenerational level (reduction of the GPI in the last 30 years).

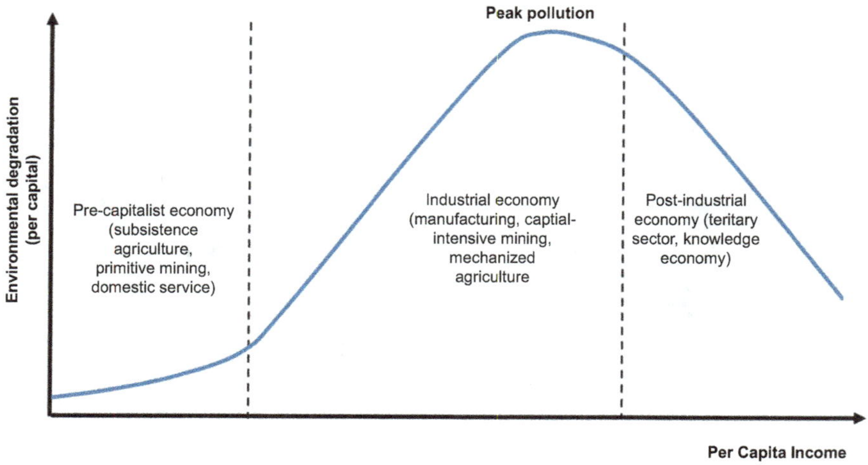

Figure 1.4 An environmental Kuznets curve [5]

Environmental aspects are closely related to the average income and consequently income has a decisive influence also on the sustainability of development. Figure 1.4 reports the environmental Kuznets curve which shows how the environmental degradation changes when the income per capita changes.

Figure 1.5 shows the development of population related to birth and death rates. In pre-industrial era, both the birth and death rates were very high and the population was stable or slowly increasing. Then, the death rate starts to decrease, thanks to the increased wellness, changes in social assets, technical progresses and progress in medicine. The consequence is a rapid increase in population. When also birth rate starts to decrease as a consequence of new social behaviours, the slope of population increasingly starts to lower. Finally, both birth and death rates tend to become stable at a low level and the population shows small changes.

Coming to current situation, Figure 1.6 reports the total primary energy consumption, the fossil fuel consumption and the climate altering gas emissions (CO_2, CH_4 and N_2O from all sources, including land-use change, and evaluated according to GWP-100, i.e. emissions measured in tonnes of carbon dioxide-equivalents over a 100-year timescale) versus increasing classes of GDP per capita in 2021. Climate altering gas emissions, though strictly related only to climate change, have been considered in a broader sense as an index of environmental degradation in the absence of more comprehensive data (e.g. data simultaneously involving climate change, pollution and other environmental degradation mechanisms). A low income (<8000$ at 2017 per capita) is generally accompanied by a very low use of energy sources and the level of environmental degradation is almost constant and not related to the income and to the energy use. As GDP increases, energy consumption, both non-renewable and renewable, also progressively increases. At the same time, emissions are also increasing when increasing GDP. However, it can be noted that a sharp increase in emissions (for GDP between 8000 and 16,000$ per

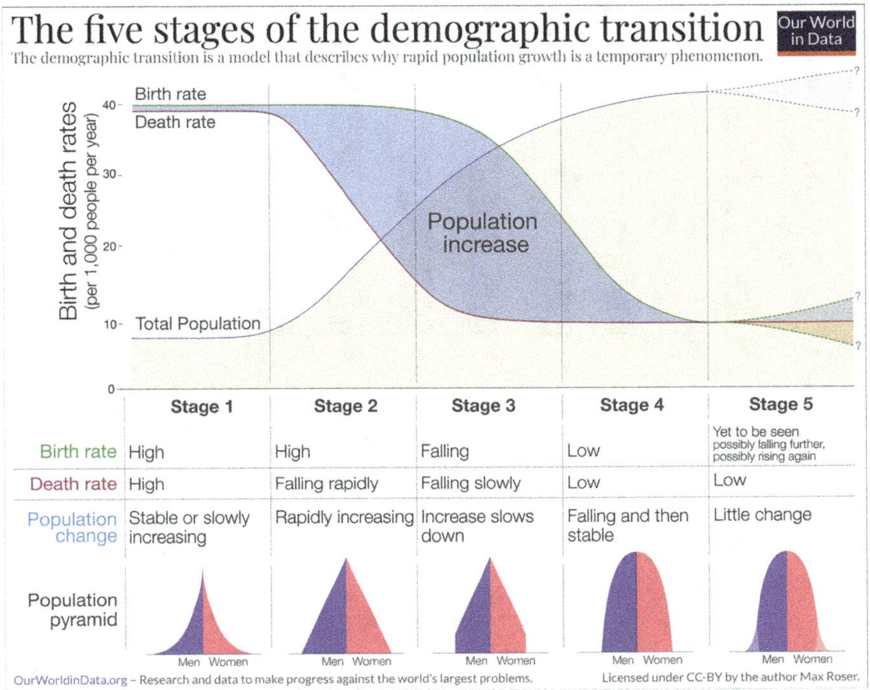

Figure 1.5 The five stages of the demographic transition (adapted from Max Roser, 2023) [6]

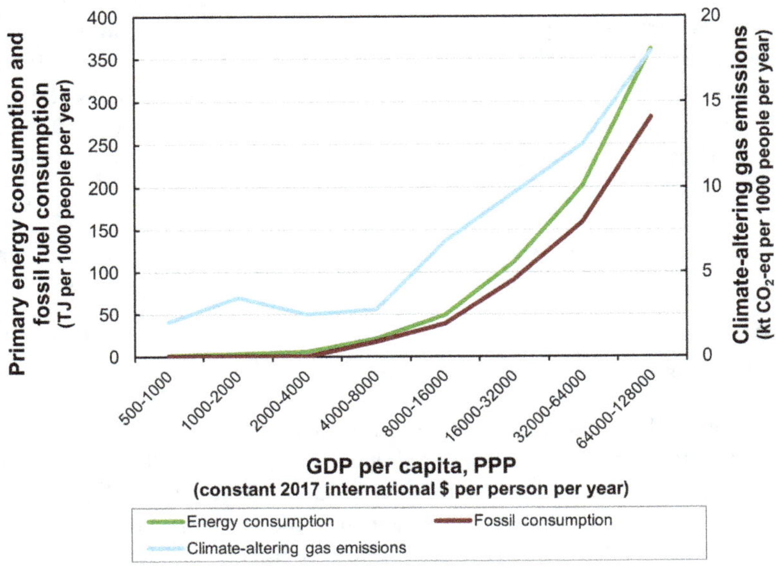

Figure 1.6 Energy consumption and climate-altering gas emissions vs GDP-PPP per capita (data from [7–11])

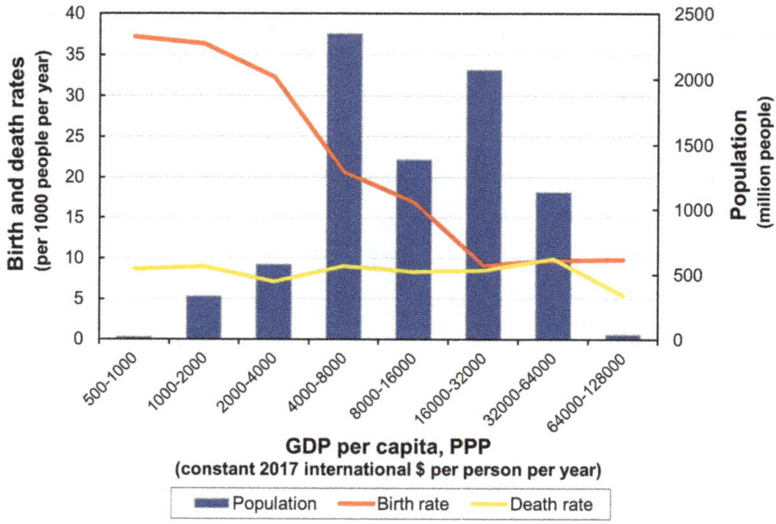

Figure 1.7 Birth and death rates and population distribution versus GDP-PPP per capita (data from [12–14])

capita) is followed by an increase in the latter practically parallel to the consumption of fossil fuels (for GPD between 16,000 and 32,000$ per capita) and finally a slightly less steep increase. This can also be explained by the fact that generally the development of low carbon and emissions abatement technologies occurs only later than the diffusion of technologies and behaviours that generate pollution.

Figure 1.7 reports the population, and the birth and death rates versus increasing classes of GDP per capita. Comparing this graph with that of Figure 1.5, it is possible to notice that the first two stages belong to the past, when GDP values were significantly lower than the current lowest values. As can be seen from Figure 1.7, the death rate has now fallen to a level almost independent of GDP, whereas the birth rate remains high for low values of GDP, and currently the world population is substantially divided in two groups located in the stages 3 and 4, with a very small minority of the world population, already at stage 5.

From the comparison of Figures 1.6 and 1.7, the following conclusions can be drawn:

- approximately 60% of the world population currently has a birth rate higher than the mortality rate and, therefore, it is destined to increase in the near future;
- this part of the population still has a very low energy consumption, which is destined to increase both due to population growth and to the desire of improving its level of well-being, and it has yet to cross the GDP range for which the emissions are growing faster than energy consumption;
- the remaining 40% of the world population is almost completely in a phase in which emissions proceed hand in hand with the consumption of energy from

fossil sources which, despite the constant increase in the use of renewable sources, is growing;

- if these correlations are not modified, the development of the poorest populations will be accompanied by a demographic boom which will be followed by industrialisation with a massive demand for energy, goods, and services and, consequently, a significant increase in environmental impact.

1.2 The environmental impact

Every activity, human or otherwise, has an impact on the surrounding environment. And it is not necessarily a negative impact, even if currently we tend to give a negative meaning to the term 'environmental impact'. In energy conversion, the negative environmental impact is generally attributable to pollution, emission of climate-altering gases and territorial occupation which can also lead to changes in the surrounding habitat.

In the most general definition, pollution is an alteration of the environment that is not in balance with existing natural cycles and which is harmful. This alteration can have a local impact or a global impact: it depends on the type of pollution and the type of pollutant.

When the type of pollution is not specified, it generally means chemical pollution, but the pollution can be of a different type, for example acoustic. Chemical pollution can affect soil, water or air. The pollutants released into the soil are dispersed more slowly and in a rather limited volume, while those released into the water, and even more into the air, will have a higher speed and radius of dispersion.

Pollutant emissions into the air (gases, vapours and particulate matter) can be classified as primary pollutants, i.e. emitted directly, or secondary pollutants, i.e. harmful substances produced starting from primary pollutants and/or from non-directly harmful substances.

Generally, primary pollutants produce the most negative effects locally because in the emission zone they have a very high and dangerous concentration, while once dispersed in the atmosphere, dilution can be such as to make them practically harmless. For example, carbon monoxide is very dangerous because it poisons the blood forming carboxyhaemoglobin, but its average concentration in the atmosphere does not cause any problems.

Conversely, secondary pollutants produce the greatest harmful effects globally since they are generated following the dispersion of the substances emitted and, therefore, the damage created is suffered by a much larger community than the one responsible for the emission. For example, nitrogen oxides, which are a primary pollutant, after being dispersed into the atmosphere give rise to photochemical reactions that lead to the generation of tropospheric ozone, which is a secondary pollutant. Sulphur oxides, however, generate acid rain and a country like Norway, which has very low emissions of sulphur oxides, is subject to acid rain due to emissions from other countries [15,16].

In the energy field, pollution is associated with the use of fossil fuels both because it refers above all to chemical pollution, and because fossil fuels are by far the most used type of primary energy.

The main problem posed by the use of fossil fuels is that of the environmental damage caused by the products of their combustion, but the whole supply chain starting from the extraction is affected by critical issues towards the environment, including those related to any accidents and/or losses. The extraction of coal from the mines has had a lasting impact on the soil, while the extraction and transport of oil and gas have caused the pollution of air and water, sometimes with disastrous effects.

In addition to the emissions of pollutants, there are those of climate-altering gases which are all the 'greenhouse gases' with the exception of water vapour. The latter is responsible for about half of the greenhouse effect but is not considered climate-altering since it is minimally of anthropogenic origin and is condensable with a fast evaporation/precipitation cycle. The main climate-altering gases are carbon dioxide, methane, dinitrogen monoxide, hydrofluorocarbons, perfluorocarbons, sulphur hexafluoride and nitrogen trifluoride [17]. Those connected to the energy sector are essentially carbon dioxide and methane which has a climate-changing capacity almost 30 times higher than that of carbon dioxide on a time horizon of 100 years [18]. The Earth energy balance is shown in Figure 1.8.

The presence of carbon dioxide in the Earth's atmosphere is fundamental: in fact, thanks to its emissions, for example from volcanoes, the Earth has reached a climatic balance that has allowed life to develop. If there were no greenhouse gases in the Earth's atmosphere, the planet's temperature would be around −18 °C, while if there were no atmosphere, the temperature would be much lower.

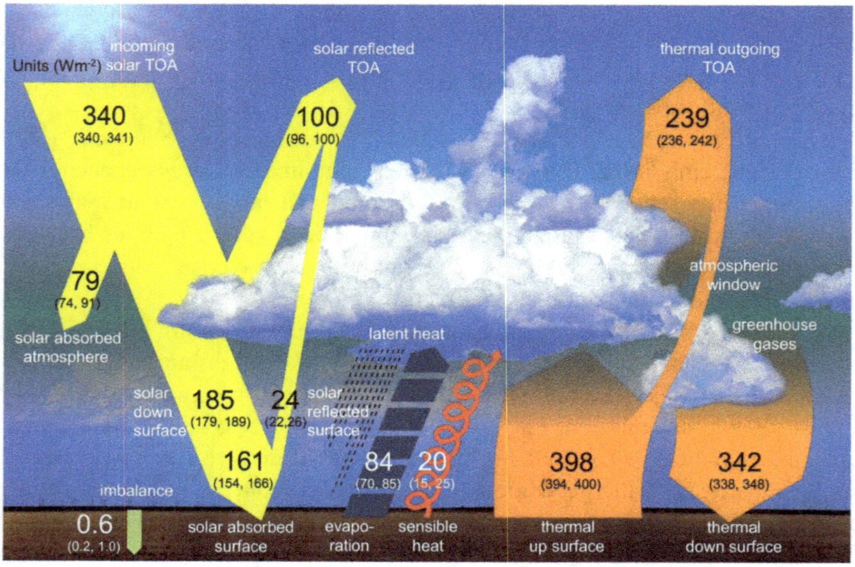

Figure 1.8 Earth energy balance. Source: IPCC Wiki (adapted from [19])

In the energy field, man began to interfere with the natural carbon cycle when he discovered fire and started burning wood. This acceleration in the production of carbon dioxide compared to the production due to natural phenomena was completely insignificant and, therefore, the natural cycle of carbon remained substantially unchanged until man put his hand to fossil fuels.

The examination carried out with cores in glaciers allowed to estimate the trend over time of the concentration of carbon dioxide (Figure 1.9) and it was estimated that the atmospheric concentration of carbon dioxide during the centuries before the industrial revolution fluctuated on average around 280 ppm. Today, the concentration has reached the value of 421 ppm [20,21], with an increase of more than 40% in the last century.

This increase in the concentration of carbon dioxide has been accompanied by an increase in the average temperature of the planet in the last century and the two trends show a good correlation. However, the scientific debate on the actual dependence of global warming on the increased concentration of carbon dioxide in the atmosphere is far from over.

What is certain is that it is more appropriate to bet in favour of the dependence of global warming on anthropogenic causes in order to be able to implement some remedy: if the increase in temperature depended on other causes, we might not have a chance to control it.

In any case, greenhouse gas emissions do not affect the climate at the local level and, therefore, this is a problem that must be addressed internationally. The initiatives of individual countries are of little value, unless they are significantly representative in terms of such emissions.

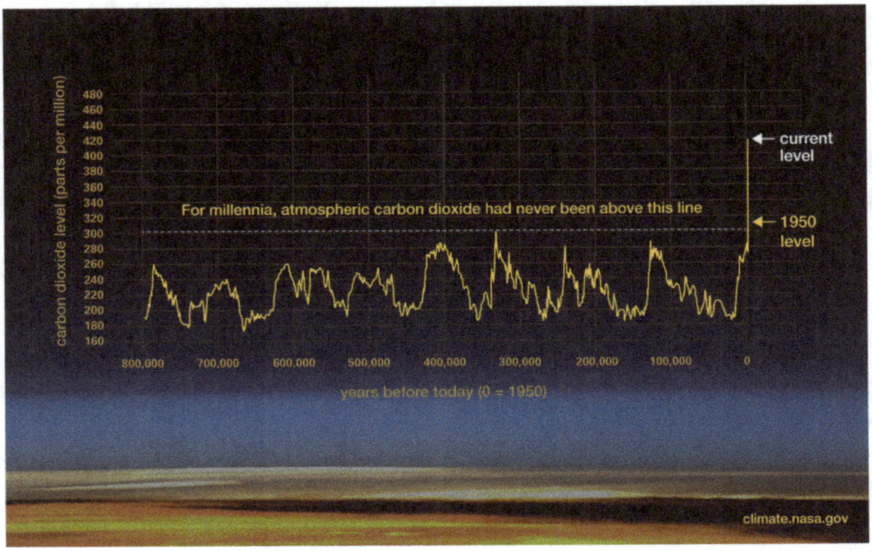

Figure 1.9 Trend of atmospheric carbon dioxide concentration [22]

1.3 Externalities

In speaking of energy conversion and environmental impact, it is good to emphasise a basic distinction between economic balances on the one hand and energy and environmental balances on the other hand.

In the case of an economic balance sheet, at each step, a data is transmitted, the cost, which summarises the entire previous history and therefore it is possible to carry out the balance regardless of this history and collecting only the costs relating to the various items that enter the balance sheet. Consider, for example, the electricity produced by a coal-fired thermoelectric power plant: the cost of coal takes into account the costs associated with the extraction process, transport and any other good/service necessary for the acquisition of the coal itself; the cost of building the plant takes into account the costs associated with the design, materials, energy, labour and all other goods/services necessary for the construction of the plant itself. In the end, by composing, possibly in a weighted way, all the costs of the goods/services necessary for the production of electricity, the cost of electricity is obtained.

This is not the case with the energy balance. Coal, for example, is considered for its calorific value, which is independent of the energy spent to extract and transport it, which, in turn, does not take into account the previous energy history of the means used for such operations. And we could still continue backwards. Then, the energy history of the plant should be considered, i.e. how much energy was needed to build it. And even from an environmental point of view, it is impossible and however useless to reconstruct the whole history of the kilowatt hour produced. The life cycle assessment partially, but not completely, reconstructs it. Thus, an overestimation of production efficiency and an underestimation of the environmental impact generated are obtained.

This fact can also be very important: in the past, the production of the first photovoltaic generators required an overall amount of energy higher than what they could generate in their entire operational life [23]! The ratio between energy produced and energy consumed is called EROEI (energy return on energy investment). Therefore, they would have been judged inefficient, but still able to provide energy from a renewable and free resource. Evidently, by consuming more ‘dirty’ energy than the amount of ‘clean’ energy they produced during their service life, they only produced a net environmental impact without producing a net amount of energy and the only valid justification to invest on this was for the development of a technology that later, as actually happens today, would have offered much better results.

Although the assessments are highly uncertain, in the energy policy choices, it would be important to also take into account external costs, that is, those costs that are also borne by subjects who do not benefit from the service that causes these costs. For example:

- effects of pollution on health;
- effects of pollution on structures;
- climate changes;
- employment effects;
- expenses for supply guarantee.

Evaluations carried out within the ExternE project [24] assigned the highest costs to health effects. And above all to the effects at the local level since the latter have a cost up to 100 times higher if the emissions occur in densely populated areas rather than in rural areas.

Finally, numerous scholars have dealt with the economic quantification of the environmental damage caused by energy sources. These costs (the so-called externalities) are currently not included in the prices of fossil fuels. They are paid by the community, directly or indirectly, through taxes, health care costs, insurance premiums and also through a decrease in the quality of life. If the respective environmental damage were economically quantified and included in the price of fuels, it would certainly resort to the introduction of 'clean' sources with significant environmental and economic benefits.

A synthetic picture of the damage caused by the use of fossil fuels is shown in Table 1.1 [25]. The estimate is from 1998 and the damage produced was, at the time, equal to 11% of the gross world product! Although the data are rather old, they prove that there is more than enough reason to move towards a more environmentally friendly energy system.

It is clear that it is better to use the less clean fuels in large centralised plants (e.g. thermoelectric power plants) using the most advanced emission abatement

Table 1.1 Estimate of the environmental damage caused, adapted from [25]

	Environmental damage 1998 $/GJ		
Type of damage	**Coal**	**Petroleum**	**Natural gas**
Effects on humans (premature deaths, medical expenses, loss of work efficiency, heat waves)	5.16	4.19	3.09
Effects on animals (loss of domestic livestock, loss of wildlife)	0.75	0.63	0.45
Effects on vegetation (reduction of crop yields due to ozone and acid rain, damage to wild flora, decline of forests also in terms of biodiversity)	1.99	1.61	1.20
Effects on the aquatic ecosystem (fuel losses, reduction of biodiversity)	0.26	1.55	0.16
Effects on man-made structures (deterioration of historic buildings and monuments, degradation of other buildings, corrosion of metal structures, soiling of clothes, etc.)	1.66	1.34	0.98
Other air pollution costs (reduced visibility, need for pollution abatement)	1.45	1.16	0.88
Effects associated with open pit mining	0.73		
Effects related to climate change (heat waves, droughts, floods, storms, hurricanes, tornadoes, agricultural and forestry losses, livestock losses, biodiversity losses, electricity generation problems)	2.04	1.66	1.22
Effects of sea level rise	0.47	0.38	0.28
Total	14.51	12.52	8.26

technologies, while for distributed uses (e.g. vehicles, homes), it is preferable to use the cleanest fuel.

1.4 The progressive depletion of fossil fuels and alternatives

Another fundamental aspect of the energy question is the availability of conventional resources. Today, over two-thirds of the world's energy needs are met by fossil resources because all the technologies involved in their supply chain are mature and cheaper than the alternatives. However, these are exhaustible resources and they are increasingly concentrated in the hands of a very small number of countries. Therefore, forecasts are made regarding the achievement of the production peak of a given primary energy source or its overall duration. For years, for example, we have been talking about reaching the peak of oil production, that is, the moment when production will begin to fall. A first estimate (Figure 1.10) was made by the American geologist Marion King Hubbert regarding the oil available in the USA, predicting the peak would have been reached in the early 1970s [26]. Later, in 1974, he also projected a peak of the global oil production for 1995 [27]. This peak has gradually moved forward and IEA (International Energy Agency) certified it was reached in 2008 for conventional oil [28]. However, the fracking technology adopted in the USA, which allows to recover other oil resources, further postponed such a peak which is currently set between 2025 and 2035 depending on the scenario [29]. The peak of natural gas is set between 2025 and 2035 in the Announced Pledges Scenario and the Net Zero Emissions, but later than 2050 in the Stated Policies Scenario. These are obviously completely hypothetical dates. In fact, they are based on consumption data collected at the time of the estimate, but the demand for energy continues to grow on average at sustained levels due to both the increase in population and the increase in per capita demand from developing countries development. Moreover, new reservoirs could be discovered and new technologies and increased sale price could make competitive the extraction from

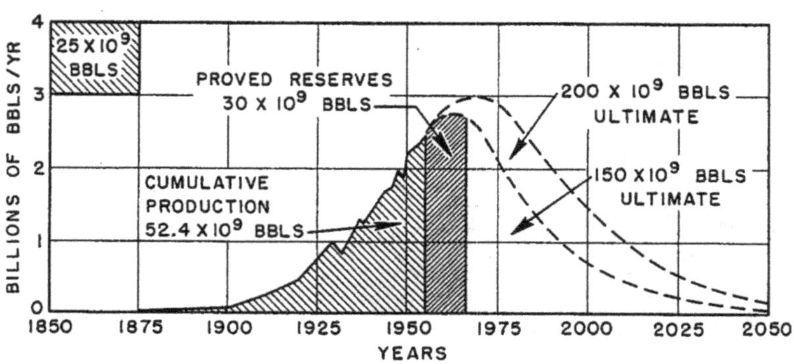

Figure 1.10 Hubbert peak for US oil production [26]

reservoirs not previously exploitable. In addition to this, technological development could lead to greater efficiency by further distancing the peak. The world population is currently around eight billion people, and although the growth rate is expected to decline, it is likely to exceed ten billion by the end of the century. Notwithstanding the greater availability, coal would peak by 2025 in all scenarios due to environmental policies, but also in this case the peak could shift forward depending on the increasing energy demand.

Observing Figures 1.11 and 1.12, it is noted that the highest energy consumption per capita (Canada, USA, Russia, Australia, Saudi Arabia and a few smaller others) is attributable to fairly modest portions of the population, whereas India, Southeast Asia, great part of Latin America and almost all Africa have a level of per capita energy consumption that is lower than the average (20 MWh) [30,31]. Obviously, countries that are below the world average will gradually increase the level of per capita consumption and therefore the total energy consumption will increase significantly.

Furthermore, the variability of prices [32] makes the economically convenient extractable quantities variable and the competitiveness of technologies variable. Finally, new explorations bring to light new fields.

Furthermore, it should be noted that human knowledge is characterised by a great uncertainty about the surrounding world and this leads different people to different considerations. The consequence is a range of positions ranging from the radical technocentrism of those who believe that man will always be able to find a

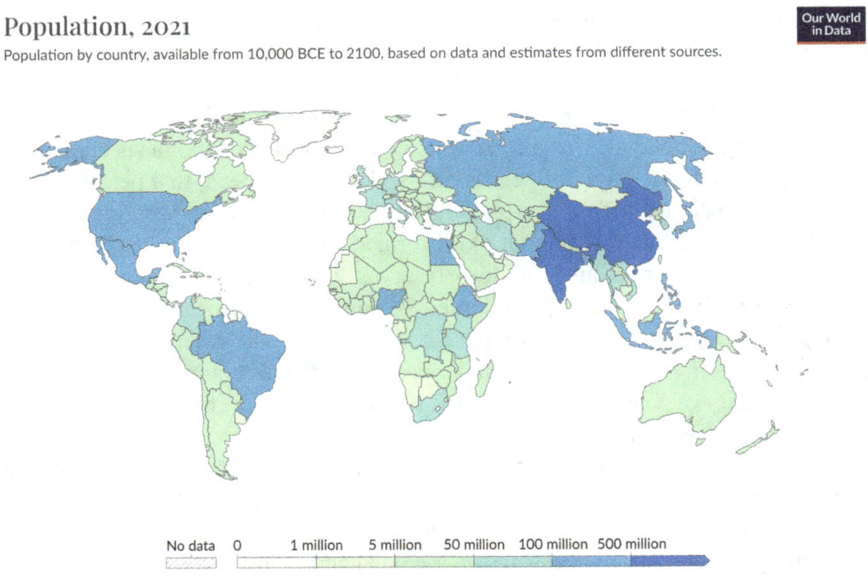

Figure 1.11 Distribution of population [30]

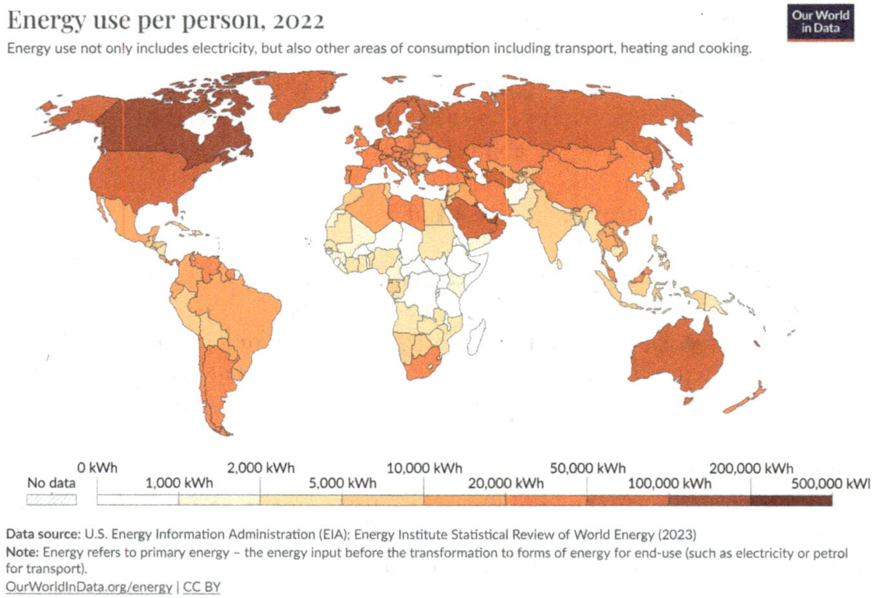

Figure 1.12 Distribution of per capita energy consumption [31]

solution to the scarcity of a resource to the radical ecocentrism of those who would like to impose drastic constraints on the use of resources and on population growth. In the energy sector, this uncertainty has given rise, for example, to the famous hypothesis about the abiotic nature of oil. This hypothesis was formulated at the beginning of the 19th century by the German naturalist Friedrich Heinrich Alexander Freiherr von Humboldt and revived with vigour in the 1950s by the Russian geologist Nikolai Alexandrovich Kudryavtsev. This hypothesis contradicts the theory of the fossil origin of petroleum by claiming, instead, that petroleum has been formed since the birth of the planet from carbon deposits at very great depths under the Earth's crust. The amount of oil formed would have been far greater than that identified up to now and the deep reserves would be able to fuel the fields already in use and to originate new ones. Among the arguments put forward in favour of this hypothesis, perhaps the most significant is the discovery of methane on other planets in the solar system. In any case, the theory commonly accepted by the vast majority of scientists remains that of the fossil origin first hypothesised by the Russian Lomonosov in the 18th century.

What is certain is that these are in any case exhaustible resources and that, therefore, despite the unpredictability of the dates, the peak of production will arrive soon or later and, without alternative energy sources, prices will begin to rise significantly. It is therefore necessary on the one hand to improve energy efficiency and on the other hand to gradually replace exhaustible resources with renewable resources. In this regard, Sheikh Yamani's phrase has remained famous: 'The stone age did not end due to a lack of stones and so it will be for the oil age too'.

Hydrogen is strongly inserted in this context, both because it can be obtained from any primary source of energy, and because its use, especially at urban level, would lead to a drastic reduction in the external costs of the energy sector.

The push towards hydrogen, however, is written in the history of the energy system. In addition to being the basis of energy conversion at a universal level, being the fuel that powers nuclear fusion in the stars, in the sequence coal → oil → natural gas there is a parameter that is constantly reduced and is the C/H ratio. Figure 1.13 shows a famous graph in this regard [33]. Natural gas consists mainly of methane and after methane (C/H = 0.25) there is only one possible value: 0, that is, hydrogen.

Hydrogen is therefore the fuel of the future, the only one that does not involve carbon dioxide emissions. It is not a primary energy source, but an energy carrier, as is the electricity that has already been used for a long time. But, being a fuel and a chemical, it is more versatile than electricity.

In the recent year, the colourless hydrogen assumed several different colours and, obviously, 'green hydrogen' is the most interesting option. However, it is very important to have clear that green is not equivalent to renewable. Hydrogen is green only if produced starting from a surplus of renewable energy not easily distributable by a more efficient system (i.e. a capable electric grid) to high demand area. If a photovoltaic system produces electricity that can be fed into the grid and instead it is decided to produce hydrogen, that will not be green hydrogen.

However, it is not possible to replace rapidly fossil fuels with green hydrogen but there could be advantages also by switching to hydrogen even if it were partly

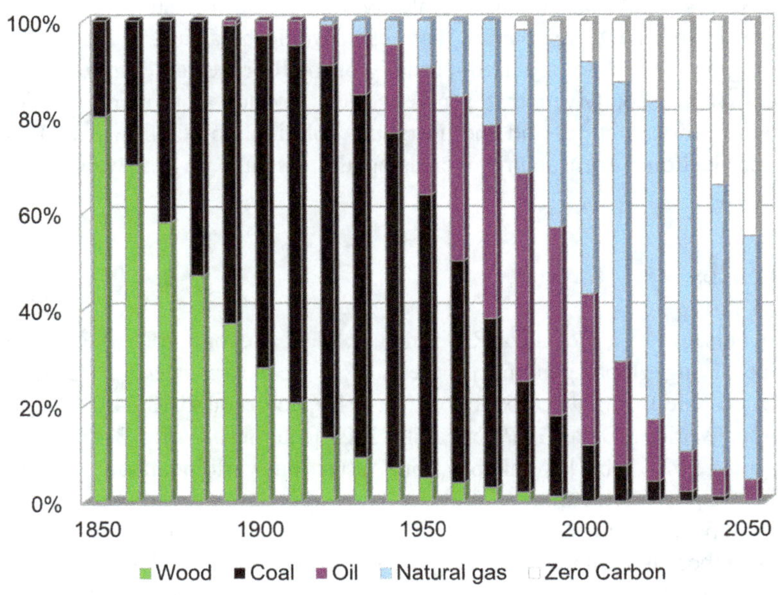

Figure 1.13 Evolution of the use of energy sources (data from [33])

produced from traditional fossil sources. Since carbon and sulphur oxides and, above all, particulate matter are pollutants very harmful to health, their emission in the urban environment is particularly harmful and represents by far the greatest social cost, as ExternE project have shown. The use of hydrogen would therefore be able to reduce this cost even if its origin were fossil, provided that the conversion from fossil fuel to hydrogen is carried out at an appropriate distance from urban areas. Centralised production would also make it possible to control pollution by reducing it to a minimum and eventually also to sequester carbon dioxide, which would otherwise not be achievable.

Considering the first step of the whole supply chain, i.e. hydrogen production, mature technologies already exist on a commercial level, both starting from fossil fuels and starting from electricity. Direct production of hydrogen from renewable sources, however, requires further research. The systems under study are photoelectric, photosynthetic and thermochemical ones.

There are no problems about how to use hydrogen. It could be used also in internal combustion engines even if the manufacturers are now concentrated on the production of vehicles equipped with fuel cells, which have the disadvantage of still high prices, although gradually decreasing. And also its use for residential or industrial applications has already been demonstrated.

The crucial issues are storage and distribution. Here, a new infrastructure is needed and it is problematic to decide to invest in the production infrastructure when the currently mature storage systems do not seem to be entirely satisfactory.

A very interesting alternative to overcome this bottleneck is the production of synthetic fuels by combining carbon and hydrogen. Carbon can derive directly or indirectly from coal and biomass, while hydrogen from the surplus of electricity generated from renewable sources. The result would be a fuel, liquid or gaseous, depending on the technology used, partially or totally renewable, in relation to the carbon source used. Given that synthetic fuels do not require interventions on distribution infrastructures and end-users, this solution would allow investing only in the production infrastructure, increasing the demand for renewable hydrogen more rapidly.

References

[1] Kubiszewski I, Costanza R, Franco C, *et al.* Beyond GDP: measuring and achieving global genuine progress. *Ecol Econ [Internet]*. 2013;93:57–68. Available from: http://dx.doi.org/10.1016/j.ecolecon.2013.04.019.

[2] Our World in Data. GDP per capita, 2021 [Internet]. World Bank; 2023 2023 [cited 2024 Feb 7]. Available from: https://ourworldindata.org/grapher/gdp-per-capita-worldbank.

[3] European Commission. Human development index | Composite indicators & scoreboards explorer [Internet]. 2022 [cited 2024 Feb 7]. Available from: https://composite-indicators.jrc.ec.europa.eu/explorer/explorer/indices/hdi/human-development-index.

[4] European Commission. Happy planet index | Composite indicators & scoreboards explorer [Internet]. 2019 [cited 2024 Feb 7]. Available from: https://composite-indicators.jrc.ec.europa.eu/explorer/explorer/indices/hpi/happy-planet-index.

[5] Sadik-Zada ER, and Ferrari M. Environmental policy stringency, technical progress and pollution haven hypothesis. *Sustainability*. 2020;12(9):3880.

[6] Our World in Data and Roser M. Demographic transition: why is rapid population growth a temporary phenomenon? [Internet]. 2019 [cited 2024 Aug 11]. Available from: https://ourworldindata.org/demographic-transition.

[7] Our World in Data. Energy use per person vs. GDP per capita, 2021 [Internet]. Washington, DC: U.S. Energy Information Administration; 2023 and other sources. 2023 [cited 2024 Feb 7]. Available from: https://our-worldindata.org/grapher/energy-use-per-person-vs-gdp-per-capita.

[8] Our World in Data. CO_2 emissions per capita vs. GDP per capita, 2018 [Internet]. Global Carbon Budget; 2023; Population based on various sources (2023); Maddison Project Database 2020 (Bolt and van Zanden, 2020). 2023 [cited 2024 Feb 7]. Available from: https://ourworldindata.org/grapher/co2-emissions-vs-gdp.

[9] Our World in Data. Carbon dioxide emissions by income level, 1980 to 2022 [Internet]. Global Carbon Budget; 2023 2023 [cited 2024 Feb 7]. Available from: https://ourworldindata.org/grapher/co2-income-level.

[10] Our World in Data. Per capita greenhouse gas emissions, 2022 [Internet]. Jones (eds.) *et al.* (2024); Population based on various sources (2023) – with major processing by Our World in Data. 2024 [cited 2024 Aug 11]. Available from: https://ourworldindata.org/grapher/per-capita-ghg-emissions.

[11] British Petroleum. BP statistical review of world energy 2022 [Internet]. London: 2022. Available from: https://www.bp.com/content/dam/bp/business-sites/en/global/corporate/pdfs/energy-economics/statistical-review/bp-stats-review-2022-full-report.pdf.

[12] Wikipedia. Income and fertility [Internet]. 2023 [cited 2024 Feb 7]. Available from: https://en.wikipedia.org/wiki/Income_and_fertility.

[13] Our World in Data. Birth rate vs. death rate, 2021 [Internet]. United Nations, World Population Prospects; 2022 [cited 2024 Feb 7]. Available from: https://ourworldindata.org/grapher/birth-rate-vs-death-rate.

[14] Our World in Data. Population by income level, 1960 to 2021 [Internet]. United Nations Population Division and others (via World Bank); 2022 [cited 2024 Feb 7]. Available from: https://ourworldindata.org/grapher/population-by-income-level.

[15] Norwegian Institute for Water Research. Acid rain still affects water quality [Internet]. 2014 [cited 2023 Apr 16]. Available from: https://partner.sciencenorway.no/environmental-surveillance-forskningno-niva/acid-rain-still-affects-water-quality/1401300#:~:text=At the peak of acidification, suffers damage from acid deposition.

[16] Garmo ØA, Skjelkvåle BL, De Wit HA, *et al.* Trends in surface water chemistry in acidified areas in Europe and North America from 1990 to 2008. *Water Air Soil Pollut.* 2014;225:1180.

[17] Eurostat. Glossary: greenhouse gas (GHG) – statistics explained [Internet]. 2016 [cited 2023 Apr 16]. Available from: https://ec.europa.eu/eurostat/statistics-explained/index.php?title=Glossary:Greenhouse_gas_(GHG).

[18] Greenhouse Gas Protocol. Global warming potential values [Internet]. 2015. Available from: https://www.ghgprotocol.org/sites/default/files/ghgp/Global-Warming-Potential-Values %28Feb 16 2016%29_1.pdf.

[19] IPCC Wiki. 152.3.1 global mean radiation budget [Internet] (adapted from Wild *et al.*, 2013). 2023 [cited 2024 Feb 7]. Available from: https://ipcc.fandom.com/wiki/152.3.1_Global_Mean_Radiation_Budget.

[20] National Oceanic and Atmospheric Administration (NOAA) U.S. Department of Commerce. Carbon dioxide now more than 50% higher than pre-industrial levels [Internet]. 2022 [cited 2023 Nov 2]. Available from: https://www.noaa.gov/news-release/carbon-dioxide-now-more-than-50-higher-than-pre-industrial-levels#:~:text=Carbon dioxide measured at NOAA's,of California San Diego announced.

[21] Scripps Institution of Oceanography at UC San Diego. The Keeling Curve [Internet]. 2022 [cited 2023 Nov 2]. Available from: https://keelingcurve.ucsd.edu/.

[22] NASA. Graphic: the relentless rise of carbon dioxide – climate change: vital signs of the planet [Internet]. 2022 [cited 2024 Feb 7]. Available from: https://climate.nasa.gov/climate_resources/24/graphic-the-relentless-rise-of-carbon-dioxide/.

[23] Odum HT. *Environmental Accounting: EMERGY and Environmental Decision Making [Internet]*. In: Odum HT (ed.). New York: Wiley; 1995 [cited 2023 Nov 2]. Available from: https://www.wiley.com/en-us/Environmental+Accounting%3A+Emergy+and+Environmental+Decision+Making-p-9780471114420.

[24] University of Stuttgart, IER Institute of Energy Economics and Rational Energy Use. ExternE – Externel Costs of Energy | IER Institute of Energy Economics and Rational Energy Use | University of Stuttgart [Internet]. 2005 [cited 2023 Apr 16]. Available from: https://www.ier.uni-stuttgart.de/en/research/projects/externe/.

[25] Veziroğlu TN, and Şahin S. 21st century's energy: hydrogen energy system. *Energy Convers Manag.* 2008;49(7):1820–31.

[26] Hubbert MK. Nuclear energy and the fossil fuels [Internet]. Publication No. 95. Drilling and Production Practice (1956). Houston, TX: American Petroleum Institute; 1956. Available from: http://www.energycrisis.com/Hubbert/1956/1956.pdf.

[27] Alimonti G. Our energy future starts from actual energy limits. *In: EPJ Web of Conferences*. 2018.

[28] Tiribuzi S, Cambi P, Celi L, *et al. Verso un sistema energetico italiano basato sulle fonti rinnovabili. Prima parte: analisi introduttiva, problematiche e scenari*

propedeutici [Internet]. Roma: CNR Edizioni; 2023. Available from: https://www.cnr.it/sites/default/files/public/media/attivita/editoria/Scetur-ASPO-Italia.pdf.

[29] IEA. *World Energy Outlook* 2021 [Internet]. Paris, France: 2021. Available from: https://iea.blob.core.windows.net/assets/4ed140c1-c3f3-4fd9-acae-789a4e14a23c/WorldEnergyOutlook2021.pdf.

[30] Our World in Data. Population, 2021 [Internet]. HYDE: 2017; Gapminder; 2022; UN: 2022. 2022 [cited 2023 Nov 2]. Available from: https://ourworldindata.org/grapher/population-unwpp?tab=map.

[31] Our World in Data. Energy use per person, 2022 [Internet]. Washington, DC: U.S. Energy Information Administration (EIA); *Energy Institute Statistical Review of World Energy*; 2023. 2023 [cited 2023 Nov 2]. Available from: https://ourworldindata.org/grapher/per-capita-energy-use.

[32] ICE. Dutch TTF natural gas futures | ICE [Internet]. [cited 2023 Apr 16]. Available from: https://www.theice.com/products/27996665/Dutch-TTF-Gas-Futures/data?marketId=5429405&span=3.

[33] Marchetti C. *On Decarbonization: Historically and Perspectively*. Interim Report IR-05-005. Laxenburg, Austria: International Institute for Applied Systems Analysis; 2005.

Chapter 2
Hydrogen characteristics

In the 16th century, Philipp Theophrast Bombast von Hohenheim (better known as Paracelsus), Turquet De Mayerne and Robert Boyle produced hydrogen by the reaction of acids (particularly sulphuric acid) with iron and other metals. In 1766, Henry Cavendish verified that hydrogen was much lighter than air and demonstrated that its combustion in air is accompanied by the formation of water according to the reaction:

$$H_2 + \frac{1}{2}O_2 \Rightarrow H_2O \qquad \Delta H^\circ = -242.0\ \text{kJ mol}^{-1} \tag{2.1}$$

Joseph Priestley called it flammable air, since the term gas was not yet used and it was introduced in the 17th century by Johann Baptista van Helmont. There are different opinions on the origin of the word gas, and in particular the one that makes it derive from the Greek $\chi\alpha o\varsigma$ = abyss and the one that makes it derive from the German gäschen = boil. In 1783, Antoine Lavoisier identified hydrogen as a chemical element and assigned it its current name deriving from the ancient Greek ($\upsilon\delta\omega\rho$ = water and $\gamma\varepsilon\nu$ – root of various words concerning the act of generating). The meaning is sometimes identical even when the root is not the Greek one: for example the Russian name is воао, from the term воа, which means water, and from the root о-, analogous to the Greek $\gamma\varepsilon\nu$-. Lavoisier also measured its heat of combustion.

Hydrogen was used for a long time to inflate balloons; but due to its flammability was replaced by helium, slightly heavier, but not flammable. Although not everyone agrees, fuel expert Addison Bain exonerates hydrogen to some extent in his book *Hindenburg: Exploring the Truth* [1], theorising his 'incendiary paint theory' in 1996. In fact, the coating of the airship was based on a highly flammable substance which did not allow the dissipation of the electric charge. The hydrogen, therefore, would have ignited only after the starting ignition caused by the fabric casing. Certainly, the airship burned faster due to the concomitant presence of hydrogen, but even if there had been helium in place of hydrogen, the whole structure would have caught fire and the airship would have crashed.

Nowadays hydrogen is a major industrial gas. It is used as a raw material in a large number of chemical processes, the most important of which is the synthesis of nitrogen trihydride (NH_3), commonly known as ammonia. It is also used in the petrochemical industry, for the hydrogenation of foods and for the manufacture of methanol and other synthetic fuels.

Hydrogen is rightly considered an excellent energy vector since it can be obtained from various primary energy sources; it is flexible with respect to applications and its use does not involve environmental impact or reduces it to a minimum. When burned with pure oxygen or used in fuel cells, it gives water as the only product. Only if burned in air, due to the simultaneous presence of oxygen and nitrogen, nitrogen oxides are also generated.

In more than 200 years since its discovery, hydrogen has had no use in the energy sector apart from its use as a propellant in aerospace missions. Liquid hydrogen and oxygen were used to propel aerospace vehicles such as the space shuttle. In addition, hydrogen and oxygen were used on board to feed fuel cells and produce the electricity required for the on-board systems and water for the crew.

This scarcity of energy uses of hydrogen in the human history is quite strange if one considers that nature has always made energy use of hydrogen and at various levels, starting from the stars (nuclear fusion) down to microorganisms (bacterial photosynthesis). The idea of using it as an energy carrier can be traced back only to the late 1960s and found its stable place with the establishment of the International Association for Hydrogen Energy in the early 1970s.

Those were the years of the oil crisis and the excessive dependence of the energy sector on oil led the industrialised countries to look for valid alternatives, but the only primary sources which in those years could offer sufficient guarantees were coal and nuclear power. Two resources that are not exactly flexible in terms of employment, especially in the transport sector. The idea of using electric vehicles equipped with accumulator batteries was joined by that of using vehicles equipped with hydrogen internal combustion engines, united to the electric ones by the problem of autonomy, but certainly capable of performances more similar to those of traditional vehicles.

Hydrogen, produced by gasification of coal or by thermal scission of water (with the use of nuclear energy as the primary source of energy), was thus proposed as a bridge between the primary sources available and the most common energy uses. The project of Dr. Cesare Marchetti's 'energy islands' [2,3] dates back to those years: atolls destined to host nuclear power plants and hydrogen production plants to be distributed to users scattered throughout the world.

Since then the scenario has changed, but still without producing large-scale effects. Today the focus is more on the production of hydrogen from renewable and partially from nuclear sources rather than fossil ones and on its use in fuel cells rather than in internal combustion engines. The aspect that still remains the most problematic is that of distribution. Huge investments have been promised and the goal of net-zero emissions has been set for the next decades by many countries.

Hydrogen is, at room temperature, a colourless, odourless and tasteless gas, practically insoluble in water. It does not cause damage to the environment. If released into the atmosphere it does not damage the ozone layer and does not directly contribute to the greenhouse effect; however, it may generate secondary greenhouse consequences [4]. Its combustion produces only water, neither carbon dioxide nor dust or other dangerous products. It is highly flammable and forms explosive mixtures with air. It is a simple asphyxiant, i.e. it is non-toxic as opposed to ammonia, methanol and others energy vectors, but if released in confined spaces, like any gas other than oxygen, it reduces oxygen concentration in the air with the risk of asphyxiation. In any case, such a risk is practically nil because access to people in environments where the concentration of hydrogen could be high is absolutely prohibited due to the risk of explosion. In fact, to reduce oxygen to less than 12%, the hydrogen must reach a concentration of at least 42%, much higher than that required to create an explosive mixture.

It is undoubtedly the most abundant chemical element in the universe. Spectral analysis of the light emitted by the Sun tells us that this star is made up of approximately 90% hydrogen atoms. And the same goes for the other stars with the result that over 75% of the mass of the universe is made up of hydrogen and the remaining part of helium for 23% and of all other species for the remaining 2% (Figure 2.1(a)). At the level of the Earth's crust, hydrogen appears significantly less abundant, only 0.14% by weight. The weight distribution of matter on the Earth's surface shows oxygen in first place (46.6%) followed by silicon (27.7%), aluminium (8.13%), and iron (5%). All other elements (calcium 3.63%, sodium 2.83%, potassium 2.59%, magnesium 2.09%, etc.) contribute the rest (Figure 2.1(b)).

Hydrogen on Earth is, however, almost exclusively present in combined form since the Earth's gravitational attraction is insufficient to hold a gas about 14 times lighter than air and it is found free only in small quantities at a height of more than 100 km. In closed environments it tends to stratify upwards. In its elemental state, it can be found in volcanic emanations and fumaroles, or trapped underground, for example in oil fields. The most important compound that hydrogen forms is water, in which it is combined with oxygen. In combination with carbon, hydrogen is present in hydrocarbons. Furthermore, together with carbon and oxygen, it forms organic compounds (carbohydrates, proteins, lipids, etc.) that are the fundamental constituents of the plant and animal world.

Table 2.1 shows the identification codes and safety data for hydrogen.

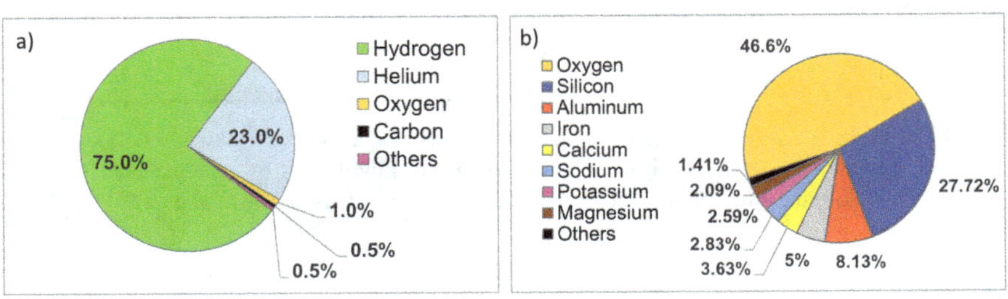

Figure 2.1 Atomic composition of universe (a) and Earth's crust (b) (data from Darling [5] and Science Notes [6])

Table 2.1 Identification codes and safety data for hydrogen

CAS (Chemical Abstracts Service) number	1333-74-0	
Substance identification number	Gas	1049
	Liquid	1966
Hazard identification number	Gas	23
	Liquid	223
Cylinder ogive colour		Red

2.1 Atomic and molecular properties

The hydrogen atom consists of a nucleus, composed of a single proton, around which a single electron moves. It is, therefore, the simplest of all elements and, therefore, it is the first element of the periodic table. When Dmitri Ivanovich Mendeleev drew the first periodic table of elements, he did not yet know the noble gases and, therefore, hydrogen was the only element in the first row and was positioned in the centre as it possessed characteristics similar not only to the elements of the alkali metal series but also to the elements of the halogen series.

The hydrogen molecule is formed by two atoms with the protons in a central position, 75 pm apart surrounded by the electron cloud that extends for about 300 pm from the centre of mass of the molecule.

Tables 2.2 and 2.3 show the main properties of hydrogen [7,8].

Compressing hydrogen at 5.5 GPa at room temperature it becomes solid.

Almost all of the mass is concentrated in the nucleus, which weighs about 1800 times more than the electron that orbits around it, while the space occupied is

Table 2.2 Atomic and molecular properties of hydrogen [9]

Formula	–	H_2
Atomic number (Z)	–	1
Electronic configuration	–	$1s^1$
Atomic weight	u	1.008
Atomic radius	pm	53
Molecular mass		2.016
Medium covalent radius	pm	31
Electronegativity (Pauling Scale)	–	2.2
Ionization energy	kJ mol^{-1}	1312

u is the symbol of the atomic mass unit: it is not a unit of measure of the International System, but it is the unit recommended by the IUPAC (International Union of Pure and Applied Chemistry) and, in any case, it is equivalent to g mol^{-1}.

Table 2.3 Gas constant and physico-chemical properties of hydrogen at normal temperature and pressure [8]

Gas constant	J kg^{-1} K^{-1}	4124.3
Specific heat at constant pressure	kJ kg^{-1} K^{-1}	14.89
Specific heat at constant volume	kJ kg^{-1} K^{-1}	10.77
Ratio between specific heats	–	1.383
Density	kg m^{-3}	83.764
Speed of sound	m s^{-1}	1294
Diffusion coefficient in air	cm^2 s^{-1}	0.61
Dynamic viscosity	g m^{-1} s^{-1}	0.00875
Thermal conductivity	W m^{-1} K^{-1}	0.1897

practically empty since the radius of the nucleus is about 100,000 times smaller than the radius of the electron's orbit. In addition to the one just described, called protium, which makes up over 99.9% of the total, hydrogen has two other isotopes (Figure 2.2) present in trace amounts on our planet: deuterium, on the one hand, whose nucleus is made up of a proton and a neutron and is stable, and tritium, which, on the other hand, has two neutrons and is an emitter of β-rays with a half-life of 12.26 years. Isotopes with a higher number of neutrons have also been observed, but these are absolutely unstable isotopes whose fundamental characteristics have not been able to be measured since they immediately decay into tritium. Deuterium and tritium are of great interest as fuels for nuclear fusion.

The hydrogen molecule consists of two atoms. There are two isomers of the hydrogen molecule: ortho-hydrogen, which has the spins of the two protons parallel, and para-hydrogen, in which they are antiparallel (Figure 2.3). At room temperature or above, hydrogen is a mixture of three parts ortho-hydrogen and one part para-hydrogen and is called normal-hydrogen. The orientation of the spins has an influence on some physical characteristics of the hydrogen molecule as shown in Tables 2.4–2.6 [10]. As the temperature drops, part of the ortho-hydrogen converts,

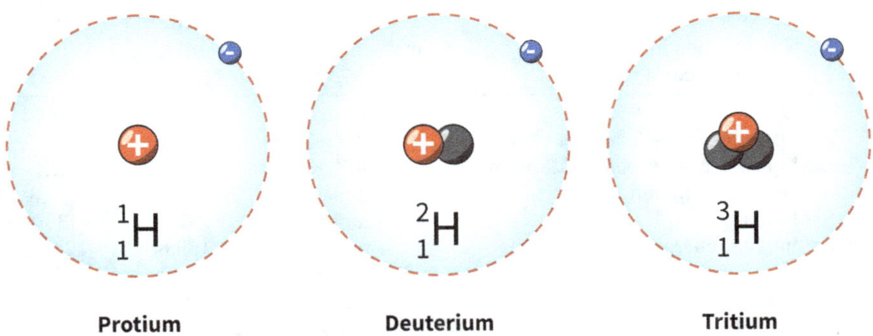

Figure 2.2 Hydrogen isotopes: protium, deuterium and tritium

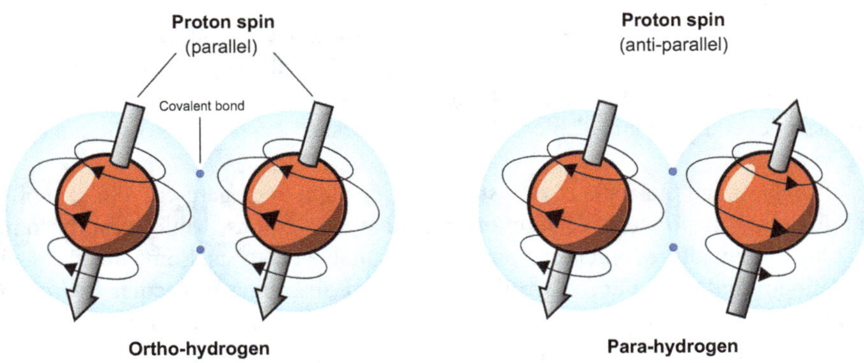

Figure 2.3 Ortho- and para-hydrogen

Table 2.4 Para- and normal- hydrogen properties at triple point

	Property	Unit	Para-hydrogen	Normal-hydrogen
	Temperature	K	13.803	13.957
	Pressure	kPa	7.042	7.204
Solid	Density	kmol m^{-3}	42.91	43.01
	Enthalpy	J mol^{-1}	−740.2	321.6
	Internal energy	J mol^{-1}	−740.4	317.9
	Entropy	J mol^{-1} K^{-1}	1.49	20.3
	Thermal conductivity	mW cm^{-1} K^{-1}	9	9
	Dielectric constant	–	1.286	1.287
Liquid	Density	kmol m^{-3}	38.207	38.3
	Specific heat at constant pressure	J mol^{-1} K^{-1}	13.13	13.23
	Specific heat at constant volume	J mol^{-1} K^{-1}	9.5	9.53
	Heat of fusion	J mol^{-1}	117.5	117.1
	Enthalpy	J mol^{-1}	−622.7	438.7
	Internal energy	J mol^{-1}	−622.9	436
	Entropy	J mol^{-1} K^{-1}	10	28.7
	Speed of sound	m s^{-1}	1273	1282
	Viscosity	µPa s	26	25.6
	Thermal conductivity	mW cm^{-1} K^{-1}	0.73	0.74
	Dielectric constant	–	1.252	1.253
Vapour	Density	kmol m^{-3}	0.0623	0.0644
	Specific heat at constant pressure	J mol^{-1} K^{-1}	21.2	21.22
	Specific heat at constant volume	J mol^{-1} K^{-1}	12.52	12.52
	Heat of evaporation	J mol^{-1}	905.5	911.3
	Enthalpy	J mol^{-1}	282.8	1350
	Internal energy	J mol^{-1}	169.8	1234.8
	Entropy	J mol^{-1} K^{-1}	45.63	93.6
	Speed of sound	m s^{-1}	305	307
	Viscosity	µPa s	0.73	0.74
	Thermal conductivity	mW cm^{-1} K^{-1}	0.104	0.105
	Dielectric constant	–	1.00038	1.00039

with a very slow spontaneous process, into para-hydrogen, tending to 100% para-hydrogen in equilibrium conditions as it approaches absolute zero (Figure 2.4). Since the energy content of para-hydrogen is lower than that of ortho-hydrogen, the process is exothermic and Table 2.7 shows the heat of conversion as the temperature varies.

The hydrogen molecule is very stable having a binding energy of 436 kJ mol^{-1} and, therefore, is not very reactive under ordinary conditions, while it can become so at high temperatures and in the presence of suitable catalysts. Some metals (such as platinum and palladium) have the property of absorbing molecular hydrogen, favouring its dissociation into atomic hydrogen which is much more reactive.

Hydrogen combines directly with most alkali and alkaline earth metals and non-metals, reacting violently with chlorine and fluorine. With sulphur it combines

Table 2.5 Para- and normal-hydrogen properties at normal boiling point

	Property	Unit	Para-hydrogen	Normal-hydrogen
	Temperature	K	20.268	20.390
	Pressure	kPa	101.325	101.325
Liquid	Density	kg m^{-3}	35.11	35.2
	Specific heat at constant pressure	J mol^{-1} K^{-1}	19.53	19.7
	Specific heat at constant volume	J mol^{-1} K^{-1}	11.57	11.6
	Enthalpy	J mol^{-1}	−516.6	548.3
	Internal energy	J mol^{-1}	−519.5	545.7
	Entropy	J mol^{-1} K^{-1}	16.08	34.92
	Speed of sound	m s^{-1}	1093	1101
	Viscosity	μPa s	13.3	13.3
	Thermal conductivity	mW cm^{-1} K^{-1}	1	1
	Dielectric constant	–	1.23	1.231
Vapour	Density	kg m^{-3}	0.6636	0.6604
	Specific heat at constant pressure	J mol^{-1} K^{-1}	24.5	24.6
	Specific heat at constant volume	J mol^{-1} K^{-1}	13.11	13.2
	Heat of evaporation	J mol^{-1}	898.3	899.1
	Enthalpy	J mol^{-1}	381.7	1447.4
	Internal energy	J mol^{-1}	229	1294
	Entropy	J mol^{-1} K^{-1}	60.41	78.94
	Speed of sound	m s^{-1}	355	357
	Viscosity	μPa s	1.1	1.11
	Thermal conductivity	mW cm^{-1} K^{-1}	0.163	0.165
	Dielectric constant	–	1.004	1.004

Table 2.6 Para- and normal-hydrogen properties at critical point

Property	Unit	Para-hydrogen	Normal-hydrogen
Temperature	K	32.98	33.18
Pressure	kPa	1.2928	1.3152
Density	kmol m^{-3}	15.59	14.94
Specific heat at constant pressure	J mol^{-1} K^{-1}	n.a.	n.a.
Specific heat at constant volume	J mol^{-1} K^{-1}	19.7	19.7
Enthalpy	J mol^{-1}	77.6	1164
Internal energy	J mol^{-1}	5.7	n.a.
Entropy	J mol^{-1} K^{-1}	35.4	54.57
Speed of sound	m s^{-1}	350	n.a.
Viscosity	μPa s	3.5	3.5
Dielectric constant	–	1.098	1.0937

spontaneously at around 250 °C. Reaction with nitrogen to form ammonia requires the use of catalysts, high temperature and high pressure.

The ideal gas law states a linear dependence between pressure and density. However, hydrogen is made up of small non-polar molecules so that the attractive

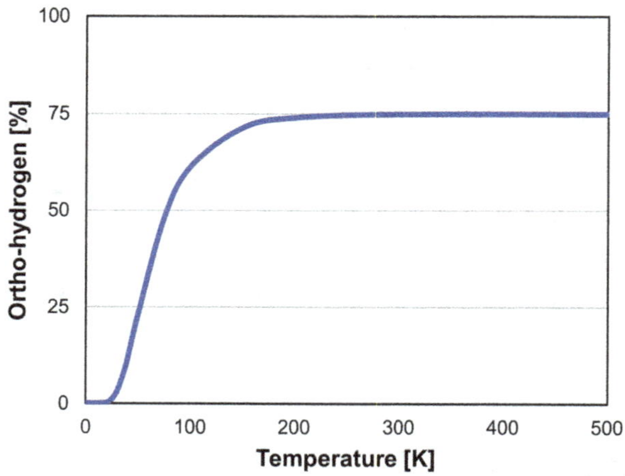

Figure 2.4 Ortho-hydrogen content vs. temperature (data from Hust and Stewart [11] and own elaboration)

Table 2.7 Heat of conversion from ortho-hydrogen to para-hydrogen (adapted from [12])

Temperature (K)	Heat (J mol^{-1})	Temperature (K)	Heat (J mol^{-1})
10	1416.90	80	1381.41
20	1416.90	90	1345.99
20.39	1416.90	100	1294.70
30	1416.90	120	1148.40
33.10	1416.90	150	866.82
40	1416.85	200	440.16
50	1416.12	250	189.58
60	1412.56	298.16	76.78
70	1402.48	300	74.09

forces between the molecules are very modest and the co-volume becomes the predominant factor in the deviation from the ideal behaviour. It follows that, for hydrogen at room temperature, the approximation is acceptable up to a pressure of the order of about 10 MPa, while for higher pressures, which are those of some interest for accumulation, the behaviour differs more and more from that of an ideal gas and the space occupied is greater. Indicating with p the pressure, R the gas constant, T the temperature and v the specific volume, the hydrogen equation of state can be well approximated by:

$$p = \frac{R \cdot T}{v - b} - \frac{a \cdot \alpha(T)}{v \cdot (v + b)} \tag{2.2}$$

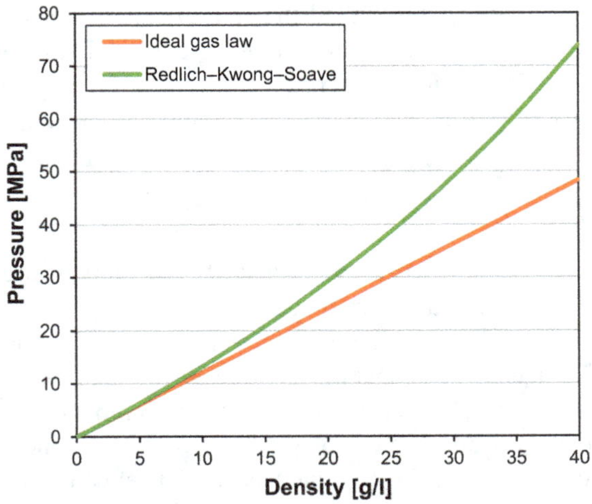

Figure 2.5 Ideal gas and real hydrogen

called the Soave-Redlich-Kwong equation of state, in which the attractive parameter a and the repulsive parameter b are functions of pressure and temperature at the critical point and are, respectively, given by:

$$a = \frac{0.42747 \cdot R^2 \cdot T_c^2}{p_c}; \; b = \frac{0.08664 \cdot R \cdot T_c}{p_c} \tag{2.3}$$

while the expression of $\alpha(T)$ for hydrogen is:

$$\alpha(T) = 1.202 \cdot e^{-0.30288 \cdot \left(\frac{T}{T_c}\right)} \tag{2.4}$$

Figure 2.5 shows the graph of the volumetric density approximated with the above equation and compared with the case of perfect gas.

2.2 Compression and expansion

The expansion of a gas implies an increase in the average distance between the molecules: it follows that the potential energy due to the forces of intermolecular attraction increases. If the expansion occurs without enthalpy change, the kinetic energy of the molecules must decrease. However, the increase in the intermolecular distance also implies a rarefaction of the collisions between molecules: it follows that the conversion from kinetic energy to potential energy undergoes a reduction.

Therefore the expansion causes two phenomena which have opposite effects on the balance between potential energy and kinetic energy and, consequently, on the final temperature of the gas. The prevalence of one or the other depends, for a given gas, on the pressure and temperature at which the gas is at the start of expansion.

This behaviour of real gases, called the Joule–Thomson effect, was observed for the first time in 1852 by James Prescott Joule and William Thomson, and the Joule–Thomson coefficient (μ_{JT}) is defined as the ratio between the temperature and the pressure variations in an isoenthalpic expansion process:

$$\mu_{JT} = \left(\frac{\partial T}{\partial p}\right)_H = \frac{\frac{2 \cdot a}{R \cdot T} - b}{c_p} \tag{2.5}$$

where a and b are the van der Waals constants, R is the universal gas constant and c_p is the molar specific heat at constant pressure.

For a given gas and a given pressure, there are two Joule–Thomson inversion temperatures defined as those temperatures value at which $\mu_{JT} = 0$ (Figure 2.6). For initial temperature values between the two inversion temperatures, the numerator assumes positive values and therefore it is $\mu_{JT} > 0$ and, since in the case of expansion $\partial p < 0$, it must be $\partial T < 0$ so the isoenthalpic expansion involves cooling of the gas. Conversely, isoenthalpic expansion involves heating the gas for values of the initial temperature external to the inversion temperatures. For ideal gases, it is always μ_{JT}=0 and therefore there is no inversion temperature and they do not heat up or cool down when subjected to isoenthalpic expansion.

For most real gases, the upper inversion temperature at atmospheric pressure is higher than ambient temperature and, therefore, such gases can be cooled by isoenthalpic expansion. Hydrogen, however, has a Joule–Thomson inversion temperature at ambient pressure of about 200 K and therefore exhibits the opposite behaviour: it does not cool down if it is subjected to isoenthalpic expansion starting from room temperature. Helium and neon have a similar behaviour [13]. After helium, it is the most difficult gas to liquefy. It is a good conductor of heat and electricity, and it is easily absorbed by some metals.

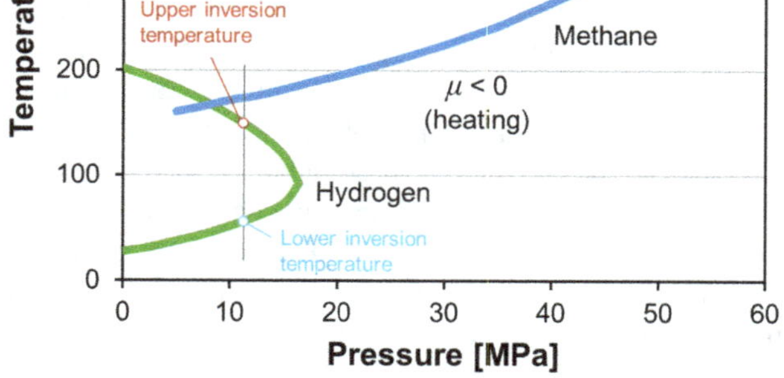

Figure 2.6 Inversion curves for hydrogen (green) and methane (light blue)

2.3 Hydrogen combustion

Hydrogen reacts not only with oxygen but also with all oxidising agents such as chlorine. The flame generated by the combustion of hydrogen and oxygen is colourless, but even combustion in air develops a flame that is almost invisible in daylight.

Table 2.8 shows the main characteristics of hydrogen as a fuel.

The heat generated by the complete combustion of a unit quantity of fuel is called heat of combustion or higher heating value (HHV). Traditionally, the study of energy systems has instead considered the lower heating value (LHV), i.e. the one evaluated by subtracting from the HHV the latent heat of evaporation of the water vapour contained in the combustion products. When using hydrogen it is possible to obtain liquid water as a product, for example when converted using a low temperature fuel cell. For such a reason, unless otherwise indicated, HHV will be considered as the reference in this text.

In practical applications, the heating value per unit of volume is more significant: the capacity of the pipes depends on their section and on the speed of the fluid.

Table 2.8 Hydrogen and combustion (data from [14] and own elaboration)

Stoichiometric air-fuel ratio (weight)	–	34.2931
Stoichiometric air-fuel ratio (volume)	–	2.3869
Flammability limits in air	%	4–75
Adiabatic flame temperature in air	°C	2045
Ignition energy in air	mJ	0.019
Self-ignition temperature in air at 1 atm	°C	585
Gas volume change after stoichiometric combustion in air	%	–14.76
Burning speed in air	$m\ s^{-1}$	2.55–3.2
Detonation limits in air	% volume	18-59
Speed of detonation in air	$m\ s^{-1}$	1480–2150
Diffusion coefficient (NTP)	$cm^2\ s^{-1}$	61
Velocity of diffusion in air (NTP)	$cm\ s^{-1}$	<2
Speed of rise in air	$m\ s^{-1}$	1.2–9
Maximum flame speed in air	$m\ s^{-1}$	2.65
Flammability limits in oxygen at 20°C and 1 atm	%	4–94
Self-ignition temperature in oxygen at 1 atm	°C	560
Adiabatic flame temperature in oxygen	°C	2830
Maximum burning speed in oxygen	$m\ s^{-1}$	11.75
Detonation limits in oxygen	% volume	15–90
Speed of detonation in oxygen	$m\ s^{-1}$	3500
Maximum flame speed in oxygen	$m\ s^{-1}$	14.36
Higher heating value	$MJ\ kg^{-1}$	141.91
	$kJ\ mol^{-1}$	286.1
Lower heating value	$MJ\ kg^{-1}$	119.89
	$kJ\ mol^{-1}$	241.7
Lower Wobbe index	$MJ\ m^{-3}$	40.898
Upper Wobbe index	$MJ\ m^{-3}$	48.340

Some of the hydrogen characteristics have significant consequences on energy conversion devices performance. For example, the flame speed affects the efficiency of an internal combustion engine. However, this kind of considerations will be made in the appropriate chapters.

The safety aspects of a fuel essentially concern the risk of fire and/or explosion and their own toxicity or that of the products of combustion. Its flammability range is very large both in air and in oxygen, and therefore, formation of flammable mixtures is very easy. Moreover ignition energy is extremely low, with the consequence that a weak electrostatic spark or a hot spot is able to ignite the mixture. Since hydrogen burns rapidly, it is preferable to let it burn completely without trying to extinguish the flame unless it is possible to intercept the flow of hydrogen: in fact, if the flame goes out and the flow continues, an explosive mixture can form.

Hydrogen is characterised also by a high diffusion coefficient: this allows a fast mixing in air. Therefore, it is difficult to reach the lower limit of the flammability range in open air, whereas it can create dangerous situations in poorly ventilated environments.

Hydrogen is lighter than air and rises quickly, and also its flame tends to go upwards. Due to its low density and low viscosity, there is a risk of leakage from circuits and materials normally impermeable to other gases. An ignited compressed hydrogen leak produces an almost invisible, very narrow and directional flame. Near the exit hole the concentration of hydrogen is too high to allow a combustion, whereas at a distance equal to a few hundred times the diameter of the hole the concentration is too low; therefore, the flammability zone is always rather close to the hole outlet, but not in direct contact with the surface.

In conclusion, a gaseous hydrogen leak becomes dangerous only if there is a limited accumulation of gas and in the presence of a trigger. However, the flame has characteristics such as to be, in some respects, less dangerous than other types of flame.

As far as safety in the use of liquid hydrogen is concerned, first of all it must be observed that contact with liquid hydrogen, with cold hydrogen that has just evaporated, with suddenly cooled and/or liquefied air and with surfaces that are not well insulated causes severe burns and hypothermia. The human body's tissues at these very low temperatures can undergo a ductile-brittle transition.

A liquid hydrogen leak can have various effects on the surrounding environment, including temporary solidification of the air. Since the gas that has just evaporated is very cold, its density is still relatively high and diffusion begins in a horizontal direction and then gradually passes to the vertical direction as the density decreases with the progressive increase in temperature. So it takes longer to disperse, using a large surface in space and generating a dangerous situation. Furthermore, solidified air may have a higher concentration of oxygen than nitrogen, due to different melting and boiling temperatures. Therefore, following re-gasification, pockets of oxygen-enriched air can be created with potentially greater risks in the event of mixing with hydrogen that has not yet been dispersed.

When dealing with hydrogen, the choice of materials is very important and depends also on the operating temperature and pressure. Embrittlement can affect materials such as rubber and plastic. Carbon steels are also subject to embrittlement. A further critical aspect is the risk of freezing of humidity which can cause blockage of certain devices such as safety valves. Not only very low temperatures but also high pressures can cause embrittlement of normally ductile materials. In general, stainless steel, copper, brass and aluminium are fine at room temperature, while cast iron is absolutely to be avoided. Even organic materials such as rubber and plastic can be used. For liquid hydrogen only resilient steel or copper must be used. As far as connections are concerned, welded joints are to be preferred because they offer a greater guarantee of tightness than screw joints.

It evaporates quickly expanding by a factor greater than 800, and therefore, the evaporation of liquid hydrogen inside a cryogenic tank causes an increase in pressure which, if the safety threshold is exceeded, triggers the safety valves with the release of hydrogen into the environment.

Particular attention must be paid to the start-up of any plant or machinery that uses hydrogen. Before putting into service, the lines must be inerted with nitrogen in order to eliminate the air from the lines themselves. The oxygen content must be $<1\%$. Furthermore, any installation must always work in over-pressure, to avoid the entry of air and before start-up it is also necessary to check all the seals by flushing first with nitrogen and then with hydrogen at the operating pressure. In the case of liquid hydrogen, helium should be used instead of nitrogen.

The low ignition energy allows catalytic combustion, that is much safer being without a flame. Furthermore hydrogen can be used for direct steam generation since its combustion with stoichiometric oxygen generates only steam which can be used for high efficiency cycles (see Section 5.1.4). Finally hydrogen can be used to produce power by means of electrochemical conversion (see Section 5.2).

2.4 Fuels comparison

In this section, hydrogen main characteristics as a fuel will be compared with correspondent characteristics of several fuels.

Fundamental requirements for combustion to take place are that there is a fuel mixed with an oxidising agent, that this mixture is within its flammability limits and that a sufficient amount of energy is available to trigger the reaction.

In general, an excess of air is used in combustion to avoid the presence of unburned products. On the one hand, this favours efficiency because all the fuel is used, but, on the other hand, it disadvantages it because the gases produced are diluted and the temperature is lower. However, the completeness of the combustion is also connected to the homogeneity of the fuel/air mixture and to the combustion speed. The homogeneity of the mixture depends in turn on the mixing rate of the components of the mixture itself, while the combustion rate depends on the reaction rate with oxygen of the fuel components.

Therefore, there are several characteristics that affect the combustion process as well as several characteristics which affect safety aspects. The next series of figures will compare hydrogen with several fuels, common and less common.

Hydrogen has a density decidedly lower than all other gaseous fuels (Figure 2.7): it is about eight times less dense than methane, which is the least dense of the others, and about 28 times less dense than butane, which is the densest of those shown in the figure.

Thanks to liquefaction it reaches a density almost 800 times higher than that of gas at atmospheric pressure. Even in this case, however, the comparison with other liquid fuels is merciless (Figure 2.8). It is about 6.5 times less dense than liquid natural gas (LNG) and about 11 times less dense than hydrazine, a fuel often used in aerospace missions.

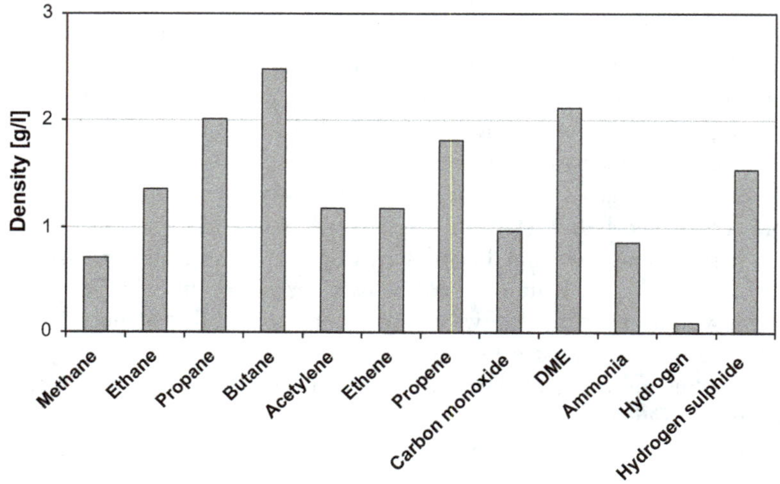

Figure 2.7 Density of some gaseous fuels (standard conditions)

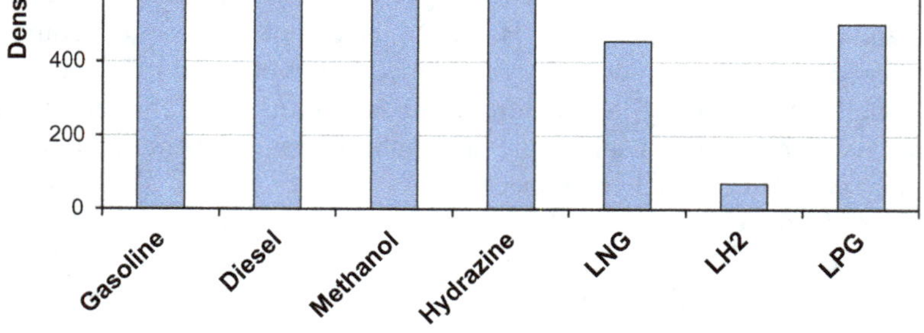

Figure 2.8 Density of some liquid fuels

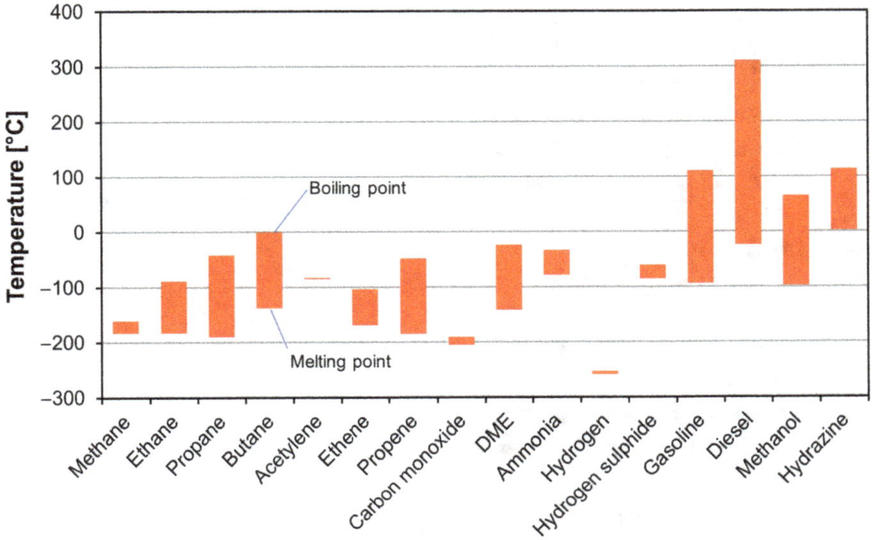

Figure 2.9 Melting and boiling points (respectively lower and upper limit of the bar) of some fuels

Unfortunately also as regards the melting and boiling temperatures, hydrogen has values that are highly misaligned with respect to other fuels (Figure 2.9). The boiling temperature, in fact, is much lower even than the melting temperatures of the other gases. This makes the liquefaction operation rather complex, as already mentioned in Section 2.2.

And the differences between hydrogen and other fuels remain enormous even if the heating values are considered (Figure 2.10). Both the LHV and the HHV of hydrogen are much more than double the corresponding values of other fuels. And since the only product of combustion is steam, the difference between HHV and LHV is also far greater than for other fuels.

However, due to its very low density, the situation is reversed when considering the heat of combustion per unit volume. Among gaseous fuels, only carbon monoxide and ammonia have comparable values (Figure 2.11). And in the case of liquid fuels, methanol has values that are approximately double, while diesel reaches values that are approximately quadruple (Figure 2.12).

The flammability range of hydrogen in air is very wide: only the ranges of acetylene and hydrazine are wider (Figure 2.13). This characteristic has very important consequences especially for the use of hydrogen in internal combustion engines (see Section 5.1.1) since commonly used fuels have flammability ranges more than ten times narrower.

Another important characteristic for the use of hydrogen in internal combustion engines is the self-ignition temperature, since a high value allows the use of a high compression ratio. From this point of view, hydrogen ranks immediately

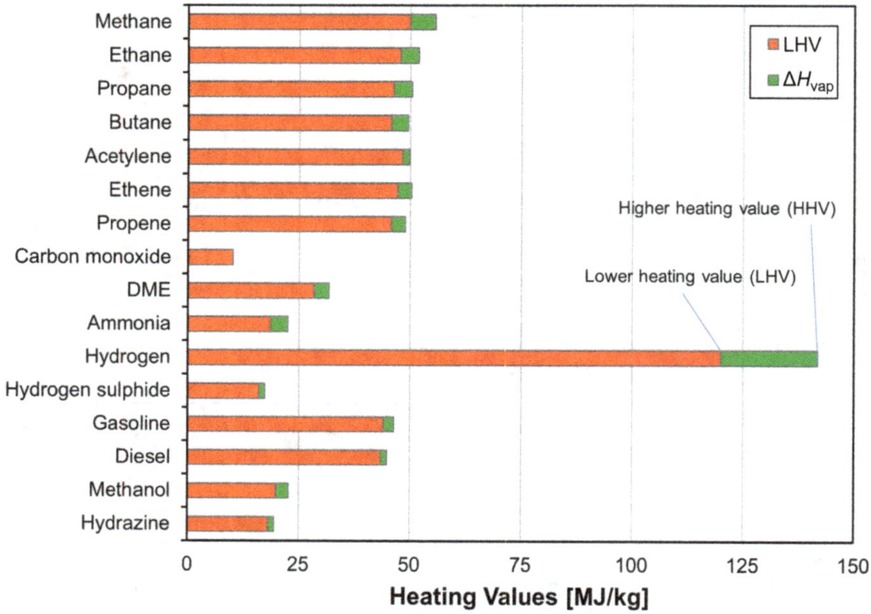

Figure 2.10 Lower and higher heating values of some fuels

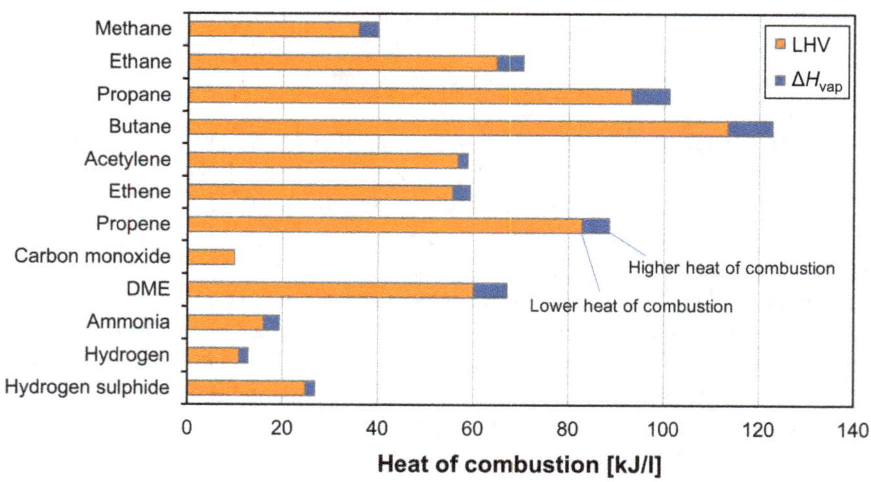

Figure 2.11 Lower and higher volumetric heat of combustion for gaseous fuels

behind ammonia and carbon monoxide (Figure 2.14). Among the fuels normally used in such engines, only natural gas shows a value close to that of hydrogen.

Like any other fuel, hydrogen also carries risks. Risks that depend on the specific physical and chemical characteristics which, being different from those of

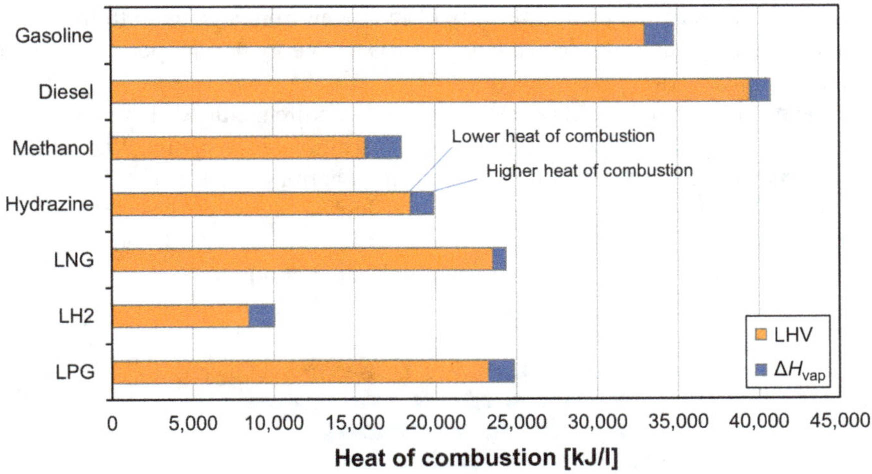

Figure 2.12 Lower and higher volumetric heat of combustion for liquid fuels

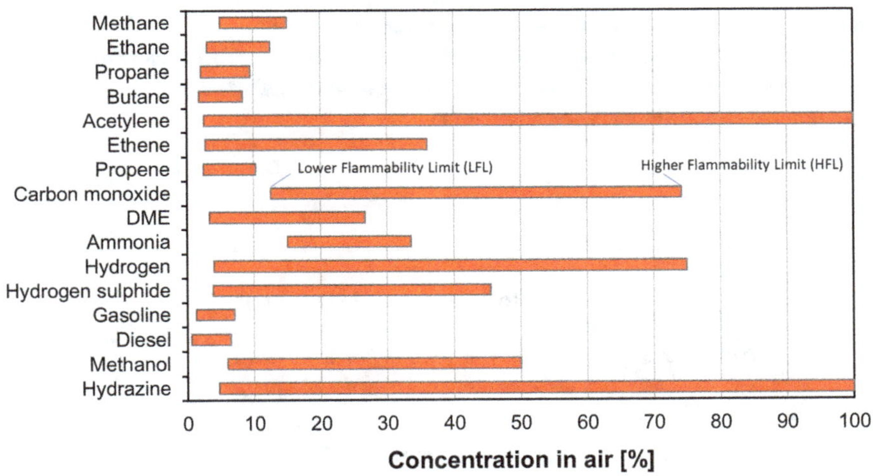

Figure 2.13 Flammability range and flammability limits in air for different fuels

other fuels, make it less safe in certain respects and safer in others. For this reason, its use may require different materials and precautions.

First of all, the hydrogen molecule is the smallest and, therefore, escapes more easily through small cracks. On the one hand, this requires greater care in the seals, and, on the other hand, it implies greater risks of leaks in the event of micro-cracks in the containers. However, the energy released by a hydrogen leak is less than the energy released by a natural gas leak. In fact, the escape velocity is influenced by density, viscosity and diffusion coefficient in air and is also limited by the sound velocity. As a result, the escape velocity of hydrogen from a crack is from 1.26 to 2.8 greater than natural gas, which however has an energy density per unit volume over three times higher than hydrogen [15].

The parameters that favour the development of a fire are the wide flammability range, the low ignition energy and the low autoignition temperature. While once a fire has started, the parameters that make it more harmful are the high flame temperature, the high explosion energy and the high emissivity of the flame. For hydrogen, methane and gasoline, they are listed in Table 2.9. All listed data come from different sources collected with many other more in [8].

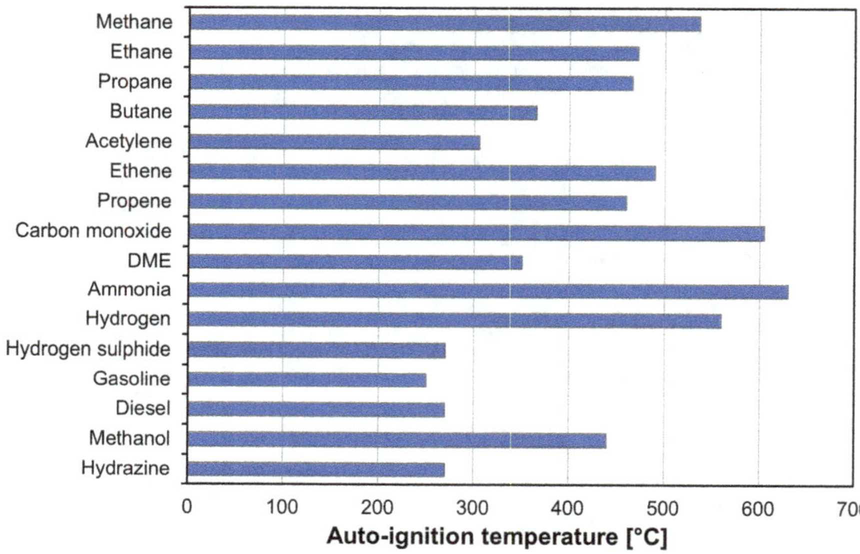

Figure 2.14 Auto-ignition temperature for different fuels

Table 2.9 Comparison of safety characteristics of gasoline, methane and hydrogen [8]

Property	Gasoline	Methane	Hydrogen
Flammability range	1.0–7.6	5.3–15.0	4.0–75.0
Minimum ignition energy in air (mJ)	0.24	0.29	0.02
Self-ignition temperature (K)	501–744	813	858
Flame temperature in air (K)	2470	2148	2318
Flame emissivity (%)	30–42	23–33	17–25
Burning velocity in air (cm s^{-1})	37–43	37–45	265–325
Theoretical explosion energy (g TNT/Sm^3)	44.22	7.03	2.02
Adiabatic sound velocity (m s^{-1})	154	448	1294
Limits of detonability in air, vol%	1.1–3.3	6.3–13.5	18.3–59.0

Compared to gasoline and methane, hydrogen has wider flammability range and lower ignition energy, characteristics that make it more dangerous. Its flame temperature is higher than that of methane, but lower than that of gasoline. However, it has a higher autoignition temperature and a less emissive flame, characteristics which mitigate its danger.

Overall, it would seem that hydrogen is more dangerous; however, some more specific considerations need to be made. In fact, if a wide flammability range constitutes a danger because it is easier to obtain flammable mixtures, it must be taken into account that the lower limit is very important: in the event of a fuel leak, a very low lower limit allows the dangerous condition to be reached quickly. Gasoline and diesel vapours are already flammable with concentrations, respectively, equal to one third and one sixth compared to hydrogen. Therefore, the risk of fire in the event of a leak is much higher for petrol and diesel.

Furthermore, hydrogen has a much lower density than other fuels, so it tends to rise very quickly. For this reason, it is extremely unlikely that it will reach ignition rates outdoors. Petrol and diesel vapours have a higher density than air, so they tend to accumulate downwards and can quickly reach the lower flammability limit.

Natural gas has a lower flammability limit that is slightly higher than that of hydrogen. However, it is eight times denser than hydrogen, which is why it is still easier for it to reach flammability conditions. Furthermore, it is not very relevant that the minimum ignition energy of a hydrogen/air mixture is only one tenth of the minimum energy required for the ignition of a methane/air mixture. Indeed, the energy of a static electric spark is sufficient to ignite methane and, however, the ignition energy depends on fuel/air ratio and is comparable for hydrogen and methane at the lower flammability limit [15].

A further safety factor for hydrogen is its high diffusion coefficient (0.61 $cm^2\ s^{-1}$) compared to the more modest ones of natural gas (0.16 $cm^2\ s^{-1}$) and gasoline (0.05 $cm^2\ s^{-1}$).

The hydrogen flame has two very important characteristics in terms of safety: it is almost invisible and has very low emissivity. This means that one can approach a hydrogen flame almost without realising it, which is very dangerous. However, the low emissivity reduces the risk of fire ignition by irradiation of the surrounding materials.

The lower fuel/air detonability ratio for hydrogen is approximately three times that of natural gas and sixteen times that of gasoline (Table 2.9). To have an explosion would require a build-up of hydrogen to about 18% without ignition occurring first. For natural gas, it would be sufficient to reach about 6% and for petrol it would be enough to reach about 1%. Furthermore, in the event of an explosion, hydrogen has the lowest explosive energy per unit of energy for the same volume (Table 2.9).

Combustion of hydrogen produces water vapour and neither hydrogen nor water vapour is toxic. However, its combustion in air also involves the production of nitrogen oxides and, therefore, the combustion products are toxic.

However, the toxicity of a fuel, as well as that of its combustion products, increases as the carbon/hydrogen ratio increases. Therefore, in terms of toxicity, the use of hydrogen is much more favourable than that of other fuels.

From an environmental point of view, the combustion of hydrogen only produces water. However, if the oxidising used is air, nitrogen oxides are also formed due to the presence of oxygen and nitrogen in the air. The amount of nitrogen oxides formed depends essentially on the temperature reached during the combustion process and can therefore be controlled by using lean mixtures. In particular cases, traces of particulate matter, unburned hydrocarbons, carbon and sulphur oxides can also be found: for example in the case of use in internal combustion engines due to the partial combustion of lubricating oil.

Although the use of hydrogen in a fuel cell avoids even the modest emissions just mentioned, it must not be forgotten that, in almost all cases, it is an energy carrier and not a primary energy source. For this reason, in evaluating the environmental aspects of its use, the entire life cycle of the fuel must be taken into consideration, starting from the primary energy source up to the final application.

References

[1] Bain A. *Hindenburg: Exploring the Truth*. Melbourne, FL: Blue Note Publications; 2014.

[2] Marchetti C. *Transport and storage of energy [Internet]*. 1975. Available from: https://pure.iiasa.ac.at/id/eprint/222/1/RR-75-038.pdf.

[3] Marchetti C. Primary energy substitution models: on the interaction between energy and society. *Technol Forecast Soc Change*. 1977;10(4):345–56.

[4] Sand M, Skeie RB, Sandstad M, *et al.* A multi-model assessment of the Global Warming Potential of hydrogen. *Commun Earth Environ*. 2023;4:203.

[5] Darling D. Terrestrial abundance of elements [Internet]. 2016 [cited 2023 Aug 16]. Available from: https://www.daviddarling.info/encyclopedia/E/elterr.html.

[6] Science Notes. Composition of the universe – element abundance [Internet]. 2022 [cited 2023 Nov 27]. Available from: https://sciencenotes.org/composition-of-the-universe-element-abundance/.

[7] Periodictable.com. Periodic table of the elements [Internet]. 2017 [cited 2024 Feb 8]. Available from: https://periodictable.com/.

[8] Hord J, U.S. Department of Commerce, and National Bureau of Standards. Is hydrogen safe? NBS technical note 690 [Internet]. Boulder, Colorado; 1976. Available from: https://www.govinfo.gov/content/pkg/GOVPUB-C13-97d5571e29c3a5137645bfe890822f26/pdf/GOVPUB-C13-97d5571e29c3a5137645bfe890822f26.pdf.

[9] Perry RH, and Green DW. *Perry's Chemical Engineers' Hanbook*. New York: McGraw-Hill; 1998. 2582 p.

[10] Roder HM, Childs GE, McCarthy RD, and Angerhofer PE. *Survey of the properties of the hydrogen isotopes below their critical temperatures – NBS technical note 641 [Internet]*. Boulder, Colorado; 1973. Available from: https://www.govinfo.gov/content/pkg/GOVPUB-C13-5119986d63f44a7b21dba4231f3465d4/pdf/GOVPUB-C13-5119986d63f44a7b21dba4231f3465d4.pdf.

[11] Hust JG, and Stewart RB. *A compilation of the property differences of ortho and para hydrogen or mixtures of ortho and para hydrogen – NBS report 8812 [Internet]*. Boulder, Colorado; 1965. Available from: https://ntrs.nasa.gov/citations/19660003795.

[12] McIntosh GE. Applications of ortho-para hydrogen catalyst. *IOP Conf Ser Mater Sci Eng*. 2015;101:012079.

[13] Atkins P, and de Paula J. *Atkins' Physical Chemistry [Internet]*. 8th edn. Atkins P, and de Paula J (eds). New York: Oxford University Press, W. H. Freeman and Company; 2006. Available from: https://www.rnlkwc.ac.in/pdf/study-material/chemistry/Peter_Atkins__Julio_de_Paula__Physical_Chemistry__1_.pdf.

[14] HySafe. Biennal report on hydrogen safety – chapter I: hydrogen fundamentals [Internet]. 2006. Available from: http://www.hysafe.org/download/1040/BRHS_Chap1_Fundamentals-version 1_0_1.pdf.

[15] Barbir F. Safety issues of hydrogen in vehicles. Energy partners [Internet]. 1999;1–5. Available from: https://courses.engr.illinois.edu/npre470/sp2019/web/readings/Hydrogen safety issues.pdf.

Chapter 3

Production of hydrogen

Hydrogen in free form is present on our planet in insignificant quantities compared to its presence in combined form. Recent discoveries of underground deposits of geological hydrogen [1], readily referred to as 'white hydrogen' or 'gold hydrogen', make us dream of a future in which it will not be necessary to extract it from water or hydrocarbons. The current prevailing theory [2] believes that there is a continuous underground production of hydrogen caused by the reaction between water and iron-rich minerals such as olivine and pyroxene. This reaction is characteristic of the geological process of serpentinisation which transforms olivine (3.1) or pyroxene (3.2) into serpentine, iron oxide (and silicon oxide in the case of pyroxene) by sequestering the oxygen from the water and releasing the hydrogen:

$$2Fe_{0.5}Mg_{1.5}SiO_4 + \frac{7}{3}H_2O \Rightarrow Mg_3Si_2O_5(OH)_4 + \frac{1}{3}Fe_3O_4 + \frac{1}{6}H_2 \tag{3.1}$$

$$2Fe_{0.5}Mg_{1.5}Si_2O_6 + \frac{7}{3}H_2O \Rightarrow Mg_3Si_2O_5(OH)_4 + \frac{1}{3}Fe_3O_4 + 2SiO_{2(aq)} + \frac{1}{6}H_2 \tag{3.2}$$

Hydrogen rises to the surface and disperses or encounters an impermeable layer and accumulates.

However, at present there is no certainty that deposits located at economically reasonable depths are sufficient to cover the entire energy requirement. Therefore, hydrogen must be produced starting from a primary source of energy and from a substance that contains it which, consequently, can release it. All primary energy sources can feed hydrogen production processes, possibly indirectly through intermediate production of other substances or electricity. And there are many substances in nature that contain hydrogen, for example water and all organic compounds.

In both cases, it is a question of checking also availability. In terms of primary energy sources, solar energy is undoubtedly the dominant one, while water certainly stands out among the substances containing hydrogen. Table 3.1 lists the main hydrogen production processes, highlighting both the primary sources of energy and the substances that make hydrogen available. As can be seen, the versatility of hydrogen in terms of production is extremely vast and this allows the primary sources of energy to be diversified based on what are the actual

Table 3.1 Main hydrogen production processes

	Energy source	Hydrogen donor
Coal gasification	Fossil	Water/coal
Hydrocarbons partial oxidation	Fossil/renewable	Water/hydrocarbons
Hydrocarbons reforming	Fossil/renewable	Water/hydrocarbons
Kvaerner process	Fossil/nuclear/renewable	Hydrocarbons
Thermolysis	Solar/nuclear	Water
Electrolysis	Fossil/nuclear/renewable	Water
Plasma-driven electrolysis	Fossil/nuclear/renewable	Organic compounds/solid waste
Photoelectrolysis	Solar	Water
Photosynthesis	Solar	Water
Biomass gasification	Solar	Water/biomass
Waste gasification	Fossil/renewable	Water/waste materials
Biomass anaerobic digestion	Solar	Water/biomass
Biomass fermentation	Solar	Water/biomass
Water/hydride reactions	Hydride	Water/hydride

availability. Some have been used on a large scale for some time, others are already commercially mature but on a small/medium scale, and still others are in the experimental phase.

Using hydrogen as an energy carrier results in its conversion into water and therefore the production of hydrogen from water originates a globally cyclical process. In reality, there may be 'disturbances' to the ideality of this process, such as the production of nitrogen oxides if air is used to burn hydrogen. However, these are side effects that are not always present (i.e. using a fuel cell) and in any case much less impactful than the extraction of hydrogen from other substances. In fact, in the event that the production of hydrogen is not carried out starting from water, or at least not from water alone, by-products and/or waste materials will be obtained during the production process. The different effects of this different hydrogen value chains on the environment are therefore evident. Starting from water to produce hydrogen and producing water to obtain energy from hydrogen without waste or by-product is a circularity behaviour feasible with hydrogen as energy carrier.

In this regard, however, this concept of cyclicality should not be confused with the concept of renewability of sources: electrolysis can generate hydrogen from water, but it is not certain that the electricity used is of renewable origin.

The renewability of the primary energy source used is, therefore, a second evaluation criterion and, in this regard, there may be non-renewable processes (e.g. coal gasification), renewable processes (e.g. photosynthesis), processes that can be renewable or not based on the source used (e.g. electrolysis) and also processes that can, under certain conditions, be hybrid (e.g. Kvaerner process).

It follows that hydrogen production processes starting from water and renewable energy sources must be considered as preferential ones. However, it must be borne in mind that drinking water is a commodity of primary importance for humanity and it will be more and more scarce in the following years. Extracting hydrogen from drinking water is a regrettable option. Therefore, it will be essential to develop those technologies that will make it possible to obtain hydrogen from soft not drinkable water or seawater that obviously is the most abundant water on the Earth. Soft not drinkable water is water from river or wells not drinkable for biological or human contamination reason or water from purification of urban, breeding or industrial sewage.

3.1 Hydrogen from hydrocarbons

As already mentioned, hydrogen is mainly produced from hydrocarbons. Hydrocarbons are compounds made up of hydrogen and carbon and, therefore, are simultaneously a primary source of energy and a source of hydrogen atoms.

A first distinction in the production of hydrogen from fossil fuels, and also from lignocellulosic biomass, is that between non-oxidative and oxidative processes.

A process is non-oxidative if there is no external supply of oxygen: hydrogen is then produced by extracting the one already present into the fuel through a decomposition process powered by an external energy source (allothermic process).

The alternative to decomposition consists in the use of substances capable of oxidising the carbon present in the fuel. Oxidative processes differ both on the basis of the nature of the fuel used, solid or fluid, and on the type of substance used to supply oxygen. The preferred source of oxygen, in general, is water vapour. More rarely, air and even oxygen and carbon dioxide are used. The reason is obvious: water, by yielding oxygen to carbon, increases the production of hydrogen.

The oxidative process that produces hydrogen starting from hydrocarbons and their oxygenated derivatives (alcohols and phenols) is generically called reforming, but substantially different processes are grouped under this name.

Oxidation with a stoichiometric quantity of oxygen would lead to an almost complete combustion reaction and, therefore, it is necessary to remain below the stoichiometric proportion to avoid oxidising all the hydrogen as well. In this case, we speak of partial oxidation or catalytic partial oxidation:

$$C_nH_m + \frac{n}{2}O_2 \Rightarrow nCO + \frac{m}{2}H_2 \tag{3.3}$$

An alternative is to oxidise the carbon using the oxygen present in the water in such a way as to add to the hydrogen present in the hydrocarbon also that present in the water. In this case, we speak of steam reforming:

$$C_nH_m + nH_2O \Rightarrow nCO + \left(\frac{m}{2} + n\right)H_2 \tag{3.4}$$

While partial oxidation is an exothermic process, steam reforming is an endothermic process: by thermally balancing the two processes, a third is obtained called autothermal reforming.

Furthermore, carbon dioxide can also be used, which is reduced to monoxide by supplying oxygen to other carbon atoms:

$$C_nH_m + nCO_2 \Rightarrow 2nCO + \frac{m}{2}H_2 \tag{3.5}$$

In this case, we speak of reforming with carbon dioxide or dry reforming.

The above reactions are ideal, but the gas obtained is a mixture that also contains carbon dioxide and water. Furthermore, in rather modest quantities, hydrocarbons and compounds of sulphur and nitrogen may be present. To obtain pure hydrogen, it is therefore necessary to subject this gas to more or less complex treatments depending on the fuel used and the pressure and temperature conditions at which the reforming process took place.

3.1.1 Steam reforming

Steam reforming (3.4) is an endothermic process that takes place at a high temperature and is mainly used for the conversion of gaseous light hydrocarbons.

The number of final moles is higher than that of the initial moles and, therefore, the process is disadvantaged by high pressures. However, the convenience of having more compact reformers is such that pressures in the 2–3 MPa range are often used.

In order to favour the conversion of hydrocarbons, it is advisable to use an overstoichiometric amount of vapour (usually in the range from 3n to 5n). This excess of steam is also useful in the subsequent phase of water gas shift (WGS) reaction:

$$CO + H_2O \Rightarrow CO_2 + H_2 \qquad \Delta H^\circ = -40.8 \text{ kJ/mol} \tag{3.6}$$

which, at lower temperature, allows to increase the hydrogen production.

Combining (3.4) and (3.6), we obtain:

$$C_nH_m + 2nH_2O \Rightarrow nCO_2 + \left(2n + \frac{m}{2}\right)H_2 \tag{3.7}$$

As can be seen from this last reaction, the steam reforming makes use of a significant contribution of the hydrogen contained in the water: in the case of methane, which is the hydrocarbon with the lowest n/m ratio (1:4), the theoretically generated hydrogen derives half from hydrocarbon and half from water.

The main reaction is endothermic and requires heat supplied from the outside. Therefore, the reformer is substantially a heat exchanger with a tube bundle which is the site of chemical reactions favoured by catalysts. The area outside the tube bundle is equipped with burners that develop the heat necessary for the process. Since there is no exchange of matter with the main reaction site, the fuel used can be the same as the feed or a different fuel. The exhaust gas, consisting mainly of nitrogen, oxygen, carbon oxides and water vapour, has a still high temperature, around 1000 °C, and is therefore used to produce and superheat steam and to preheat the reactants and the combustion air. The final exhaust gas temperature can be lowered to around 150 °C.

Unlike what happens in a normal heat exchanger, the tubes that are inside the reformer are internally coated with a catalyst. Such a catalyst contains from 5% to 25% nickel present in the form of nickel oxide. The oxide is generally supported on a refractory material such as aluminium oxide or calcium aluminate. The catalyst also contains potassium and calcium to inhibit the formation of solid carbon and its consequent deposition on the catalyst itself. In continuous operation, the catalyst can remain active for about 50,000 h, while its duration is reduced if it is subject to frequent switching on and off due to thermal stress.

Approximately half of the hydrogen produced today comes from the steam reforming of natural gas. Natural gas is a gas coming from spontaneous emission, drilling or physicochemical underground processing (i.e. fracking) and mainly made up of methane. However it also contains variable, and sometimes substantial, quantities of other hydrocarbons (mainly light), inert gases (especially carbon dioxide) and sulphur compounds. The composition of natural gas depends on the region of extraction, and in Table 3.2 the composition ranges for the main components are indicated [3].

The sulphur compounds must be eliminated upstream of the steam reforming process as the sulphur poisons the nickel catalysts used to promote the kinetics of the reaction. There is therefore a pre-treatment with hydrogen at 300 °C–400 °C in the presence of a cobalt–molybdenum catalyst in such a way that all sulphur compounds present are converted into dihydrogen sulphide. The latter is then removed at temperatures of 340 °C–390 °C by adsorption on a bed of zinc oxide:

$$ZnO + H_2S \Rightarrow ZnS + H_2O \tag{3.8}$$

The heavier hydrocarbons present in natural gas could lead to the production of solid carbon, also capable of deactivating the nickel catalysts. For this reason, the reforming process can be preceded by a pre-reforming that converts these hydrocarbons, extending the life of the catalysts and allowing greater flexibility in relation to the power supply (e.g. naphtha and petrol can also be treated). Since non-methane hydrocarbons are more reactive, pre-reforming can be performed at lower temperatures, around 500 °C, and this allows heat to be recovered within the plant itself to generate steam suitable for feeding the pre-reforming reaction. This opportunity to increase heat recovery within the plant itself allows the reformer to reduce fuel demand. At such a temperature, the only residual hydrocarbon is methane and, precisely for this reason, steam reforming of natural gas is often referred to simply as steam methane reforming.

Table 3.2 Composition of natural gas

CH_4	C_2H_6	C_3H_8+C_4H_{10}	H_2S	CO_2
60%–90%	0%–20%	0%–20%	0%–5%	0%–8%

Once the non-methane hydrocarbons have been eliminated, there are two reactions that could lead to the formation and deposition of solid carbon, namely the Boudouard reaction:

$$C + CO_2 \Rightarrow 2CO \quad \Delta H^\circ = 159.7\ \text{kJ/mol} \tag{3.9}$$

and the decomposition reaction of methane:

$$CH_{4(g)} \Rightarrow C_{(s)} + 2H_{2(g)} \quad \Delta H^\circ = 74.95\ \text{kJ/mol} \tag{3.10}$$

for which, however, the temperature is too low.

The first reaction is favoured at temperatures below about 700 °C, while for higher temperatures, the reverse reaction is favoured. Therefore, it is disadvantaged inside the reformer, which in the case of methane generally operates at 800–900°C. However it could occur downstream of it when the gas is cooled: in this case, however, it is in competition with the WGS reaction (3.6) and the latter is favoured by the presence of a considerable quantity of water vapour.

In the case of methane, the generic steam reforming reactions [(3.4) and (3.7)] specialise, respectively, in the following:

$$CH_4 + H_2O \Rightarrow CO + 3H_2 \quad \Delta H^\circ = 206.3\ \text{kJ/mol} \tag{3.11}$$

$$CH_4 + 2H_2O \Rightarrow CO_2 + 4H_2 \quad \Delta H^\circ = 216.5\ \text{kJ/mol} \tag{3.12}$$

These two reactions determine the conversion of methane, which is completed at a temperature between 800 °C and 900 °C, obviously in the presence of a suitable quantity of steam. The gas produced by the reformer is therefore composed of hydrogen, water vapour and carbon oxides and the composition is regulated by the WGS reaction (3.6), i.e. by the excess of vapour introduced with respect to the stoichiometric quantity. At these temperatures, the equilibrium of the WGS reaction does not change significantly; therefore, for temperatures higher than that of completion of the methane conversion, there is only a slight increase in the molar fractions of water and carbon monoxide at the expense of those of hydrogen and carbon dioxide.

To convert the entire chemical energy content into hydrogen, it is necessary to convert the carbon monoxide by shifting the equilibrium of (3.6) to the right. This is achieved by cooling the gas and accelerating the reaction with suitable catalysts. Usually, the WGS process is carried out in two separate units that operate at different temperatures and with different catalysts: the first unit at 350 °C–400 °C with iron oxide catalyst and the second unit at 200 °C–215 °C with catalyst at zinc–copper supported on an aluminium oxide matrix. A typical (dry) gas composition at the end of the process is as follows: 77% H_2, 18% CO_2, 0.3% CO and 4.7% CH_4.

At this point, if the carbon monoxide and methane levels are acceptable for end use, all that is needed is to remove carbon dioxide (see Section 3.3).

Figure 3.1 shows the scheme of a steam reforming plant for hydrogen production.

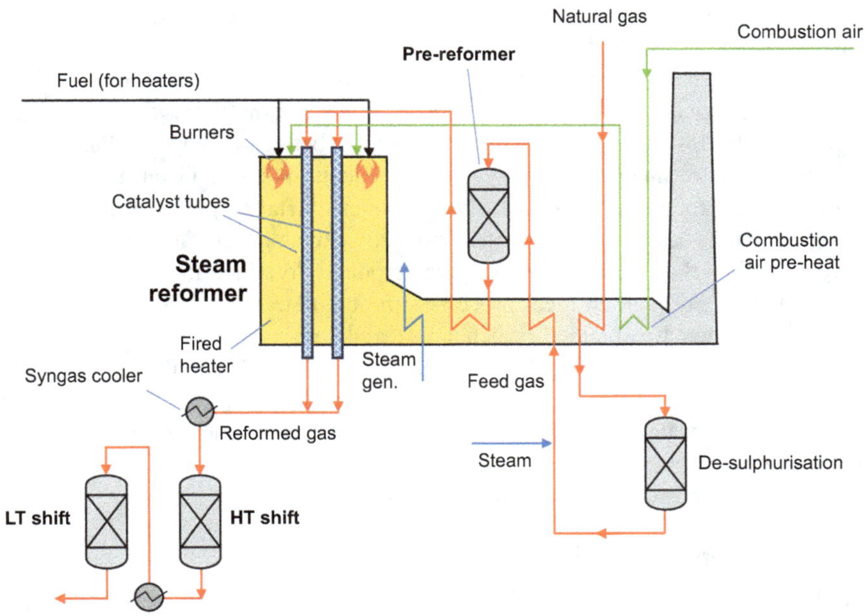

Figure 3.1 Steam reforming plant for hydrogen production

3.1.2 Partial oxidation

Partial oxidation (3.3) is mainly used for the conversion of liquid heavy hydrocarbons. In the ideal process with oxygen in a stoichiometric ratio, if the hydrocarbon is oxygenated the amount of oxygen required is reduced but the final products remain unchanged:

$$C_nH_mO_k + \frac{n-k}{2}O_2 \Rightarrow nCO + \frac{m}{2}H_2 \tag{3.13}$$

In reality, the process is much more complicated, both due to the formation of intermediate compounds and because the feed does not consist of a single pure hydrocarbon. A better schematisation, albeit also simplified, can be the one that also considers the other main reactions: oxidation of carbon monoxide (3.14), oxidation of hydrogen (3.15), WGS reaction (3.6), steam reforming (3.16) and dry reforming (3.17):

$$CO + \frac{1}{2}O_2 \Rightarrow CO_2 \quad \Delta H^\circ = -282.8\ \mathrm{kJ/mol} \tag{3.14}$$

$$H_2 + \frac{1}{2}O_2 \Rightarrow H_2O \quad \Delta H^\circ = -242.0\ \mathrm{kJ/mol} \tag{3.15}$$

$$C_nH_mO_k + (n-k)H_2O \Rightarrow nCO + \left(n-k+\frac{m}{2}\right)H_2 \tag{3.16}$$

$$C_nH_mO_k + (n-k)CO_2 \Rightarrow (2n-k)CO + \frac{m}{2}H_2 \tag{3.17}$$

The combustion reactions are strongly exothermic, while the WGS reaction [6] is weakly exothermic and the steam and dry reforming reactions are strongly endothermic. Overall, the process is strongly exothermic. In the case of a catalytic process, the temperature is in the range 600 °C–800 °C; otherwise, it operates in the range 1100 °C–1500 °C. The pressure is generally between 2 and 6 MPa.

Compared to steam reforming, partial oxidation has a faster kinetics and therefore does not require catalysts and combustion takes place entirely in the reactor which is more compact since no heat exchange surfaces are required. However, it has the disadvantage of lower productivity since the steam supply is lacking: for example, in an ideal process with complete oxidation of carbon, four moles of hydrogen are obtained from a single mole of methane in the case of steam reforming and two moles of hydrogen in the case of partial oxidation.

Therefore, thanks to the greater flexibility towards the fuel and the greater compactness, partial oxidation could be more suitable for mobile applications in comparison with steam reforming, while the greater efficiency of steam reforming makes the latter preferable for stationary applications.

3.1.3 Autothermal reforming

Autothermal reforming consists of a reforming process in which exothermic reactions occurring inside the reactor provide the whole amount of heat required by the contemporaneous endothermic reactions. In other words, by adjusting the quantities of oxygen and steam supplied, the partial oxidation is made in thermal equilibrium with the steam reforming. The overall effect is that of a compromise between the advantages of the two processes discussed above. In fact, since the heat is developed internally it is more compact, but requires less molecular oxygen than partial oxidation, being able to extract oxygen even from the steam. It is particularly suitable for small sizes applications.

3.1.4 Pyrolysis

Pyrolysis is a non-oxidative endothermic process which consists of a thermochemical decomposition carried out through a series of rather complex radical reactions which progressively lead to increasingly lighter radicals and hydrocarbons. As an example, we report one of the countless possible pyrolysis paths of pentane (the three points indicate that it is a radical):

$$2C_5H_{12} \Rightarrow 2C_3H_7\cdots + 2C_2H_5\cdots \Rightarrow 2CH_2\cdots + 2C_2H_5\cdots + C_4H_{10}$$

$$\Rightarrow 2C_2H_4 + C_2H_6 + C_4H_{10}$$

Hydrocarbons consist only of carbon and hydrogen and therefore in the ideal pyrolysis process for hydrogen production there is no formation of carbon oxides, but only the production of gaseous hydrogen and solid carbon:

$$C_nH_m \Rightarrow nC + \frac{m}{2}H_2 \tag{3.18}$$

Since the reaction is endothermic, an increase in temperature favours the conversion of hydrocarbons. However, an increase in pressure is unfavourable since the decomposition of the hydrocarbons themselves involves an increase in moles and, in particular, in the number of moles in the gaseous phase.

The best-known pyrolytic technology, called CB&H (carbon black & hydrogen), is the one developed in the 1980s by the Norwegian company Kvaerner (today Aker Solutions) using a plasma torch at around 1600 °C. A plant using this process was built in Bergen (Norway) for the decomposition of methane (3.10).

The plant essentially consists of a reactor in which hydrogen is ionised (plasma) by means of an electric arc so that the reaction receives the required heat. Once started, the process is kept in place by supplying methane and electricity in sufficient quantities and by recirculating an adequate flow of hydrogen for the formation of the plasma [4]. The methane is preheated at the expense of the heat of the products and the excess heat is used to generate steam. The process is very efficient since about 98% of the energy used is recovered in the form of hydrogen (48%), carbon (40%) and superheated steam (10%). Obviously, taking into account the fact that the carbon obtained should not be used as a fuel, but as a raw material, the energy available at the end of the process is reduced to less than 60%. However, the sale of carbon, used mainly in the production of tires, but also in other industrial sectors, contributes significantly to the reduction of the cost of hydrogen.

The CB&H process has the following advantages over other hydrogen production processes:

- low costs for the production of hydrogen (even competitive with the steam reforming of natural gas depending on the price of carbon black);
- low emissions because the impurities other than any residual hydrocarbons are of the order of ppm and there are no carbon dioxide emissions;
- compactness of the system thanks to the high energy density associated with the plasma;
- high flexibility with regard to power supply as all hydrocarbons and their mixtures can be used with the sole consequence of a different ratio between hydrogen and carbon produced.

3.2 Gasification of solids

Coal and lignocellulosic biomass are composed by the main chemical species. Coal is composed mainly of carbon (50%–95%), oxygen (3%–40%), hydrogen (2%–6%) with chlorine, nitrogen sulphur and silicon; similarly biomass is generally composed of carbon (around 50%), oxygen (around 40%), hydrogen (around 6%) with chlorine, nitrogen sulphur and silicon [5].

Gasification has been a known process for at least a couple of centuries and has been used in gas generators to produce the so-called ‘air gas’, consisting essentially of carbon monoxide and nitrogen, and ‘water gas’, essentially consisting of monoxide carbon and hydrogen. It is used in the chemical industry (e.g. production of ammonia), in the production of gaseous and liquid fuels (e.g. methanol) and in the generation of electricity.

The use of natural gas has reduced its importance, but the availability of coal in quantities greater than other fossil fuels, combined with the need to limit the environmental local impact deriving from the use of coal itself, has awakened interest in this technology. This happened not only with the aim of generating electricity but also to produce hydrogen for use in sectors where the use of a fuel is preferable to that of electricity. In 2016, there were 117 coal and/or solid waste gasification plants in the petrochemical industry in the world and about 35 additional facilities in various stages of development, design, and construction [6]. Gasification is a thermochemical conversion process of a solid fuel into a combustible gas. The reactor in which this occurs is called a gasifier. The solid fuel fed to the gasifier is converted into gas reacting with water vapour thanks to the heat supplied by incomplete oxidation with air or oxygen or coming from an external heat source.

In the case of biomass, the interest in gasification was first born to make up for the lack of fossil fuels, but currently it is mainly connected to its neutrality towards climate change. In fact, the carbon dioxide emitted in the conversion of biomass generally does not contribute to the greenhouse effect if the time of biomass growth is comparable with the time of its consumption (around a few years). In this condition using biomass is part of the natural cycle and the emission of carbon dioxide is equal to the amount removed from the atmosphere during the growth of the biomass itself, although it is important to consider the time for growing in comparison to the time of consumption [7]. In other words, the use of biomass does not automatically imply that its production, conversion and use are sustainable. Using biomass from a century-old forest is not a sustainable mode. Another interest regarding the gasification of biomass is the realisation of a negative carbon emission (CO_2 capture and storage) by production of biochar that can be used as soil amendment or other application as construction material [8,9]. However, if the ultimate goal is to obtain hydrogen, this is a relative advantage since the entire synthesis gas purification process, including the separation of carbon dioxide, would have to be carried out anyway. It follows that, especially for small-sized gasifiers, the direct use of synthesis gas seems convenient compared to the production of hydrogen.

The gasification process, schematised in Figure 3.2, is quite complex, but the sub-processes it comprises are essentially the following:

- dehumidification;
- pyrolysis;
- proper gasification;
- combustion.

In case the gasifier is a counter-current type, the hot syngas is able to dehumidify the fuel before exiting from the reactor. Otherwise, it could be necessary to adjust the humidity of the fuel before supplying it to the gasifier.

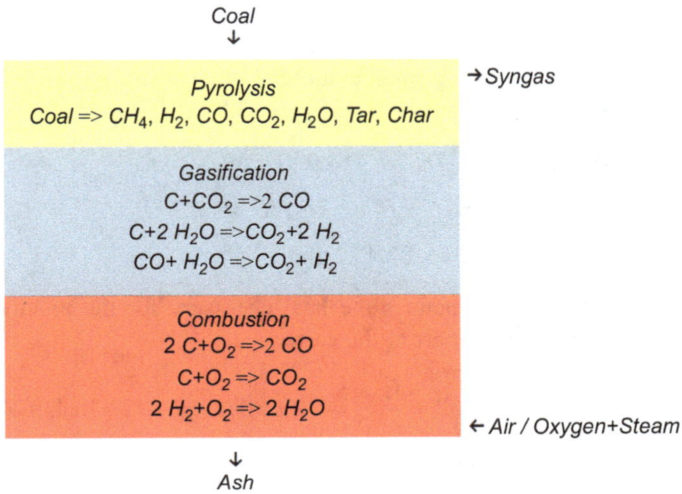

Figure 3.2 Flows and processes in a gasifier

In the case of solid fuels (coal and biomass), the final product of pyrolysis consists of:

- a solid carbonaceous residue, poor in hydrogen and oxygen, called char;
- a liquid phase at room temperature, but gaseous at the process temperature, called tar;
- a gaseous phase consisting of carbon oxides, hydrogen and light hydrocarbons.

The distribution among the three phases essentially depends on the speed and temperature of the process. In fact, we speak of:

- slow pyrolysis, at a temperature of 550 °C–950 °C and a residence time of the order of minutes, through which heavy hydrocarbons are generated and the size of the solid particles can be in the range of 5–50 mm;
- fast pyrolysis, at a temperature of 850 °C–1250 °C and a residence time of the order of seconds, which generates lighter hydrocarbons with solid particle generally smaller than 1 mm;
- flash pyrolysis, at a temperature of 1050 °C–1300 °C and a residence time of the order of tenths of a second, which generates light hydrocarbons with solid particle generally smaller than 0.2 mm.

As the speed increases, the solid and liquid fractions are reduced, while the gaseous fraction increases. In the case of gasification, pyrolysis takes place in an anoxygenic zone and mainly provides for the separation of the volatile material and the cracking of heavy hydrocarbons.

In the proper gasification phase, solid carbon reacts with steam (steam gasification) and carbon dioxide (3.9) to produce mainly hydrogen and carbon monoxide. Furthermore, numerous reactions occur as a consequence of the presence of different hydrocarbons that are formed during the process due to the reactions between carbon and hydrogen (hydrogasification). However, the hydrocarbons are reformed by the steam and only a small amount of methane remains. Therefore, the main reactions that define the whole process are (3.9) as follows:

$$C + H_2O \Rightarrow CO + H_2 \quad \Delta H^\circ = 118.9\ \mathrm{kJ/mol} \tag{3.19}$$

$$C + 2H_2O \Rightarrow CO_2 + 2H_2 \quad \Delta H^\circ = 78.1\ \mathrm{kJ/mol} \tag{3.20}$$

$$C + 2H_2 \Rightarrow CH_4 \quad \Delta H^\circ = -87.4\ \mathrm{kJ/mol} \tag{3.21}$$

The combustion phase generates the heat necessary for the previous phases mainly through oxidation of a part of the carbon:

$$\text{Carbon partial oxidation } C + \frac{1}{2}O_2 \Rightarrow CO \quad \Delta H^\circ = -123.1\ \mathrm{kJ/mol} \tag{3.22}$$

$$\text{Carbon direct oxidation } C + O_2 \Rightarrow CO_2 \quad \Delta H^\circ = -405.9\ \mathrm{kJ/mol} \tag{3.23}$$

and of the hydrogen (3.15) present into the fuel.

In the reactor zones at a lower temperature, again in the presence of an adequate amount of water vapour, the WGS reaction (3.6) takes place.

All the listed reactions involve only carbon, hydrogen and oxygen. The presence in coal or biomass of other substances gives rise to other reactions of secondary importance from an energy point of view. However, these reactions lead to the formation of undesirable products: for example sulphur and nitrogen cause the production of oxides (SO_x, NO_x), hydrogen cyanide (HCN), sulphides (H_2S and COS) and ammonia (NH_3). The contribution to heat generation is significant only if the sulphur content is high.

The gasification process is very versatile with respect to fuel: although most of the existing plants are fuelled by coal, plants fuelled with coke, biomass of various kinds, asphalt, tar, refining residues or mixtures of these fuels are in operation.

The classification of the gas produced can be carried out on the basis of the higher heating value (HHV) and therefore, indirectly, on the basis of the composition of the gas itself in relation to methane (39.130 MJ/Nm^3), carbon monoxide (12.600 MJ/Nm^3), hydrogen (12.742 MJ/Nm^3) and diluents:

- gas with high higher heating value (HHV=33.500–37.500 MJ/Nm^3), a product very similar to natural gas being essentially composed of methane with low percentages of hydrogen, carbon oxides and nitrogen;
- gas with medium higher heating value (HHV=9.400–20.600 MJ/Nm^3), characterised by a lower percentage of methane and a prevalent presence of hydrogen and carbon monoxide;
- gas with low higher heating value (HHV=3.500–9.400 MJ/Nm^3), characterised by a strong dilution with nitrogen and carbon dioxide.

In Figure 3.3, the full gasification process is schematised, highlighting the components and the type of gas obtained as the final product and in the intermediate stages.

<table>
<tr><td>fuel+air+steam</td><td rowspan="4">Gasification</td><td>raw gas
low heating value
CO - H_2 - CO_2 - N_2 - CH_4 - H_2S</td><td rowspan="4">Cleaning</td><td>clean gas
low heating value
CO - H_2 - CO_2 - N_2 - CH_4</td><td>No treatment</td><td></td></tr>
<tr><td rowspan="3">fuel+air+steam
or
fuel+steam+heat
or
fuel+hydrogen</td><td rowspan="3">raw gas
medium heating value
CO - H_2 - CO_2 - CH_4 - H_2S</td><td rowspan="3">clean gas
medium heating value
CO - H_2 - CH_4</td><td>No treatment</td><td>syngas</td></tr>
<tr><td>Methanation</td><td>CH_4</td></tr>
<tr><td>Reforming+
Water Gas Shift</td><td>H_2</td></tr>
</table>

Figure 3.3 Gasification paths and products

The main substances contained in the raw gas are carbon oxides, hydrogen and methane. Based on the sulphur content in the starting fuel, the presence of dihydrogen sulphide can be significant. If air is used as an oxidant, the raw gas produced will contain a high percentage of nitrogen and will have a modest heating value which makes it suitable, mainly, to be used as a fuel directly on site, possibly after a purification treatment. The use of pure oxygen, however, guarantees, albeit with high costs, a raw gas with medium calorific value which, after purification, is suitable for distribution or to be subjected to further treatments.

For the purposes of hydrogen production, it is not advisable to use air as an oxidant as it would then be necessary to separate a large amount of nitrogen. It is also preferable to limit the oxygen/vapour ratio: in fact the oxidation of a mole of carbon obtained using oxygen does not produce hydrogen (3.23) whereas using steam, two moles of hydrogen are obtained (3.20). The ideal would be to zero such a ratio by using an external heat source.

The operating temperature depends above all on the melting point of the ashes and is regulated by the amount of oxygen supplied, since it is on this quantity that the progress of the combustion depends. Using air, the temperature is usually around 800 °C–1000 °C, whereas with oxygen it is in the 1000 °C–1400 °C range. High temperatures favour the pyrolysis reactions, and therefore the conversion of methane and other hydrocarbons, and disfavour the WGS reaction (3.6), and therefore reduce the quantity of hydrogen in favour of carbon monoxide. The high pressure, however, favours the reduction of the number of molecules, i.e. the formation of methane and other hydrocarbons at the expense of hydrogen and carbon monoxide.

The biomass gasification technologies derive from those developed for the gasification of coal suitably modified to take into account the different characteristics of the fuel. Therefore, regardless of the fuel used, most gasifiers can be classified as indicated and characterised in Table 3.3 [10–12].

The fixed bed gasifiers operate usually with a discontinuous feeding of the fuel which can have very different granulometry. The discontinuity of the supply creates some problems in the case of pressurised gasifiers and limits the potential. Therefore, generally, they are used for small and medium-sized plants. The pressure can reach 10 MPa. Some initiatives have been done to create continuous mode fixed bed gasifiers [13,14].

Table 3.3 Kind of gasifiers and their main characteristics

Kind	Fixed bed	Fluidised bed	Entrained bed
Size of coal/biomass	Coarse (cm)	Average (mm)	Fine (<mm)
Presence of fine particles	Limited	Good	Unlimited
Residence time	Long (hours)	Medium (minutes)	Short (seconds)
Temperature of syngas	300 °C–700 °C	840 °C–1100 °C	1250 °C–1600 °C
Oxygen consumption	Low	Moderate	High
Steam flow required	High[a]/low[b]	Moderate	Low
Char and tar	High	Low	Trace
Ash condition	Dry/slagging	Dry/agglomerate	Slagging
Carbon conversion efficiency	89%–93%	90%–95%	98%–99.5%
Commercial examples	Lurgi BGL	HTW KBR U-Gas	GE Energy Shell E-Gas Siemens

[a]dry ash.
[b]slagging ash.

The fuel always enters from the top and the gasifiers are then classified according to the relative direction of the oxidant:

- counter-current fixed bed gasifiers (updraft gasifier);
- co-current fixed bed gasifiers (downdraft gasifier).

In the first case, therefore, the oxidant rises from below and the synthesis gas exits from the upper part of the gasifier (Figure 3.4). The fuel then encounters the outgoing hot gas and is first preheated and dehumidified and subsequently subjected to pyrolysis. The gas that emerges from the pyrolysis zone also carries tar vapours with it, part of which, due to the lowering of the gas temperature before leaving the gasifier, condenses and returns downwards to undergo a new pyrolysis process (tar-craking) and produce more gas. The char that leaves the pyrolysis zone descends into the gasification zone in which it encounters reducing gases at high temperatures that come from the underlying combustion zone. These gasifiers are constructively simple and produce a gas with a high heating value due to the presence of tar vapours.

In the case of co-current gasifiers, the oxidiser also goes downwards. Therefore, the incoming fuel is not dehumidified and must have a low humidity level at the start. The temperature of the gas generated is higher than in the counter-current case and, therefore, on the one hand the heating value is reduced, on the other hand the tar-cracking is more extreme and the gas cleaner.

The fluidised bed gasifiers are fed with small size fuel (1–5 mm) which is kept in suspension by the oxidant stream (steam mixed with air or oxygen) supplied from below. The motion inside the reactor mixes the reactants throughout the volume, eliminating temperature gradients. These gasifiers can operate at pressures up to a few MPa.

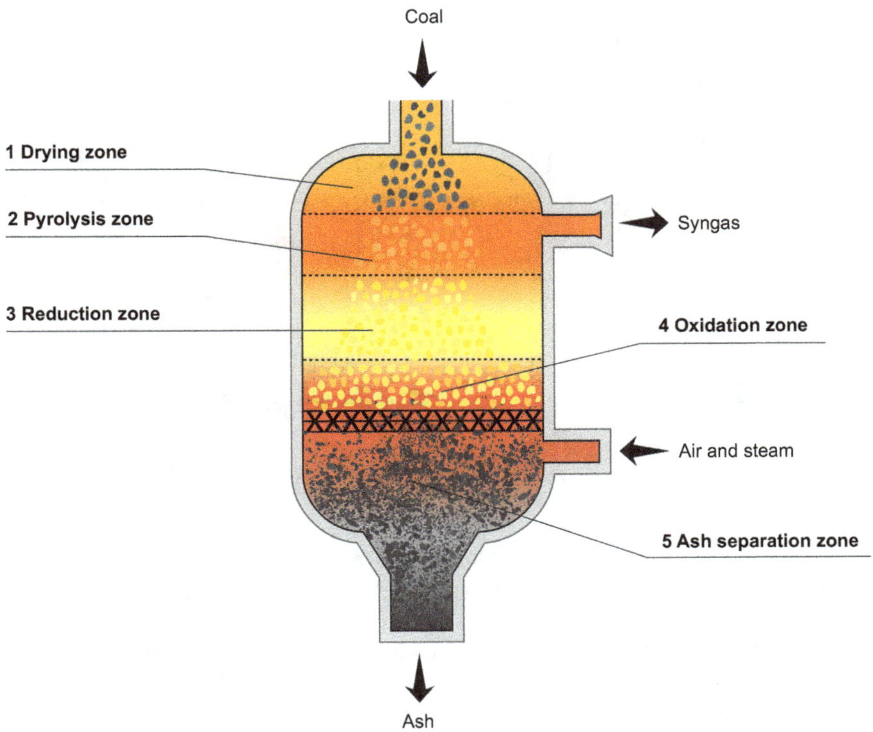

Figure 3.4 Schematic of counter-current fixed bed gasifier

In relation to the size of the fuel and the speed of the oxidiser, fluidised bed gasifiers are divided into:

- boiling fluidised bed gasifiers;
- circulating fluidised bed gasifiers.

In boiling fluidised bed gasifiers, the speed of the oxidant is between 0.7 and 2 m/s and the bed takes on the appearance of a boiling liquid overwhelmed by the gas generated. It is possible to introduce particular sorbents into the bed (generally limestone and dolomite) with the aim of directly desulphurising the gas:

$$CaCO_3 + SO_3 \Rightarrow CaSO_4 + CO_2 \quad (3.24)$$

$$MgCO_3 + SO_3 \Rightarrow MgSO_4 + CO_2 \quad (3.25)$$

The circulating fluidised bed technology is characterised by a fuel with a smaller grain size and higher oxidant speeds (3–5 m/s) so that there is no clear division between phases with different densities, but rather a reduction in progressive density. The solid particles transported by the gas are in greater quantities and therefore a partial recirculation of the gas is advisable in order to bring the

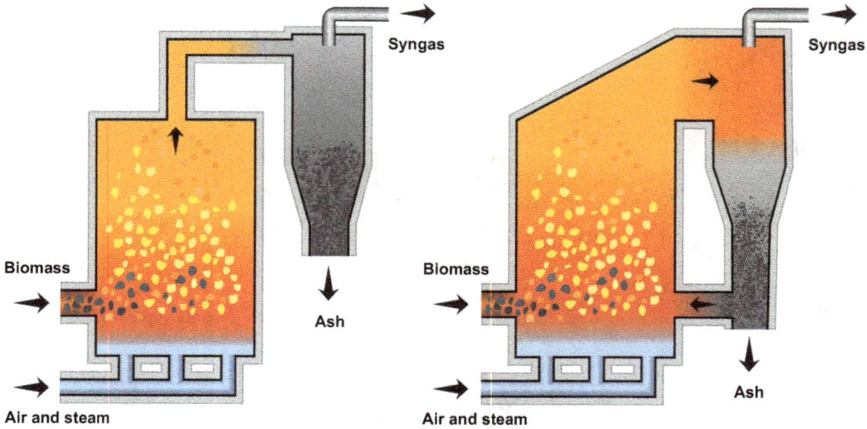

Figure 3.5 Schematics of boiling and circulating fluidised bed gasifier

entrained solid back into the bed. Due to the recirculation and the higher temperature, the cracking of the tar is more complete and therefore the degree of purity of the synthesis gas increases.

Figure 3.5 shows the schemes of boiling and circulating fluidised bed gasifiers.

The entrained bed gasifiers require a very finely pulverised fuel (<0.1 mm) which is fed into the reaction zone together with the oxidiser. There are two different types of feeding: wet feeding (e.g. Texaco) and dry feeding (e.g. Shell). In the first case, the fuel is pulverised, mixed with water (slurry) and pumped to the gasifier, while in the second case, the pulverised fuel is pressurised in a special hopper. A further distinction can be made between gasifiers with top feeding and gas extracted from below (e.g. Texaco and Shell) and gasifiers with bottom feeding and gas extracted from above (e.g. Dow Syngas).

The speed is very high and therefore the residence time is reduced. The pyrolysis phase occurs simultaneously with the gasification and combustion phases, that is at a very high temperature (>1200 °C): the synthesis gas is therefore poor in hydrocarbons and almost entirely made up of hydrogen, water, carbon monoxide and dioxide carbon.

Although any type of gasifier can be used for the production of hydrogen, the entrained bed technology is more suitable because it allows to minimise the production of tar and methane, while maximising those of hydrogen and carbon monoxide. By way of example, the composition of the dry gas produced is shown in Table 3.4 for different kinds of gasifiers [11,15,16]. As you can see, the difference in methane production between the first and the other two is considerable. Generally, methane is not converted in the treatments following gasification and therefore the most suitable technology for the production of hydrogen is the entrained bed technology. Comparing the two entrained bed gasifiers, Dow Syngas would seem more suitable for the production of hydrogen, but this is not the case. In fact, as the temperature drops, the WGS reaction (3.6) shifts its equilibrium

Table 3.4 Main components of a dry syngas generated by different gasifiers

Gasifier	H_2	CO	CH_4	CO_2	N_2	H_2S
Fixed bed air blown	16.2%–23.3%	17.1%–27.8%	1.7%–5.8%	5.9%–14.8%	38.5%–50.5%	0%–0.2%
Fixed bed oxygen blown	28.1%–42.3%	15.1%–61.2%	5.0%–12.1%	2.4%–32.4%	0.4%–3.3%	0.1%–1.3%
Fluidised bed air blown	12.6%–28.56%	12.54%–30.7%	0.2%–3.59%	6.4%–14.47%	47.0%–55.7%	(n.a.)
Fluidised bed oxygen blown	32.8%–40.0%	31.0%–53.0%	0.3%–3.5%	5.3%–19.5%	0.3%–1.7%	0.44%
Entrained bed air blown	18.4%–18.6%	29.7%–30.5%	<0.1%	10.1%–10.5%	36.4%–37.0%	<1 ppm
Entrained bed oxygen blown	25%–28%	47%–49%	1%–2%	20%–24%	–	–

towards the consumption of carbon monoxide and the production of hydrogen. Therefore, the potential production of the Texaco gasifier is higher and this advantage is added to an extremely low methane production.

In the case of biomass, entrained bed gasifiers are still at a modest stage of technological development [17,18] and, therefore, the most suitable gasifiers for the production of hydrogen are the circulating fluidised bed gasifiers which have higher operating temperatures and therefore produce a syngas with lower hydrocarbon content.

Hydrogasification is a particular gasification process that uses hydrogen as a gasifying agent. Its use is specifically aimed at the production of a syngas with a high methane content. It is obviously a process that consumes hydrogen and in general is not interesting for the purposes of hydrogen production. Nevertheless it can be used in some complex systems that also provide for the production of hydrogen and which will be mentioned in Section 3.4 or in power to methane systems (Section 5.3.1).

Next years will be characterised by a continuous growth in human population: IEA (International Energy Agency) estimated a rate around 1.18% each year [19]. Because of this by 2025, a waste production of six million tonnes/day is estimated [20]. The amount of energy lost through incinerators and landfill in 2012 was 893,000 GWh only in the EU [21]. Landfills create a double problem: soil degradation and methane dispersion with huge greenhouse effect.

Waste to hydrogen technologies are a way to manage the waste (municipal and industrial one) with thermochemical and biochemical methods, to produce directly hydrogen without use of combustion and thermodynamic cycle to produce electricity [22].

Thermochemical techniques include pyrolysis (350 °C–550 °C) and gasification (500 °C–1200 °C) with or without use of plasma. These techniques are based on coal and biomass technologies described above. Hydrogen yield is very dependent on feedstock and process parameters. Very close to commercialisation phase are Raven SR and Nextchem that have planned municipal waste plants in both USA and Europe to produce hydrogen for road transportation and industrial uses.

Researchers at Rice University have devised a way to produce hydrogen from mixed plastic waste while also obtaining graphene. The process consists of exposing the waste to rapid heating which takes it to 3100 K in about four seconds. For example, up to 68% of the hydrogen content can be recovered from polyethylene with a purity of 94%. Given the high cost of graphene, researchers have declared that the cost of producing hydrogen could even be eliminated [23,24].

Biochemical processes include the use of microorganism to decompose waste to produce hydrogen such as photo or dark fermentation. In this field, the systems are in the demonstrative stage.

Through a pyrolysis process, waste material, as well as biomasses, can be thermally decomposed at 400 °C–450 °C in an inert atmosphere, generating vapours that can be condensed into a bio-oil consisting of oxygenated organic substances (85%) and water (15%). These vapours can be subjected to steam reforming to obtain hydrogen. An interesting option is to install small pyrolysis units distributed throughout the territory and a single centralised steam reforming plant also equipped with infrastructures for the storage and distribution of hydrogen.

3.3 Cleaning technologies

Hydrogen production from fossil fuels, biomass and waste materials also generates undesired products and a cleaning process is required.

The gasification process also produces solid residues: ashes and particulates.

The ashes are eliminated directly inside the gasification unit and consist of different oxides, mainly SiO_2, Al_2O_3, Fe_2O_3, CaO, MgO. Iron oxide has a catalytic effect on the WGS reaction (3.6) and therefore favours the formation of hydrogen and carbon dioxide.

The combustion reactions are strongly exothermic and can cause the ashes to melt. The melting point of the ashes is the higher the higher the ratio:

$$\frac{m_{SiO_2} + m_{Al_2O_3}}{m_{Fe_2O_3} + m_{CaO} + m_{MgO}}$$

The particulate is instead entrained by the synthesis gas and its presence determines phenomena of deposit, erosion and corrosion in the equipment subsequently crossed by the gas. Therefore, it is necessary to carry out a separation immediately downstream of the gasifier. This separation is usually done by cyclones coupled with scrubbers [25,26].

Cyclones (Figure 3.6) exploit the action of centrifugal force which drives the solid particles to collide against the side surface of the cyclone and to slide downwards due to the gravity to be collected in a hopper. Their advantages are the absence of moving elements, the compactness, the low cost of installation and

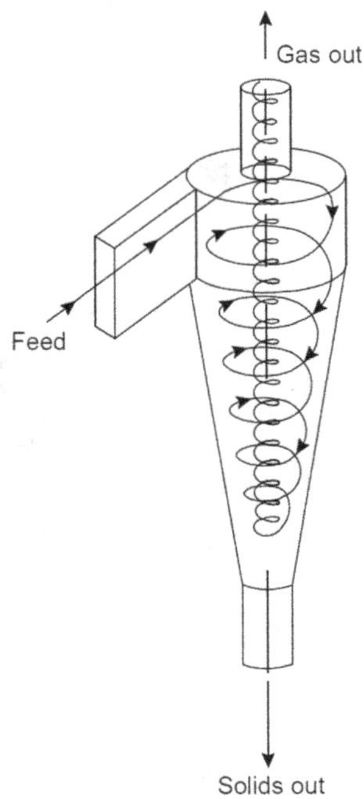

Figure 3.6 Schematic of a cyclone [27]

management, the continuous operation and the ability to operate within wide ranges of pressures and temperatures. The big disadvantage is related to the low removal efficiency (85%), which is strongly influenced by the dimensions of the particulate itself.

Scrubbers can be of various types and can also remove some gaseous impurities. First of all, they distinguish in wet scrubbers and dry scrubbers, depending on whether to operate the separation make use of water (or another liquid) or dry reactants (possibly in suspension). For particulate removal, wet scrubbers are used and, in particular, those based on the Venturi effect (Figure 3.7) which gives high speeds (60–100 m/s) to the gaseous stream and to the liquid. The liquid can be introduced together with the gas or injected near the throat, and is divided into small drops that can capture also solid particles of very small size. Scrubbers are characterised by very high particulate removal efficiency (99%) and they also achieve cooling and saturation of the gas.

Bag filters are a complementary technology. They are made with propylene or Teflon and reinforced with glass wool. The syngas is forced to pass through the filters which are capable to capture very small particles. They normally work with operating temperatures below 150 °C and with an efficiency that exceeds 99%. However, a high humidity level could make the dust too compact and make it difficult to clean the filters. This is a quite expensive technology both because of capital cost and operating costs (due to the limited life of the filters).

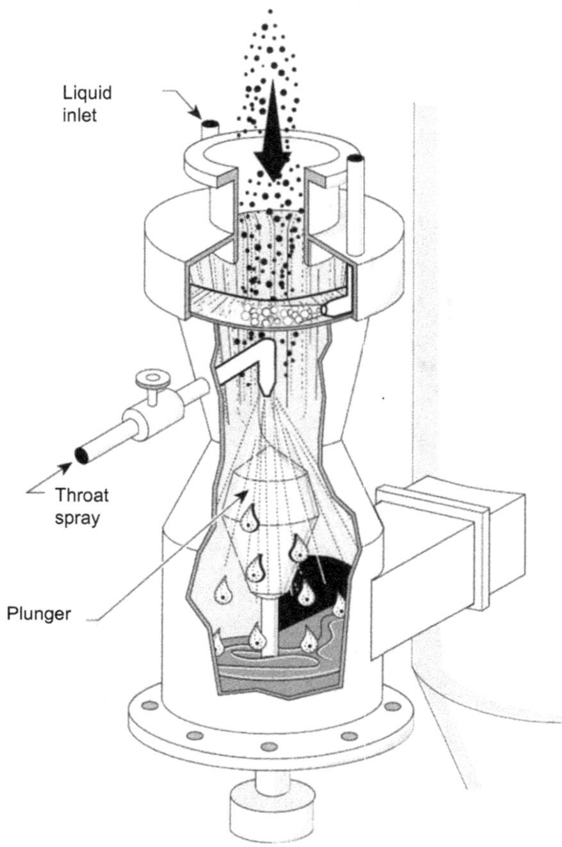

Figure 3.7 Schematic of a wet Venturi scrubber. This specific case shows an adjustable-throat Venturi scrubber with plunger [28].

The synthesis gas generated by the previous mentioned processes contains:

- a useful fraction, consisting of combustible substances; generally these are hydrogen, carbon monoxide and methane;
- a diluting fraction, consisting of nitrogen and oxidised non-polluting substances, such as water vapour and carbon dioxide;
- a contaminating fraction, consisting of those substances that can damage some components downstream of the main reactor or that are polluting for the environment.

This classification can vary depending on the final application: for example methane becomes a diluent for low temperature fuel cells and carbon monoxide even becomes a contaminant for proton exchange membrane fuel cells as it is capable of poisoning the platinum used as a catalyst (see Section 5.2).

The purification of the synthesis gas essentially consists in the elimination of the contaminating fraction which is present in very modest quantities and, therefore, does not involve a significant variation of the heating value. The unique important contaminant affecting also the heating value is tar which, condensing, could deposit and cause problems along the purification line. However, its presence varies with the gasification technology used. The most important contaminants are sulphur oxides, dihydrogen sulphide and carbonyl sulphide. Such acid gases can cause corrosion phenomena on other plant components and their emission constitutes a danger to the environment.

If the goal is the production of hydrogen, also methane, carbon monoxide and the diluting fraction must be eliminated. In this case, the purification of synthesis gas can be considered a real process of hydrogen production as it is a process that does not limit itself to increasing the concentration of hydrogen, but significantly increases its quantity thanks to the conversion of methane and carbon monoxide.

As regards the diluents, the water vapour is easily eliminated by condensation, while the separation of carbon dioxide can be achieved by various technologies. However, it is preferable to eliminate nitrogen upstream by feeding the gasifier with oxygen rather than with air. The scarce nitrogen eventually present in the fuel remains mainly at the molecular level, i.e. as a diluent, while its contaminating compounds are present only in traces.

Converting carbon monoxide is important for the production of hydrogen. The WGS reaction (3.6) takes care of this step. This reaction takes place at a low temperature and therefore requires the use of catalysts. Most suitable catalysts are sensitive to the presence of sulphur and this is why the WGS reaction (3.6) is not usually carried out upstream of the removal of acid gases. By employing sulphur-tolerant catalysts, it would be possible to convert the monoxide into carbon dioxide upstream of the removal of the acid gases and thus simultaneously remove the dihydrogen sulphide and carbon dioxide.

Gas separation can take place by means of hot systems or cold systems. Cold separation technology is a mature technology, and also hot systems reached a commercial phase [29,30]. Therefore, it could be necessary to cool the synthesis gas, an operation that entails considerable complications at the plant level and consequently increases costs. Figure 3.8 shows different techniques to obtain, for example, such a cooling downstream of an entrained bed gasifier:

- convective exchange preceded by radiant exchange;
- convective exchange preceded by quenching by direct mixing with water or cooled synthesis gas;
- simple quenching for direct mixing with water.

The first of these techniques allows for the production of high-pressure steam which can be used to generate electricity. The latter technique is instead less efficient, but more reliable and economical and particularly suitable for the production of hydrogen since the vapour produced inside the synthesis gas is subsequently used for the WGS reaction (3.6).

Figure 3.9 shows different options for gas separation.

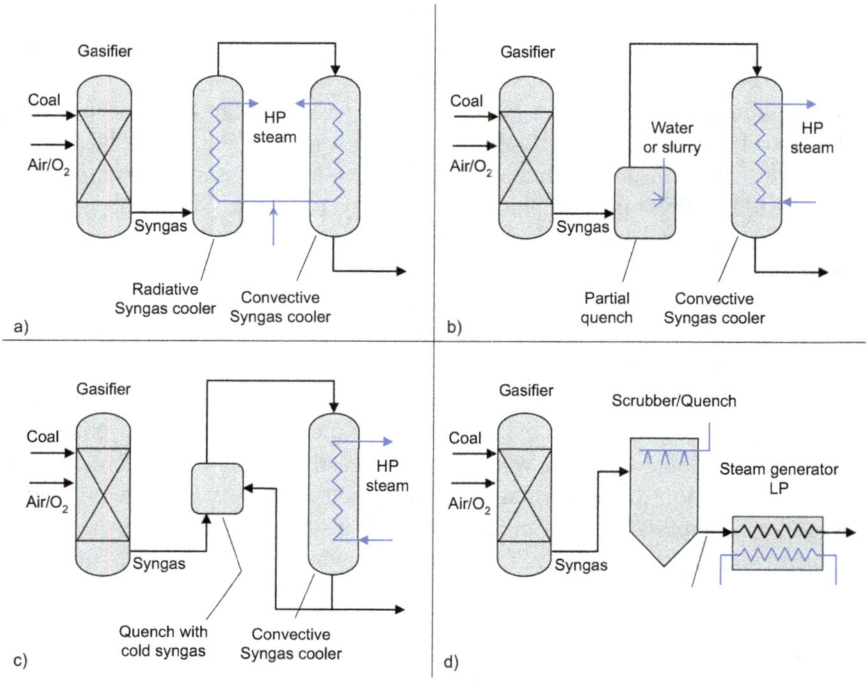

Figure 3.8 Syngas cooling technologies

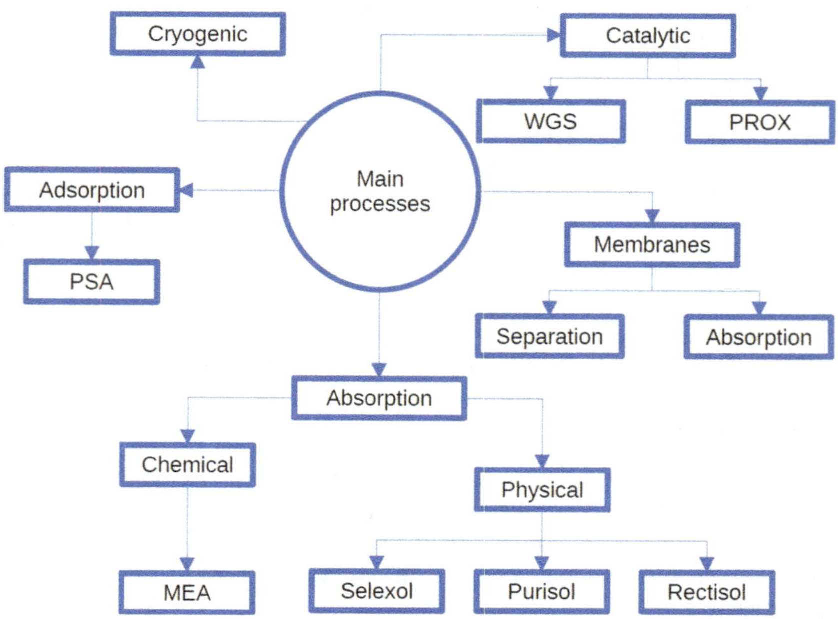

Figure 3.9 Main gas separation methods

Cryogenic technology is the main used for large-scale air separation plants to produce oxygen and nitrogen. It is based on the different boiling points of the air components (nitrogen, oxygen, argon and a few other substances) which allow their separation by liquefaction at progressively lower temperature.

Cryogenic technology is not used for hydrogen separation, but can be used for hydrogen production when oxygen is preferred to air. Moreover, it can be used for hydrogen storage (see Section 4.2).

Carbon dioxide can be captured from the syngas by chemical or physical processes or by separation through membranes. Capture is only an economic issue, what to do with captured CO_2 is the real issue. A good perspective, although not a definitive solution, is given in Section 5.3.

Dihydrogen sulphide can be separated from the syngas by chemical processes. However, the sulphur eventually contained in the primary fuel is found in the synthesis gas not only as dihydrogen sulphide but also as carbonyl sulphide. Therefore, it is necessary to convert COS to H_2S and CO_2 through an hydrolysis reaction:

$$COS + H_2O \Rightarrow H_2S + CO_2 \tag{3.26}$$

Absorption is a chemical or physical assimilation of molecular species within the mass of a solid or liquid.

The solvents used for chemical absorption are generally aqueous solutions of amines and alkaline salts which react with the substance to be removed generating other compounds. The characteristic of amines is to establish a weak bond with acid gases. They have a different selectivity towards different acid gases: for example methyldietanolamine removes the dihydrogen sulphide, but not the carbon dioxide, while monoetanolamine (MEA) removes both gases. The absorption phase takes place inside a plate column or a packed column, in which the solvent, coming into contact with the gas, reacts chemically with the substance to be removed. A plate column is a column in which the contact between the solvent and the acid gas is favoured by the presence of plates that force the two fluids to pass in restricted sections such as holes. A packed column is a column filled with material, for example ceramic cylinders, designed to increase the contact surface between solvent and acid gas. Temperatures of 20 °C–50 °C and pressure slightly higher than atmospheric favour the reactions. The purified gas is extracted from the top of the column, whereas the saturated solvent is extracted from the bottom, and then heated and sent to the regeneration section. Regeneration takes place at a temperature of 100 °C–120 °C using low pressure steam and releasing the removed substance. The removal efficiency of sulphur compounds is high: 99%, but the regeneration process absorbs a lot of heat for the production of steam, about 10 MJ/kg_{H2S}. The removal efficiency of carbon dioxide is 85% and the heat required is around 3–4 MJ/kg_{CO2} [31,32]. The residual carbon oxides are removed by a methanation process in the presence of a 35% nickel catalyst on a silicon support.

Cost is mainly related to the energy needs: the electric power to operate the ancillary equipment and the heat for solvent regeneration. Heat is generally supplied through low pressure steam either produced by heat recovery from the syngas or extracted from the plant. Once the solvent to be used has been chosen, the amount of steam to be tapped depends on the quantity of H_2S and CO_2 to be separated and on the initial characteristics of the steam. The concentration of sulphur compounds is of 0%–2% on average.

The sulphur compounds removal system consists essentially of two other steps after desulphurisation of the syngas:

- recovery of elemental sulphur from the gases leaving the desulphurisation section;
- treatment of tail gases of the process before they are released into the atmosphere.

Sulphur recovery is generally obtained using Claus process. The dihydrogen sulphide, removed and subsequently recovered during the regeneration phase, is subjected to oxidation with under-stoichiometric air (one mole of oxygen for every two moles of sulphide) in a furnace up to temperatures of 1400 °C:

$$3H_2S + \frac{3}{2}O_2 \Rightarrow 2H_2S + SO_2 + H_2O \tag{3.27}$$

The residual dihydrogen sulphide and sulphur dioxide are separated from the water and reacted at a temperature of about 230 °C in the presence of catalysts such as bauxite and vanadium, providing elemental sulphur in the liquid state:

$$2H_2S + SO_2 \Rightarrow 3S + 2H_2O \tag{3.28}$$

The conversion does not exceed 96%–97% and the 'tail gas' leaving the plant still contains H_2S and SO_2 and is sent to the SCOT process (Shell Claus off-gas treatment) in which the sulphur dioxide is reduced to H_2S by adding hydrogen

$$SO_2 + 3H_2 \Rightarrow H_2S + 2H_2O \tag{3.29}$$

and then recirculated to the Claus plant in such a way that the final abatement reaches about 99.8%.

In case of physical absorption, the compound to be removed goes into solution in a liquid solvent until a balance is established between the pressures exerted on the solvent and in the solvent.

The solubility of any gas in a liquid is governed by Henry's law:

$$x_i = \frac{p_i}{H_i} \tag{3.30}$$

where x is the concentration of the species i, p is its partial pressure and H its Henry constant, increasing with temperature. Therefore, by adopting high pressures and low temperatures, the absorption of the species by the solution is facilitated. Solvents characterised by a high absorption capacity towards dihydrogen sulphide are organic and the most used processes are Selexol, Purisol and Rectisol (Table 3.5) [33].

Table 3.5 Properties and relative solubilities of main processes

Process	Selexol	Purisol	Rectisol
Solvent	DMEPG	NMP	MeOH
Molecular mass (kg $kmol^{-1}$)	280	99	32
Viscosity, @25 °C (cP)	5.8	1.65	0.6
Density, @25 °C (kg m^{-3})	1030	1027	785
Melting point (°C)	−28	−24	−92
Boiling point, @1 atm (°C)	275	240	65
Solubility (l/l), Selexol and Purisol @ 25 °C, Rectisol @ −25 °C			
CO_2	3.628	3.568	3.179
H_2S	31.999	36.396	22.445
COS	8.345	9.706	12.463
SO_2	334.144	–	–
N_2	0.073	–	0.038
CO	0.102	0.075	0.064
CH_4	0.239	0.257	0.162
O_2	–	0.125	0.064
H_2	0.047	0.023	0.017

The solvent used by Selexol process is glycol-polyethylene dimethyl ether diluted with water. This process is characterised by H_2S removal efficiencies in the order of 97% and can also remove about 10% of the CO_2 contained in the syngas.

N-methylpyrrolidone is the solvent used by the Purisol process. Operating temperatures are close to room temperature and sulphur compounds removal efficiencies is around 99%.

The Rectisol process uses methanol as a solvent, operating at high pressures (22–140 bar) and at very low temperatures (from −73 °C to −18 °C), which involves significant operating costs. It is characterised by an efficiency of removal of sulphur compounds higher than 99%.

The process is carried out in two distinct stages: absorption of gas to be removed by an absorbent solution inside a column operating at high pressure and gas release in a series of cascading flash chambers operating at ever lower pressures. The pressure in the last flash chamber is generally fixed at a value of 1.05 bar in such a way that the whole system operates at pressures higher than atmospheric pressure and therefore the risk of air re-entry be avoided. Alternatively thermal regeneration can be used.

In general, more energy is required to regenerate a chemical solvent than a physical solvent, while the convenience in terms of effectiveness depends on the concentration of the species to be removed. Figure 3.10 qualitatively shows the trends of the substance removal capacity as a function of its partial pressure: as can be seen, chemical processes are more effective for low concentrations, while physical ones are preferable at higher concentrations.

Operating at high temperature allows a lower plant cost and a plant engineering simplification, thanks to the reduction of thermal regeneration equipment. For

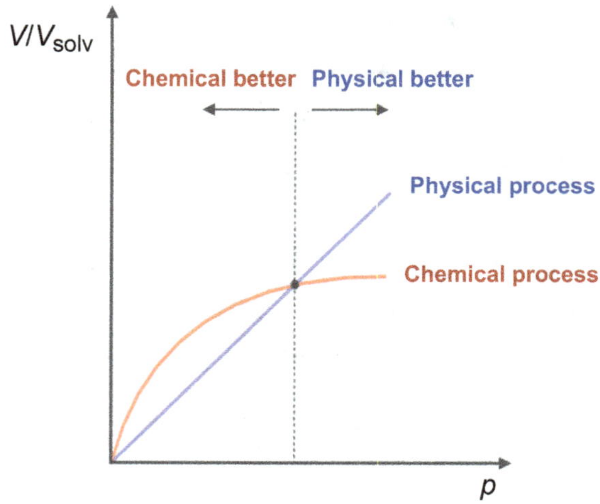

Figure 3.10 Chemical process vs. physical process for gas separation

such a reason, there is a growing interest in the development of systems for the hot treatment (450 °C–550 °C) of syngas. There is no problem to use a cyclone for particulate removal at high temperatures, but it is not able to guarantee a sufficient removal efficiency. The most interesting solution for an additional and more effective dedusting is represented by hot gas filters. They are filters in porous ceramic materials, resistant to high temperatures and also to the strongly aggressive environment such as those of coal and biomass syngas. In particular, the most efficient systems are ceramic candle filters, whose active element consists of a hollow cylinder made of porous material. Each cylinder is closed at one end, whereas on the other end it is connected to a perforated metal plate. The gas passes through the filter elements from the outside to the inside and the solid particles, which are deposited on the external surface of each element, are periodically removed by means of pressurised air jets. This type of filters can operate an efficient dedusting even at temperatures of 850 °C–950 °C.

Internal desulphurisation can be obtained through the absorption of sulphur at high temperature inside the gasifier introducing sorbents based on calcium (limestone, $CaCO_3$) or calcium–magnesium (dolomite, $CaCO_3$–$MgCO_3$), extremely abundant in nature and therefore quite inexpensive:

$$CaCO_3 + H_2S \Rightarrow CaO + CO_2 + H_2S \Rightarrow CaS + H_2O + CO_2 \tag{3.31}$$

Calcium sulphide (CaS), produced by desulphurisation reactions, must be converted into calcium tetraoxosulphate ($CaSO_4$), which is practically inert. This treatment is carried out by oxidation inside an air or oxygen fed furnace. When the ash still contains high amounts of carbonaceous residues, a fluid-bed combustor is used.

The main experiments on this technology were conducted by Westinghouse, IGT (Institute of Gas Turbines) and METC (Morgantown Energy Technology Center). Tests carried out at temperatures of 1000 °C have led to sulphur removal efficiencies in the range 80%–90% and therefore an external desulphurisation stage must always be provided.

Adsorption is a process in which a substance in the form of atoms, ions or molecules adheres to a solid surface.

Suitable materials must have high selectivity towards the species to be separated, high load capacity, low cost, ease of disposal or regeneration. They can be zeolites or silica gel or activated carbon, obtained by carbonisation of materials or by partial combustion of substances like methane, fats, oils and wood. The choice of the adsorbent must be made taking into account both the adsorption phase and the desorption (regeneration) phase, since the technical feasibility of the process is linked to the adsorption phase, whereas the desorption phase controls the cost effectiveness of the process itself.

Different adsorption processes exist: pressure swing adsorption (PSA), temperature swing adsorption, electrical swing adsorption, pressure and temperature swing adsorption. Generally, the PSA is used. It allows a selective adsorption of different substances on special adsorbent beds at high pressures. The adsorbed compounds are removed cyclically by pressure variations and inversions, and by using multiple adsorbent beds, it is possible to obtain a very high level of purity. It is a process developed in four phases that globally form a complete cycle:

- pressurisation phase, in which the gas is introduced (at about 30 bar) into a high-pressure reactor (e.g. a zeolite bed);
- adsorption phase, in which the undesired gases are adsorbed, while the purified gas is concentrated in the upper part of the bed and is extracted;
- desorption phase, in which the reactor is depressurised to be able to expel part of the gases trapped in the crystalline structure;
- cleaning phase, in which part of the purified gas (generally 10%–20%) is re-introduced into the bed in order to favour the removal of the gases remained into the pores of the adsorbent material. This fraction of the gas is lost in the tail gas stream (purge gas).

In order to obtain a continuous production, several beds work in parallel and alternate their operations, so that the four phases of the cycle are carried out simultaneously inside the different beds. Costs are relatively insensitive to changes of the flow to be treated and this makes PSA particularly suitable for large volumes of gas to be treated. PSA systems represent a good compromise between high purity, moderate costs and ease of integration into hydrogen production facilities.

Table 3.6 shows how a PSA system is capable to remove different substances from a gas stream. It is evident that PSA is very suitable to obtain pure hydrogen from a syngas since it can remove easily dihydrogen sulphide and moderately carbon oxides and light hydrocarbons, whereas hydrogen cannot be removed. However, not all hydrogen can be recovered due to the bed cleaning phase.

Table 3.6 Ease of removal of substances (courtesy: La Termotecnica [34])

Easy	Moderate	Difficult	Non-removable
C_3H_6	CO	O_2	H_2
C_4H_{10}	CH_4	N_2	He
C_5+	CO_2	Ar	
H_2S	C_2H_6		
NH_3	C_3H_8		
C_6H_6	C_2H_4		
C_7H_8			
C_8H_{10}			
H_2O			

Table 3.7 Example values for a PSA unit

	Inlet	Outlet	Residual
H_2	88%	99.97%	60.67%
N_2	6%	<0.03%	19.64%
CO	3%	~0%	9.85%
CO_2	2%	~0%	6.57%
Ar	1%	~0%	3.28%

Table 3.7 shows an example of the composition of the incoming and outgoing flows from a PSA unit. In this example, 25% of the hydrogen supplied to the PSA unit is not recovered and is found into the tail gas, which can be used to generate heat or steam for the reformer or the gasifier [35].

Currently, the high temperature external desulphurisation processes based on adsorption are in pre-commercial phase, and they are extremely promising compared to traditional low temperature processes. Such processes use a sorbent which reacts with the dihydrogen sulphide. The saturated sorbent is then regenerated by means of suitable oxidation processes producing, at the exhaust, a gas rich in H_2S or SO_2 which can subsequently be used for the production of elemental sulphur or sulphuric acid. The process generally involves two identical reactors operating alternately in the adsorption and regeneration phases.

The most studied process is based on adsorption on metal oxides. By indicating the generic metal used with M, the adsorption and regeneration reactions are respectively:

$$MO + H_2S \Rightarrow MS + H_2O \tag{3.32}$$

$$MS + \frac{3}{2}O_2 \Rightarrow MO + SO_2 \tag{3.33}$$

Sorbent based on zinc oxide (ZnO) or zinc–ferrite ($ZnO–Fe_2O_3$) allows to operate at temperatures of 450 °C–600 °C, whereas the sorbent regeneration takes place at temperatures in the range 500 °C–700 °C, with a remarkable release of thermal energy. Also iron can be used as a sorbent at temperatures below 500 °C. The residual H_2S concentration is of the order of 100 ppm in volume.

Membranes are solid, porous or semi-porous structures which allow permeation, that is the passage of molecules under the action of a driving force. They are characterised by separation selectivity: only some species present in the gaseous stream are allowed to permeate, while others are retained and leave the unit at a pressure close to that of the supply. The quality of the separation depends on the selectivity of the membrane with respect to the species to be separated and on two process parameters:

- the ratio between permeate gas flow and input flow to the separator;
- the ratio between permeate gas pressure and input gas pressure.

Membrane separation is driven by Sievert's law:

$$J = \frac{P_e}{s} \cdot \left(\sqrt{p_a} - \sqrt{p_p}\right) \cdot A \tag{3.34}$$

where:

- J ($m^3\ s^{-1}$), permeate flow,
- P_e ($m^3\ m^{-1}\ s^{-1}\ Pa^{-0.5}$), permeability coefficient,
- s (m), membrane thickness,
- p_a and p_p (Pa), partial pressures of the species to be separated on the feeding side and on the permeation side, respectively,
- A (m^2), permeation area.

Membrane units do not contain moving parts and potentially have a very high reliability. To increase the purity of the permeate, multistage plants can be used, with a consequent increase of energy needs for compression and of capital costs. Moreover, the membranes are realised in comparatively small modules and several modules are used to treat large volumes of gas. Costs therefore vary linearly with the flow of gas to be treated and this makes the membranes less competitive compared to other processes in the event of a large flow.

Another kind of membranes are absorption membranes which are microporous and water-repellent. Their task is only to separate the gas and a liquid absorbent (generally amines). Therefore, the gas to be separated is diffused on the solid membrane and is then absorbed and removed by the absorbent liquid which is selective towards a gaseous species with respect to the others. The membrane is not selective and serves only as a contact area, preventing the gas and the liquid from mixing physically. A pressure gradient is not necessary and therefore the energy costs is limited to the liquid regeneration treatment.

Another interesting external desulphurisation process is the catalytic reactions using zeolites (aluminosilicates, crystalline polymers with tetrahedral structure) as catalysts. It is a low temperature process (150 °C–200 °C) which allows to obtain directly sulphur:

$$H_2S + \frac{1}{2}O_2 \Rightarrow S + H_2O \tag{3.35}$$

In the event that the power plant is intolerant to the presence of carbon monoxide, it is also necessary to insert a preferential oxidation unit (PROX). A 100% selectivity is obviously not possible, so other reactions are generally possible. In the specific case, the reaction in competition with the oxidation of carbon monoxide (3.14) is that of oxidation of hydrogen (3.15), obviously unwanted, but in any case less favoured and limited to a loss of about 1%. One problem is the very strong exothermic reaction that requires the cooling of the reactor: usually the process is divided into two stages with an intermediate cooling. The oxygen excess fed to the first reactor is 100% and is such as to convert about 90% of the carbon monoxide, whereas the oxygen excess fed to the second reactor is 300% to reach a final concentration of carbon monoxide less than 10 ppm.

3.4 Co-generation of power and hydrogen

Similar to co-generation of heat and power, co-generation of power and hydrogen could allow to obtain a better efficiency [36,37].

As shown in the example of Figure 3.11, in a gasification-only plant, the heat recovered from the refrigeration of the syngas can be used to produce the steam that feeds the gasifier and to heat the nitrogen from the air separation unit (ASU) before its expansion.

Recovering heat to generate steam allows to increase the steam to carbon ratio S/C. This leads to a larger demand for heat generation inside the gasifier in order to maintain the desired temperature. Therefore, a greater flow of oxygen, and therefore of air, is needed with two effects on the work: on the one hand increase of air

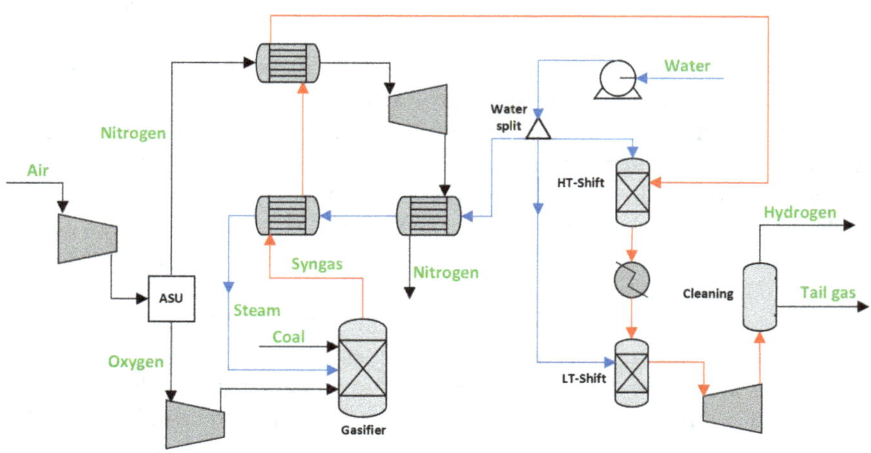

Figure 3.11 Schematics of a gasification plant for hydrogen production

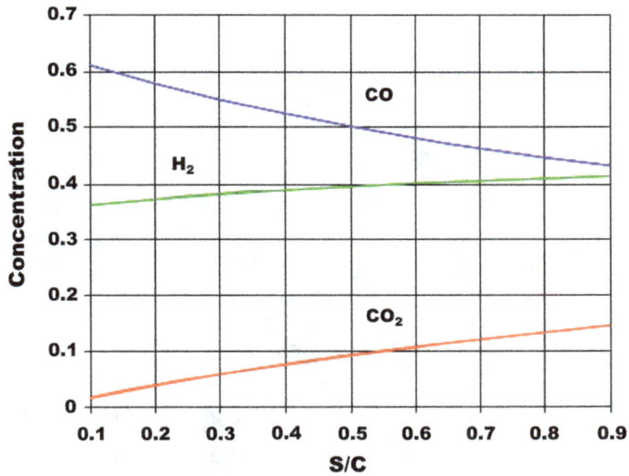

Figure 3.12 Main components of the syngas exiting the gasifier versus S/C ratio

and oxygen compression work, and on the other hand increase in nitrogen expansion work. The greater oxygen flow favours the advancement of the combustion reactions with a larger consumption of H_2 and CO. At the same time, the greater availability of steam favours the reaction of WGS (3.6), further increasing the consumption of CO and the production of H_2 and CO_2. The overall result is favourable and results in a higher hydrogen concentration at the exit of the gasifier (Figure 3.12).

A better use of high temperature steam generated from heat recovery could be the power production. In this case, the steam is generated at a pressure higher than the operating pressure of the gasifier in such a way that it is possible to expand it before entering the gasifier itself. As shown in the example of Figure 3.13, it could be possible to expand the steam in a high- and an intermediate-pressure turbines and then realise a final expansion with the steam exceeding the amount required by the gasifier.

While the power obtained is always increasing when S/C decreases, hydrogen production is increasing–decreasing with the maximum is reached at S/C=1.1 (Figure 3.14).

To appreciate the advantages of the co-generation of electricity and hydrogen, the marginal efficiency can be evaluated. Marginal efficiency is defined as the ratio between the change of energy output and the change of energy input both referred to the reference plant.

In this case, maintaining the same amount of hydrogen produced, marginal efficiency is defined as the ratio between the increase of electricity generated and the increase of chemical energy consumed with reference to the plant for hydrogen production from coal. With reference to Figure 3.15, it can be defined by the following equation:

$$\eta_{marg} = \frac{\Delta E_e}{\Delta E_{c,coal}} = \frac{E_{e,CPH} - E_{e,IGHP}}{\left(E_{c,CPH} - E_{c,IGHP}\right)_{coal}} \tag{3.36}$$

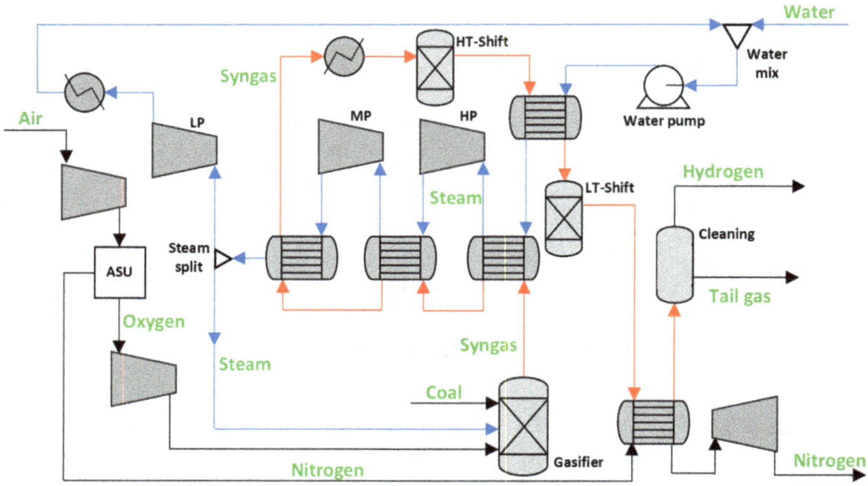

Figure 3.13 Schematics of a gasification plant for cogeneration of power and hydrogen

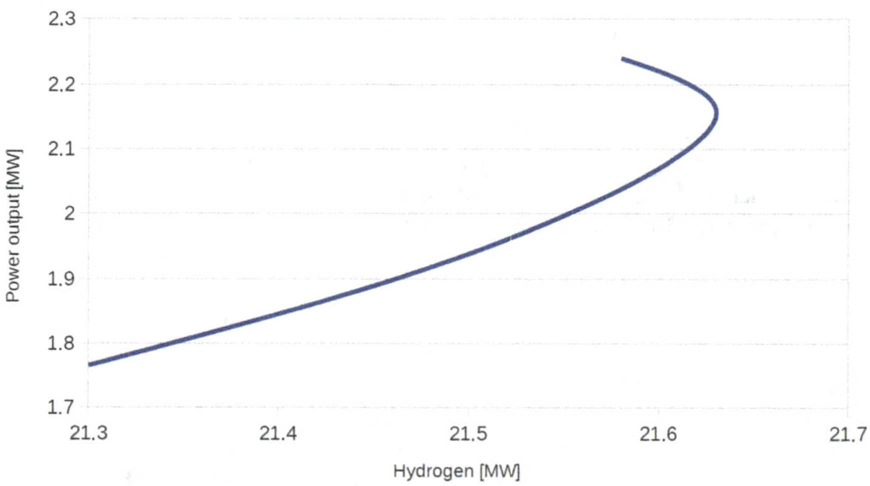

Figure 3.14 Power generation vs hydrogen production (adapted from [38])

where:

- IGHP is integrated gasifier hydrogen production
- CPH is co-generation of power and hydrogen
- E is energy, with subscripts c=chemical, e=electric and t=thermal

This efficiency can be compared to the efficiency of an integrated gasifier combined cycle, which is the best current technology for coal to power conversion.

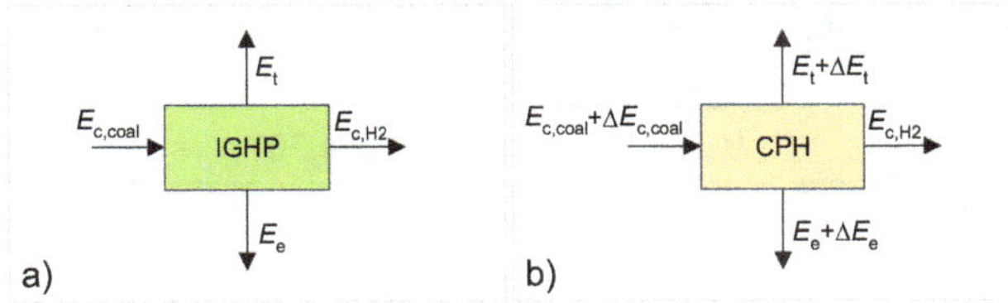

Figure 3.15 Energy flows to define a marginal efficiency

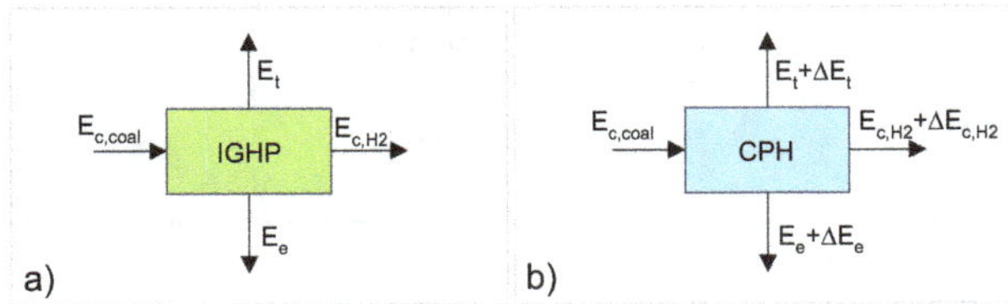

Figure 3.16 Energy flows to define an apparent efficiency

There is also another interesting interpretation which we called apparent efficiency. In this case, it is the same chemical energy input to be maintained. The apparent efficiency is defined as the ratio between the increase in electric energy generated and the decrease (in absolute value) of the produced hydrogen. With reference to Figure 3.16, it can be defined by the following equation:

$$\eta_{app} = \frac{\Delta E_e}{-\Delta E_{c,H_2}} = \frac{E_{e,CPH} - E_{e,IGHP}}{\left(E_{c,IGHP} - E_{c,CPH}\right)_{H_2}} \tag{3.37}$$

The least amount of hydrogen produced represents the chemical energy apparently consumed in the CPH plant compared to the IGHP plant. Therefore, this parameter allows to evaluate the conversion efficiency of the amount of hydrogen apparently consumed. This efficiency can be compared to the efficiency of systems that use hydrogen as a fuel, e.g. fuel cells.

Table 3.8 shows a comparison among some CPH layouts at different operating pressures and an IGHP at 70 bar, which was found to be the optimal working pressure. The reported results correspond also to the optimal steam to carbon ratio (S/C) either for hydrogen or electricity production.

Although the global output is higher for IGHP, it is interesting the result in terms of marginal and apparent efficiencies. Actually, marginal efficiency is much higher than the conversion efficiency of an integrated gasifier combined cycle, whereas the apparent efficiency is much higher than the conversion efficiency of a fuel cell.

The hydrogasification process can be combined with the steam reforming process. A particular example of hydrogen and electricity cogeneration based on this combination is provided by the ZECOMIX system (Zero Emission of CarbOn with MIXed Technologies) [39,40]. It initially uses a coal hydrogasification

Table 3.8 Comparison among IGHP and CPH layouts results [38]

	IGHP,70	**CPH,30**		**CPH,70**	
S_{shift}/C	1.2	–	–	–	–
S/C	0.2	1	1.1	1	1.1
$p_{GASIFIER}$ (bar)	70	30		70	
Coal input (MW)	27.13	27.13	27.13	27.13	27.13
W_{NET} (MW)	0.87	2.27	2.17	2.24	2.15
H_2 output (MW)	23.06	21.52	21.57	21.58	21.64
Co-generation ratio (%)	3.8	10.5	10.1	10.4	9.9
η_{H2} (%)	85.0	79.3	79.5	79.5	79.8
η_e (%)	3.2	8.4	8.0	8.2	7.9
η_{marg} (%)	–	80.6	77.7	81.9	79.8
η_{app} (%)	–	91.1	87.6	92.6	90.1

process, that is, consuming part of the hydrogen generated. The generated syngas is treated with calcium oxide in order to separate the sulphur as calcium sulphide:

$$H_2S + CaO \Rightarrow CaS + H_2O \tag{3.38}$$

while the carbon is converted into calcium carbonate inside a special reactor which also provides for the steam reforming of methane:

$$CH_4 + 2H_2O + CaO \Rightarrow CaCO_3 + 4H_2 \tag{3.39}$$

The calcium carbonate is then sent to a high-temperature calcinator in which the calcium oxide is regenerated and the carbon dioxide is released, suitable for compression and use or storage:

$$CaCO_3 \Rightarrow CaO + CO_2 \tag{3.40}$$

3.5 Water splitting

Among the substances that contain hydrogen on Earth, water is by far the most abundant. Furthermore, the combustion of hydrogen with air or oxygen produces water. Therefore, the ideal process to obtain hydrogen is certainly the splitting of water into its constituent elements:

$$H_2O \Rightarrow H_2 + \frac{1}{2}O_2 \quad \Delta H^\circ = 242.0\ \text{kJ/mol} \tag{3.41}$$

To split the water molecule, it is necessary to supply energy and this can be supplied as thermal energy, electric energy or a combination of the two.

3.5.1 Thermolysis

The process of thermolysis (from the ancient Greek θερμς = hot and λυσις = dissolution) of water consists in the splitting of the water molecule into hydrogen and oxygen by means of high temperature heat.

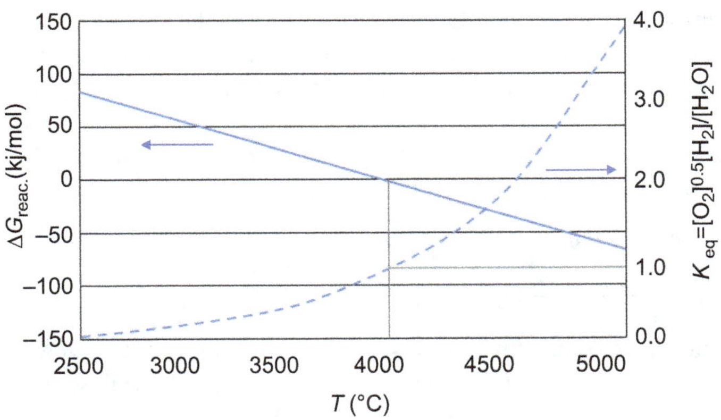

Figure 3.17 Thermodynamics of water thermolysis [41]

The aptitude of a reaction to occur spontaneously or not is indicated by the Gibbs free energy variation:

$$\Delta G = \Delta H - T\Delta S \tag{3.42}$$

Spontaneous reactions are those in which heat is subtracted from the system with a decrease of Gibbs free energy ($\Delta G < 0$), while non-spontaneous reactions are characterised by a positive variation of Gibbs free energy ($\Delta G > 0$).

Figure 3.17 shows the trends of the free energy variation and of the equilibrium constant (K_{eq}) for the water molecule splitting reaction.

The free energy goes to zero at about 4000 °C: this means that below this threshold, the splitting of the water molecule into its elementary components cannot be obtained by using only heat. It follows that direct thermolysis of water requires temperatures not compatible with conventional materials.

To overcome this drawback, starting from the 1960s, the study of thermochemical cycles at lower temperatures began and more than 100 were designed. The term 'cycle' means that all the reactants used are continuously regenerated: therefore, the set of reactions of a thermochemical cycle only carries out the thermolysis of the water and the overall result therefore coincides with the reaction (3.41). The maximum temperature required is lower than that of the direct thermolysis process and, in general, decreases as the number of reactions involved increases. The initial idea was to couple these cycles to nuclear reactors, but, over time, it was thought that they could be powered by solar energy. Below are presented, in order of increasing number of reactions, some thermochemical cycles currently considered promising.

Zinc–zinc oxide cycle

The thermochemical cycles that use metal oxides for the splitting of water generally consist of two reactions:

$$MO_{(ox)} \rightarrow MO_{(red)} + \frac{1}{2}O_{2(g)} \tag{3.43}$$

$$MO_{(red)} + H_2O_{(g)} \rightarrow H_{2(g)} + MO_{(ox)} \tag{3.44}$$

The thermal energy supplied to the reactor causes the release of oxygen by the metal oxide which passes from the oxidised to the reduced form (3.43). The latter is therefore able to return to the oxidised form by subtracting oxygen from the water and releasing hydrogen (3.44).

A cycle of this type is the one based on zinc oxide and was developed by researchers at the PSI (Paul Sherrer Institut, Switzerland):

$$ZnO \rightarrow Zn + \tfrac{1}{2}O_2 \qquad 1700\,^\circ C \tag{3.45}$$

$$Zn + H_2O \rightarrow ZnO + H_2 \qquad 700\,^\circ C \tag{3.46}$$

In reaction (3.45), the zinc oxide splits into zinc vapour. Oxygen is extracted, while zinc is subjected to the exothermic reaction of hydrolysis (3.46) which releases hydrogen by regenerating zinc oxide [42].

The zinc oxide dissociation reaction takes place inside a solar receiver which, unlike the receivers used in concentrating solar power plants, is not crossed by any heat transfer fluid. In Figure 3.18, it is possible to see a cross-section of the experimental receiver made at the PSI.

The solar energy captured and reflected by the mirror field is concentrated on a quartz lens through which it reaches and hits a sample of zinc oxide directly. In order to monitor the degree of decomposition of the zinc oxide, the above sample is placed on a precision balance which determines the weight of oxygen released. Due to the high operating temperature, the receiver is coated with an insulating material based on aluminium oxide and zirconium.

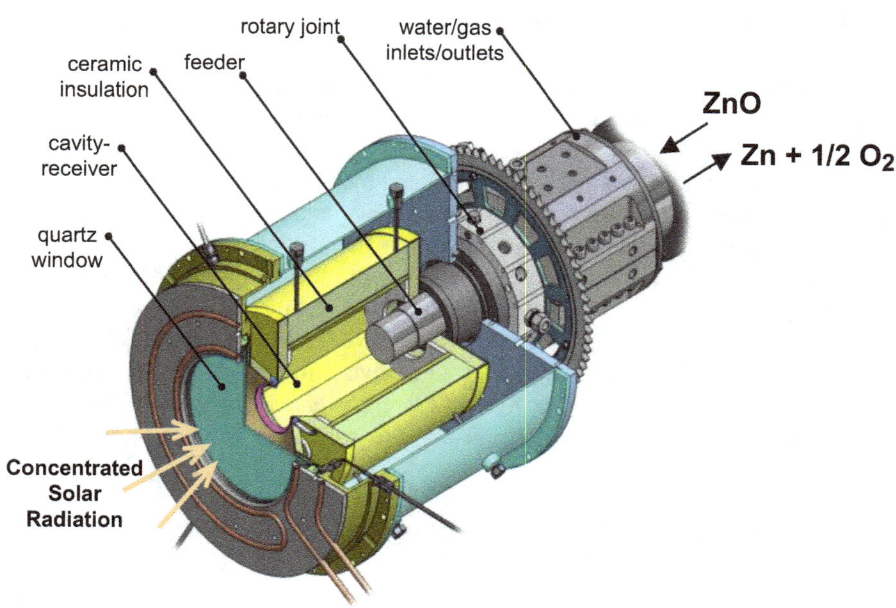

Figure 3.18 Cutaway of a solar reactor for the dissociation of zinc oxide [43]

The zinc vapour and oxygen resulting from the decomposition of zinc oxide are separated and cooled. The zinc vapour solidifies and is reacted with water vapour according to an exothermic hydrolysis reaction (3.46) which does not need external energy to trigger and sustain itself. In this way, hydrogen is released and zinc oxide is obtained which is recycled into the plant so as to close the cycle [42].

Mixed ferrites cycle

Another cycle based on metal oxides, also consisting of only two reactions, is the mixed ferrites. It was developed at the level of basic chemistry in the 1990s at the Tokyo Institute of Technology. Using chemically very reactive materials such as nanoparticulated manganese ferrites and sodium carbonate, a temperature of about 1000 °C is sufficient to obtain the release of hydrogen, whereas the release of oxygen requires a lower temperature (750 °C).

The reactions that make up the cycle are the following:

$$6Na(Mn_{1/3}Fe_{2/3})O_2 + 3CO_2 \rightarrow 2MnFe_2O_{4(s)} + 3Na_2CO_3 + \tfrac{1}{2}O_2 \qquad 750\ ^\circ C \tag{3.47}$$

$$2MnFe_2O_{4(s)} + 3Na_2CO_3 + H_2O \rightarrow 6Na(Mn_{1/3}Fe_{2/3})O_2 + 3CO_2 + H_2 \qquad 1000\ ^\circ C \tag{3.48}$$

Like the previous cycle, this one has the advantage of simplicity. It also operates at much lower temperatures, compatible with those of some concentrating solar plants. In addition to the low technological maturity, it has the disadvantage

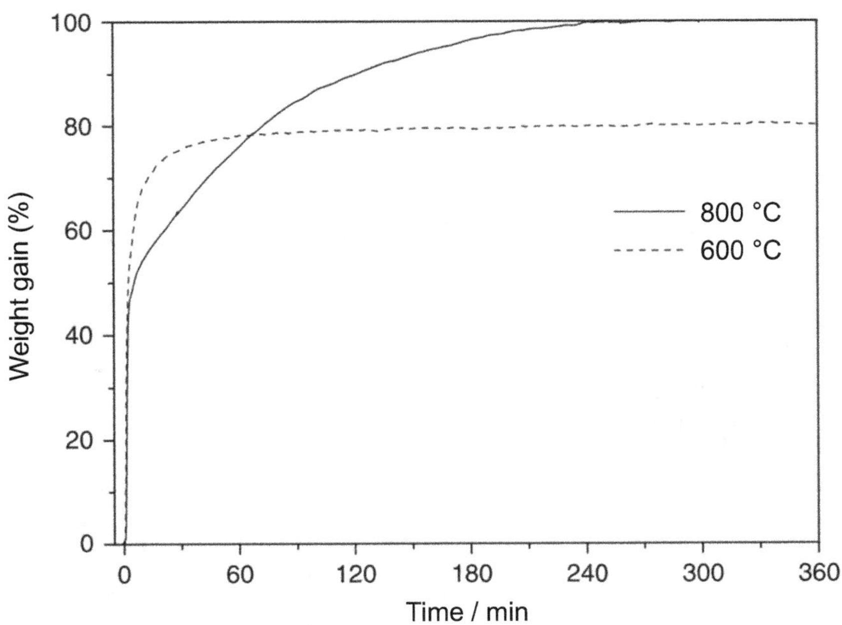

Figure 3.19 Time trend of the ferrite regeneration reaction [45]

of the slowness and incompleteness of the reaction (3.47). As shown in Figure 3.19, for a reaction maintained isothermal at 600 °C, a quasi-asymptotic level of regeneration of the ferrites equal to 45% is reached after almost seven hours [44].

Sulphur–iodine cycle

The first studies relating to this cycle date back to the 1970s by General Atomics. After ups and downs, mainly due to the difficult handling of the reagents, it has remained among the most promising and is, perhaps, the most studied. Although the feasibility of the process has been demonstrated, on an industrial scale it has limitations due to problems related to the materials that are subjected to contact with acids and to high thermal stresses as well as to the quality of the hydrogen separation membranes.

The cycle (Figure 3.20) is composed of three reactions with a maximum temperature of 850 °C:

$$H_2SO_{4(g)} \rightarrow SO_{2(g)} + H_2O_{(g)} + \tfrac{1}{2}O_{2(g)} \qquad 850\ ^\circ C \qquad (3.49)$$

$$I_{2(aq)} + SO_{2(aq)} + 2H_2O_{(aq)} \rightarrow 2HI_{(aq)} + H_2SO_{4(aq)} \qquad 120\ ^\circ C \qquad (3.50)$$

$$2HI_{(aq)} \rightarrow I_{2(aq)} + H_{2(g)} \qquad 450\ ^\circ C \qquad (3.51)$$

Due to the heating and in the presence of a catalyst, in reaction (3.49) the sulphuric acid flux is dissociated into sulphur dioxide, water and oxygen. The latter is removed while the dioxide is bubbled in a Bunsen reactor containing iodine and water at low temperature in order to regenerate sulphuric acid and hydrogen iodide.

The separation of hydrogen iodide from sulphuric acid by distillation is hindered by the fact that the reaction (3.50) is disadvantaged by temperature.

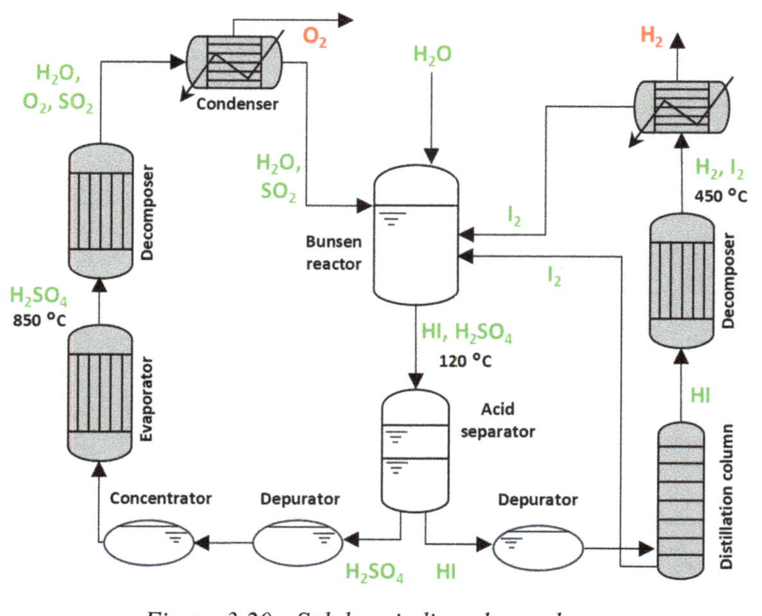

Figure 3.20 Sulphur–iodine plant scheme

To overcome this drawback, excess water and iodine are supplied so as to obtain an aqueous solution in which hydrogen iodide, iodine and part of the water segregate forming a denser phase (iodide solution) which stratifies at the bottom. The sulphuric acid, dissolved in water with a ratio between 1:4 and 1:6, instead stratifies in the upper part and the two phases are separated by decanting.

The sulphuric acid, after being purified and concentrated, is recirculated, while the iodide solution is sent to the distillation section where hydrogen and iodine are separated (3.51). Hydrogen is removed, while iodine, condensed, is recirculated.

The main advantages of this cycle consist in the high thermochemical efficiency (close to 52%) and in the high reaction rate. This is made possible by the fact that all the reagents and products are in the fluid phase, which favours the contact between reagents and impacts between molecules. Furthermore, hydrogen is produced at high pressure (50 bar), thus reducing the compression expense downstream of the process. However, the materials are subjected to corrosion by the acids present in the cycle and the recovery of the acids requires a large amount of heat [46,47].

Cerium–chlorine cycle

Another cycle based on three reactions is the cerium–chlorine cycle (Figure 3.21):

$$2CeO_{2(s)} + 8HCl_{(g)} \rightarrow 2CeCl_{3(s)} + Cl_{2(g)} + 4H_2O_{(g)} \quad 110\ °C \quad (3.52)$$

$$2CeCl_{3(s)} + 4H_2O_{(g)} \rightarrow 2CeO_{2(s)} + 6HCl_{(g)} + H_{2(g)} \quad 750\ °C \quad (3.53)$$

$$Cl_{2(g)} + H_2O_{(g)} \rightarrow 2HCl_{(g)} + ½O_{2(g)} \quad 700\ °C \quad (3.54)$$

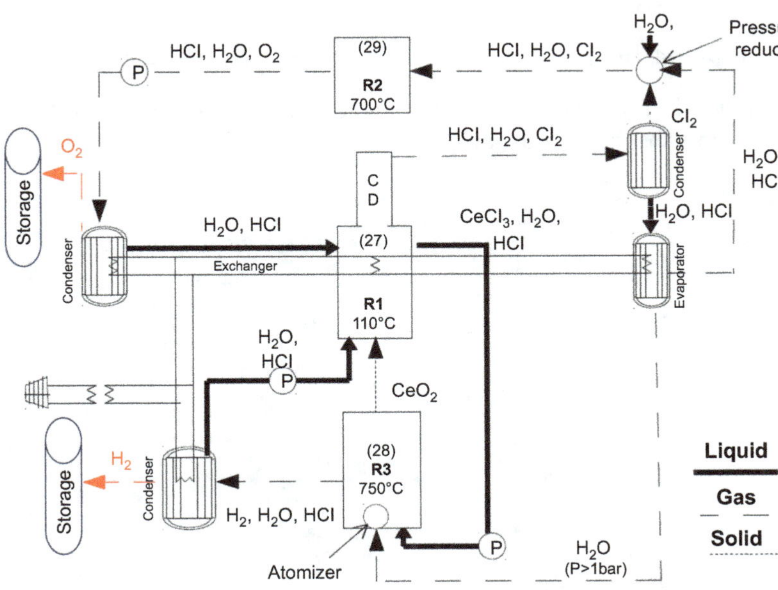

Figure 3.21 Cerium–chlorine plant scheme

An aqueous solution of hydrochloric acid is vaporised at 110 °C and, in a proportion of one to four times with respect to the stoichiometry, reacted with solid cerium oxide so as to produce cerium chloride, chlorine and water. Therefore two flows exit from the distillation column of this reactor:

- a liquid flow consisting of an aqueous solution of cerium chloride also containing part of the excess hydrochloric acid;
- a gaseous flow consisting of chlorine and water vapour also containing the remaining part of the excess hydrochloric acid.

The first of the two flows is atomised and subjected to hydrolysis at 750 °C in order to release hydrogen by regenerating cerium oxide and hydrochloric acid. The second flow is heated to 700 °C in order to release oxygen by regenerating hydrochloric acid. Both hydrogen and oxygen are separated from their respective streams by condensation of hydrochloric acid and water vapour.

This cycle has the advantage of separating the gases by simple condensation. However, it has a low yield, about 20%, and a modest reaction speed due to the presence of solid reactants.

Adiabatic UT-3 cycle

This cycle is an evolution of a cycle originally developed in the 1970s at the University of Tokyo, hence the initials UT. There are four reactions, all endothermic, with the exception of (3.57), and the maximum temperature is 760 °C:

$$CaBr_{2(s)} + H_2O_{(g)} \rightarrow CaO_{(s)} + 2HBr_{(g)} \qquad 760\,°C \qquad (3.55)$$

$$CaO_{(s)} + Br_{2(g)} \rightarrow CaBr_{2(s)} + ½O_{2(g)} \qquad 560\,°C \qquad (3.56)$$

$$Fe_3O_{4(s)} + 8HBr_{(g)} \rightarrow 3FeBr_{2(s)} + 4H_2O_{(g)} + Br_{2(g)} \qquad 220\,°C \qquad (3.57)$$

$$3FeBr_{(s)} + 4H_2O_{(g)} \rightarrow FeO_{4(s)} + 6HBr_{(g)} + H_{2(g)} \qquad 572\,°C \qquad (3.58)$$

The scheme of the cycle is shown in Figure 3.22. The system consists of two pairs of reactors that contain the solid phase of reactants/products:

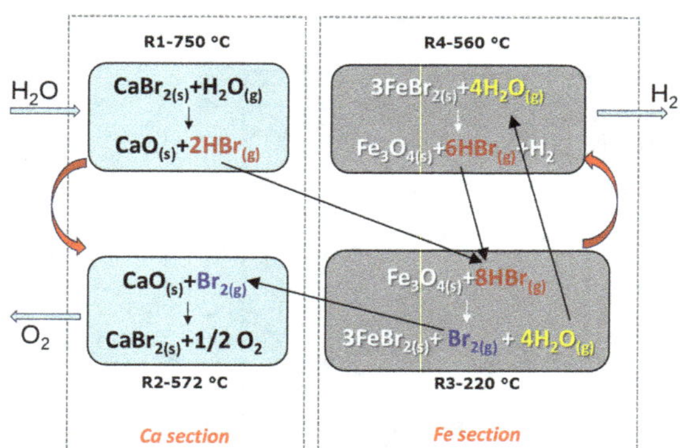

Figure 3.22 Scheme of the adiabatic UT3 cycle

- one pair (R1–R2) containing the calcium compounds ($CaBr_2$–CaO)
- the other (R3–R4) containing the iron compounds (Fe_3O_4–$FeBr_2$).

The characteristic of this cycle is in fact that it is a batch process, i.e. discontinuous, which lasts about one hour and at the end of which the reactors of each pair are reversed. The subsequent cycle will therefore be reversed.

The heat necessary for the process is supplied by sending superheated steam to the hydrolysis reaction of calcium bromide (3.55) which laps the solid bed generating calcium oxide and hydrobromic acid. From R1, calcium oxide is transferred to R2 which also receives bromine from R3 regenerating calcium bromide and releasing oxygen (3.56). Hydrobromic acid, together with those generated in R4, is transferred to R3 where reacts with tri-iron tetraoxide to generate iron bromide as well as bromine for R2 and steam for R4 (3.57). The latter reacts in R4 with iron bromide to release hydrogen, regenerate tri-iron tetraoxide and produce hydrobromic acid (3.58). The hydrogen released by this hydrolysis of iron bromide is separated from the steam and hydrobromic acid by means of a membrane separator. A second membrane separator is used for oxygen in R2.

The expected cycle efficiency is higher than 50% [48]. However this efficiency depends from several aspects, for example the efficiency of membrane separators. The use of these separators involves the need to compress the gases extracted at low pressure and, therefore, an energy expenditure which is reflected in the final cost of the hydrogen produced. There are also problems of corrosion and resistance of the materials due to the need to transport bromine and oxygen at high temperatures. Wang and Macián-Juan estimated the efficiency in a very wide range, 15.8%–49.8%, and after adopting come measures, confirmed a promising thermal efficiency around 36.7% [49].

Hybrid copper–chlorine cycle and sulphuric acid–sulphur oxide cycle

Finally, the copper–chlorine cycle (Figure 3.23) is a four-reaction cycle characterised by a maximum operating temperature of 530 °C, a value already easily reached with existing concentrating solar systems. However, it is hybrid, i.e. it requires the supply of electricity since it includes an electrolytic reaction [50]:

$$2CuCl_{2(s)} + H_2O_{(g)} \rightarrow Cu_2OCl_{2(s)} + 2HCl_{(g)} \qquad 400\,^{\circ}C \tag{3.59}$$

$$2Cu_{(s)} + 2HCl_{(g)} \rightarrow 2CuCl_{(l)} + H_{2(g)} \qquad 425\,^{\circ}C - 475\,^{\circ}C \tag{3.60}$$

$$Cu_2OCl_{2(s)} \rightarrow 2CuCl_{(l)} + \tfrac{1}{2}O_{2(g)} \qquad 530\,^{\circ}C \tag{3.61}$$

$$4CuCl_{(l)} \rightarrow 2CuCl_{2(aq)} + 2Cu_{(s)} \qquad 25\,^{\circ}C - 80\,^{\circ}C \tag{3.62}$$

Each reaction identifies a phase, respectively, hydrolysis, hydrogen production, oxygen production and electrolysis. The copper dichloride particles acquire properties typical of fluids when they are reached by a steam flow and, therefore, the hydrolysis of the copper dichloride is carried out in a fluidised bed reactor. This reaction generates hydrochloric acid and cupric oxychloride (3.59). Hydrochloric

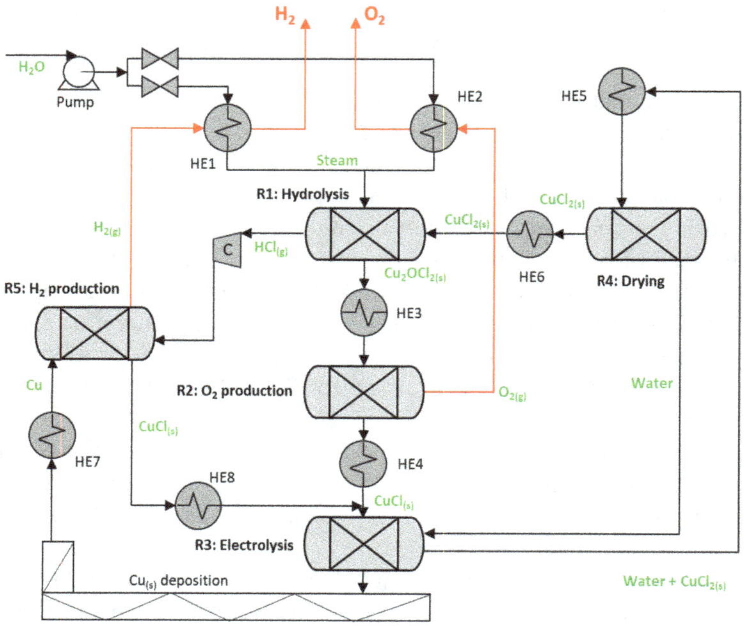

Figure 3.23 Copper–chlorine hybrid cycle process scheme

acid, reacting with copper (3.60), generates copper chloride and hydrogen, which can be easily separated as they are in different phases. Cupric oxychloride, however, is heated in a fixed bed reactor and decomposes into oxygen and copper chloride (3.61). The regeneration of copper dichloride starting from copper chloride, crystallised after cooling, is carried out by electrolysis which also regenerates the copper (3.62). Copper dichloride is separated from copper as an aqueous solution which is then heated to a temperature higher than 100 °C to evaporate the water and obtain copper dichloride.

Some variations to this cycle have been proposed. In particular, there is one that avoids the production of solid copper [51]. In this case, the reactions (3.60) and (3.62) are replaced by the following two:

$$2CuCl_{(aq)} + 2HCl_{(aq)} \rightarrow 2CuCl_{2(aq)} + H_{2(g)} \qquad < 100\ ^\circ C \tag{3.60b}$$

$$2CuCl_{2(aq)} \rightarrow 2CuCl_{2(s)} \qquad < 100\ ^\circ C \tag{3.62b}$$

This cycle has several advantages: a maximum temperature that slightly exceeds 500 °C, inexpensive chemical reagents based on chlorine and copper that are abundant and easily available and a yield that can reach values close to 43% [52]. The biggest problem consists in the corrosion of the materials by the acids that circulate at high temperatures.

Another interesting hybrid thermochemical cycle is the one involving sulphur and which consists of only two reactions [53]. The heat, of solar or nuclear origin, is used in a reactor in which, at 850 °C, sulphuric acid is decomposed into sulphur dioxide, water and oxygen (3.63). The sulphur dioxide produced is used in the second phase which consists of water electrolysis with regeneration of the sulphuric acid at the anode and releasing of hydrogen at the cathode (3.64).

$$H_2SO_4 \rightarrow SO_2 + H_2O + O_2 \tag{3.63}$$

$$SO_2 + 2H_2O \rightarrow H_2SO_4 + H_2 \tag{3.64}$$

3.5.2 Electrolysis

Electrolysis (from the ancient Greek *ελεκτρον* = amber and *λυσις* = dissolution) is the process in which a direct electric current is used to dissociate a substance into its components. When a substance is decomposed into ions and subjected to an electric field, positive ions (cations) are attracted from the negative electrode (cathode), while negative ions (anions) are attracted from the positive electrode (anode). In contact with the electrode, the former acquire electrons and are reduced, while the latter release electrons by oxidising.

To obtain hydrogen by electrolysis, it is therefore necessary to have a substance that contains it and electricity. The existing water on the Earth's surface is estimated at almost 1.4 Mm^3, of which 97% is located in the oceans and 2%, in solid form, in the polar regions. Therefore, the raw material to produce hydrogen is not lacking. As regards the electricity required, there are different production methods and, of course, those based on renewable sources would be preferable. However, it is not always true that electrolytic hydrogen from renewable sources is 'green': the real difference is not the renewability of the source but the availability of electricity in excess of the demand from the grid to which the generator is connected. Only by using a surplus from a renewable source can we speak of green hydrogen.

The first news of water splitting using electricity dates back to 1789 when Jan Rudolph Deiman and Adriaan Paets van Troostwijk succeeded by generating sparks with the help of pure gold threads. However, we must wait until 1800 to reach a continuous process of electrolysis. In that year, Alessandro Volta announced the invention of the battery and the chemist William Nicholson, after having built a specimen together with Anthony Carlisle, accidentally decomposed water during an experiment. A few months later, Johann Wilhelm Ritter built an apparatus that allowed him to collect the developed hydrogen and oxygen separately.

The basic device for separating the water molecule into its constituents by means of electricity is called the electrolytic cell. It is schematically represented in Figure 3.24. It consists of a container filled with an electrolyte consisting of water made conductive by adding salts, acids or bases. Two electrodes, one positive and one negative, are immersed into the electrolyte and separated by a diaphragm that divides the cell into two compartments avoiding mixing of products.

A single electrolytic cell has no industrial application: by placing several electrolytic cells in series and/or in parallel, an electrolyser is therefore obtained.

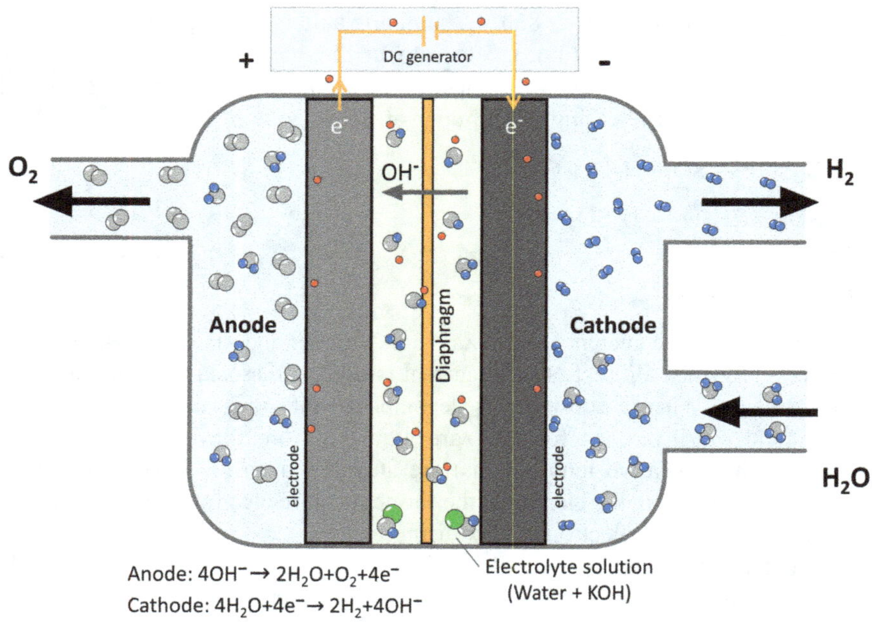

Figure 3.24 Alkaline electrolytic cell scheme

According to the connection between the cells, the electrolysers can be of the monopolar or bipolar type.

The first, as the word itself says, consists of electrodes having only one polarity. Therefore, all the cathodes are connected to each other and the same happens for the anodes. And cathodes and anodes are alternated within a single container and are separated by diaphragms. The power supply voltage is equal to that of the single cell.

The bipolar electrolyser is, on the other hand, formed by electrodes having both polarities: on one side the cathode half reactions take place and on the other the anode half reactions.

Therefore, the monopolar electrolyser is characterised by a connection of the cells in parallel which allows the disconnection of the single cell without stopping the entire electrolyser, an advantageous feature in case of maintenance of a single cell. The bipolar electrolyser, on the other hand, is characterised by a connection of the cells in series and is more compact and more suitable for operating at high pressure. Almost all the major manufacturers on the market produce bipolar electrolysers.

From the point of view of the technology used, a classification is made based on the electrolyte used on which the transferred ion depends:

- Alkaline Electrolyser Cell (AEC);
- Proton Exchange Membrane Electrolyser Cell (PEMEC);

Table 3.9 Electrode half reactions

Technology	Cathode	Anode	
AEC and AEMEC	$2H_2O + 2e^- \Rightarrow H_2 + 2OH^-$	$2OH^- \Rightarrow \frac{1}{2}O_2 + H_2O + 2e^-$	(3.65)
PEMEC	$2H^+ + 2e^- \Rightarrow H_2$	$H_2O \Rightarrow \frac{1}{2}O_2 + 2H^+ + 2e^-$	(3.66)
SOEC	$H_2O + 2e^- \Rightarrow H_2 + O^{--}$	$O^{--} - \Rightarrow \frac{1}{2}O_2 + 2e^-$	(3.67)

- Anionic Exchange Membrane Electrolyser Cell (AEMEC);
- Solid Oxide Electrolyser Cell (SOEC).

Table 3.9 shows the electrode half reactions that occur in the three cases.

The thermodynamics of a reversible electrochemical system is quite similar whether it is the dissociation of a substance into its components or whether it is a combination of substances. In Section 5.2.1, such a theoretical discussion will be made for fuel cells. In the case of water electrolysis, with reference to an alkaline electrolytic cell, the equilibrium potentials at the cathode (E_c) and at the anode (E_a) are defined by the relations:

$$E_c = E^\circ_{H_2} + \frac{R \cdot T}{2 \cdot F} \cdot \ln \frac{a^2_{H_2O}}{a^2_{OH^-} \cdot a_{H_2}} \tag{3.68}$$

$$E_a = E^\circ_{O_2} + \frac{R \cdot T}{2 \cdot F} \cdot \ln \frac{\sqrt{a_{O_2}} \cdot a_{H_2O}}{a^2_{OH^-}} \tag{3.69}$$

in which:

- E° is the standard potential of the electrode;
- R is the universal gas constant (equal to 8.314472 J · mol^{-1}·K^{-1})
- T is the absolute temperature;
- F is the Faraday constant (corresponding to the charge of 1 kmol of electrons, 96478 C);
- a is the activity.

The standard conditions according to the IUPAC (International Union of Pure and Applied Chemistry) definition correspond to 1 bar and 273.15 K, with unitary activities of all substances. Therefore, under standard conditions, the activity of the hydroxyl ion is the only that can assume values different from the unit and depends on the acidity of the electrolyte. It follows that the logarithmic term present in the two potentials is identical and, therefore, while varying the electrode potentials with the acidity of the electrolyte, the value of the reversible voltage remains constant (Figure 3.25):

$$E = E_a - E_c = E^\circ_{O_2} - E^\circ_{H_2} = 1.229 \text{ V} \tag{3.70}$$

The reversible voltage corresponds to an energy equal to 236.96 kJ, while the quantity of heat generated by the combustion of the mass of hydrogen produced is 285.58 kJ (at 25 °C). It follows that an additional energy of 48.66 kJ must be provided.

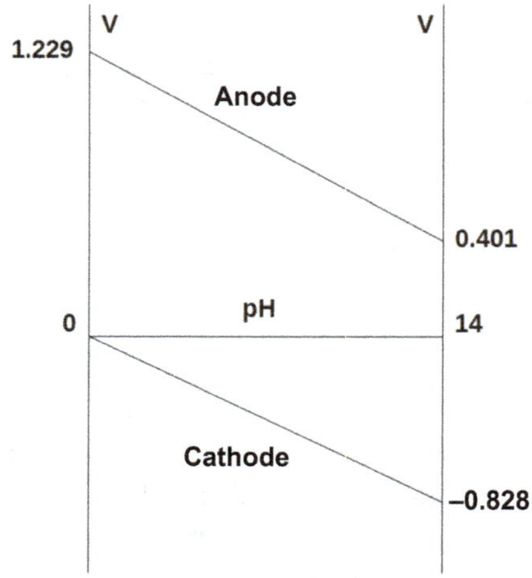

Figure 3.25 Electrodes potential versus pH

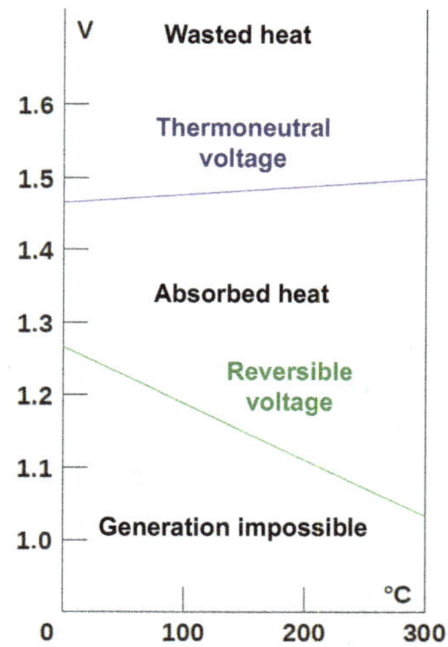

Figure 3.26 Total potential versus temperature

A voltage of 1.481 V corresponds to the value 285.58 kJ: by applying this voltage (called thermoneutral potential) to an electrolyser operating at 25 °C, there is an isothermal production of hydrogen and oxygen and there is no exchange of energy with the environment.

Therefore, for values of the voltage applied to the cell equal to that of the thermoneutral potential, it is not necessary to provide heat from the outside. For lower values, but greater than the value of the reversible voltage, it is necessary to provide heat from the outside. For values greater than the thermoneutral potential, it is instead necessary to remove heat from the cell. The reversible potential and the thermoneutral potential vary with the temperature (Figure 3.26), but while the thermoneutral potential varies modestly, the reversible one strongly decreases with the temperature.

Due to the irreversibility related to the water decomposition process, in order for electrolysis to take place, it is necessary to apply a higher voltage than the theoretical one. This over-voltage must compensate for the losses generated by the electrolytic process that occur in the cell and which are of a different nature:

$$\Delta V = \Delta V_a + \Delta V_o + \Delta V_c \tag{3.71}$$

The first over-voltage (ΔV_a) is that related to activation. In order for a reaction to take place, a certain activation energy is required which can be supplied in the form of heat, but in the case of electrolysis it is generally supplied by increasing the voltage.

The ohmic over-voltage (ΔV_o) is caused by the ohmic drop in the cell whose value is given by the product of the electric current (I) by the electrical resistance of the cell (R_e):

$$\Delta V_o = I \cdot R_e \tag{3.72}$$

The third component (ΔV_c) is the concentrated over-voltage: due to the difference in concentration of the ions in the areas in contact with the electrodes, an ion concentration gradient is created. This generates an internal battery that opposes the voltage applied to the electrodes. The slower the migration of the ions, the greater the difference in concentration and, consequently, the greater the counter-electromotive force that characterises the concentration battery.

The voltage necessary for the electrolysis is therefore:

$$V = E + \Delta V_a + \Delta V_c + \Delta V_o \tag{3.73}$$

To minimise the three components of the over-voltage to be applied: catalysts are used on the surface of the electrodes, the electrodes are brought as close as possible and, for AEC, the solution is kept agitated.

The relationship between the real voltage and the reversible one is called voltage efficiency:

$$\eta_V = \frac{V}{E} \tag{3.74}$$

Electrolysis is governed by two laws enunciated in 1832 by the English chemist Michael Faraday. The first of these laws states that the mass of a substance produced at an electrode is directly proportional to the amount of charge transferred to that electrode. The second law states that the electric charge required to produce one mole of substance at an electrode is equal to the number of elementary charges in excess of that ion.

Faraday's two laws on electrolysis can be expressed by:

$$\dot{m} = \frac{M}{n \cdot F} \cdot \int I(t)dt \tag{3.75}$$

where:

- $\dot{m}$ is the mass flow of the substance produced at the electrode;
- M is the molar mass of the substance;
- n is the valence number of the ions of the substance (electrons transferred per ion);
- F is the Faraday constant
- I is the electric current flowing through the solution;
- t is the time.

If the electric current is constant in a given time interval Δt, then the number of moles produced in that interval is:

$$N = \frac{I \cdot \Delta t}{n \cdot F} \tag{3.76}$$

A parameter generally used to define the efficiency of the electrolytic cell is the conversion factor or thermal efficiency, defined as the ratio between the amount of energy theoretically required for the electrolytic process and the amount of energy actually consumed. Since energy is generally supplied in the form of electricity, the thermal efficiency can be determined by means of the following relationship:

$$\eta_t = \frac{\Delta H}{N \cdot F \cdot V} \tag{3.77}$$

where ΔH is the enthalpy of dissociation of liquid water.

Moreover, the amount of hydrogen produced is lower than the theoretical value due to the possible formation of secondary products through secondary reactions as well as to current losses. A current efficiency is therefore defined as the ratio between the real flow rate of hydrogen and the theoretical one:

$$\eta_I = \frac{\dot{m}_{H_2}}{\dot{m}_{H_2,theor}} = \frac{\dot{m}_{H_2}}{I \cdot \vartheta} \tag{3.78}$$

where ϑ is the electrochemical equivalent, i.e. the mass of an element that is deposited on one of the electrodes of an electrolytic cell upon the passage of the charge of one Coulomb. It is obtained by dividing the gram-equivalent by the charge of one Faraday.

To obtain the maximum efficiency, the electrolytic cell should therefore be operated at a voltage just above the minimum necessary for the formation of hydrogen. However, under these conditions, a little electric current flows and the flow rate of hydrogen produced per electrode surface unit is very low. To produce interesting quantities of hydrogen, the cells should be very large, and therefore expensive. By increasing the applied voltage, the production per cell surface unit increases, but the efficiency decreases. As the temperature increases, the voltage–current density curve shifts downwards (Figure 3.27) and, therefore, it is possible to apply a lower voltage while leaving the current density unchanged.

Pressure variation, on the other hand, does not have a significant influence on the voltage to be applied to the electrolyser. This is, at least in part, due to the

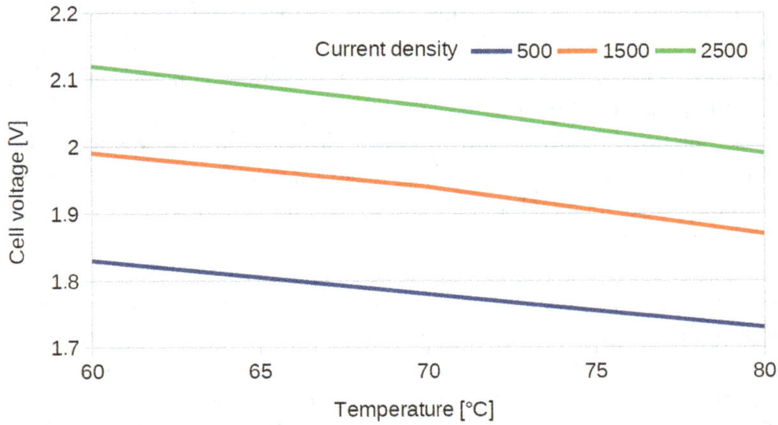

Figure 3.27 Voltage–current characteristic curves of an electrolytic cell

decrease of the size of the gas bubbles formed on the electrodes and generating leaks. This is a very interesting aspect since it allows to operate at high pressure, reducing the need to compress the gases downstream of the electrolyser. Despite this advantage, until the 1970s only one electrolyser at 30 bar (Lurgi) was on the market due to greater difficulty in controlling the pressure, greater corrosion of the materials and a higher cost associated with safety aspects. In the meantime, the technology has improved and electrolysers operating up to 200 bar will be probably manufactured and marketed in next years at grid scale. Considering that the work of pumping the water is completely negligible compared to that required for compressing the gases, the advantage is quite evident [54,55].

Alkaline electrolysers

They are by far the most common and as an electrolyte they use an aqueous solution containing 25%–35% by weight of potassium hydroxide (KOH) which, at the same concentration and temperature, is much more conductive than sodium hydroxide. The transferred ion is the hydroxyl ion that migrates from the cathode to the anode.

High values of temperature and concentration and a small distance between the electrodes severely limit the ohmic resistance. For each additional degree Celsius, the electrical conductivity of the electrolyte increases by 2%–3%, but the operating temperature is usually between 70 °C and 85 °C to avoid corrosion phenomena. As regards the concentration, the limit is imposed by the precipitation of potassium which, in ambient temperature conditions, occurs for concentrations higher than 47%.

Generally, the container is made of nickel-plated steel, a material also used for the electrodes which are then coated with a catalyst. The diaphragm consists of a microporous membrane permeable only to ions: in the past asbestos was used, but, due to its harmfulness, it has been replaced with ceramic oxides reinforced with a metal or polymer mesh. Zirfon is one of the most used nowadays.

Nickel is used as a cathode catalyst. Fine-grained and very porous catalysts have the advantage of a high contact surface, but they recrystallise at high temperatures and lose their efficiency. An alternative with a high contact surface may be nickel filaments and foams.

For the anode in the 1980s, mixtures of oxides such as perovskites and spinels were studied, also containing transition metals such as Ni, Co, Mn and Ag. The most used materials are cobalt oxides and Raney nickel, a very fine-grained nickel and aluminium alloy [56].

While conventional alkaline electrolysers operate at a pressure close to atmospheric pressure, there are 'advanced' alkaline electrolysers that allow higher pressures operations, usually up to 30 bar. These electrolysers are equipped with a diaphragm that has a specific resistance of 0.2–0.3 $\Omega\ cm^{-2}$ and better catalysts: thanks to these characteristics they are able to operate with a current density of 0.4–0.5 A/cm^2 and temperatures up to 120 °C–150 °C with consequent advantages in terms of both efficiency and overall dimensions. The cell voltage drops from 1.84 to 2.25 V (at 0.13–0.25 A/cm^2) of conventional electrolysers to 1.6–1.9 V with an increase in efficiency from about 74% to over 80%. The operating temperature is limited by the stability of the materials used; however, there is the hope to reach 200 °C [57].

This type of electrolysis allows to reach a high purity of hydrogen, even over 99.8%.

A recent article [58] proposed a new concept for low temperature electrolysis cells based on the supply of water to the electrodes which develop hydrogen through capillaries with the result that the formation of gas bubbles on the electrodes is avoided. The performance of an electrolytic cell of this type is largely superior to that of a traditional cell: with a voltage of 1.51 V at 0.5 A/cm^2 and 85 °C, the energy consumption of 40.4 kWh/kg of hydrogen, corresponding to an energy efficiency of 98%. By adopting this technology, Hysata [59] claims to produce electrolysers with only slightly lower efficiency (95%).

Another new approach to generate hydrogen by water electrolysis is based on the amphoteric behaviour of the latter. In reality, it is not possible to place this type of electrolytic cell in this position, nor anywhere else because it is neither alkaline nor acidic. Water, in fact, behaves differently depending on whether it is in an acidic or alkaline solution. It is therefore possible to create an electrolytic cell in which a membrane separates two environments, an acidic one, in which the development of hydrogen is obtained, and an alkaline one, in which the development of oxygen is obtained. This allows to reduce the overvoltage at the electrodes with a reduction of approximately 30% in energy consumption at a current density of 0.2 A/cm^2 [60].

Proton exchange membrane electrolysers

Acid electrolysis of water has made a comeback in recent decades thanks to polymer membranes. This technology was developed in the 1960s for fuel cells that were used in aerospace missions and was later transferred to electrolytic cells as well. The compactness and the absence of liquid electrolyte offer significant

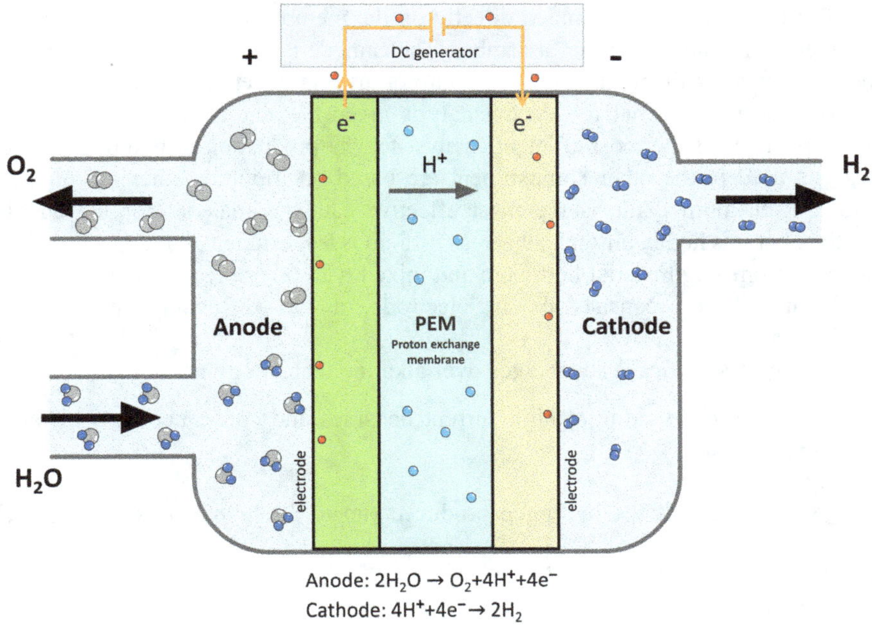

Figure 3.28 Proton exchange membrane electrolytic cell scheme

advantages over alkaline technology. They allow high pressures with a fast response to transients. They are all bipolar type and, for the most part, have a different pressurisation of the two electrode compartments with the hydrogen compartment operating at higher pressure. Although the cost of the materials used, especially membranes and catalysts, has limited in the past their use to niche sectors, PEMECs are now a mature industrial technology.

Since this is acid electrolysis, the ion that is exchanged across the membrane is the proton (Figure 3.28). The membrane is a perfluorosulfonic acid (PFSA) polymer with a thickness of 50–200 μm. It is characterised by a high proton conductivity and a low electronic conductivity. It must also have a high chemical and mechanical stability and must be slightly permeable to gases so that explosive mixtures are avoided. PFSA membranes can only be used at temperatures below 100 °C: at higher temperatures, in fact, the ionomer is subject to dehydration, with consequent loss of mechanical stability and proton conductivity [61]. Operation at higher temperatures can be achieved through the use of inorganic additives that improve water retention within the PFSA membrane (e.g. silicon and titanium oxides). Solvay claims that the membrane Aquivion has an excellent thermal stability up to 230 °C [62]. However, the historically most used polymer is Nafion, developed by DuPont Corporation and currently produced by Chemours, which can be produced in thin sheets and resists to gas penetration well, thus avoiding the use of a diaphragm and obtaining very thin electrolytic cells.

The electrodes are assembled directly on the membrane and together form the so-called membrane electrodes assembly. The cathode is made of carbon, while the anode is made of titanium. They are both porous and covered with catalysts with precious metals, essential for electrocatalysis given the acidic nature of the membrane: platinum is the most efficient catalyst for the production of hydrogen, while catalysts used in the other compartment are based on iridium, ruthenium or their oxides. Ruthenium oxide is the most effective catalytic material for the anodic reaction, but it is highly unstable, whereas iridium is less efficient, but it is considered the most promising material both pure and mixed with other precious metals.

Each half-cell consists of an electrode, a gas diffusion layer, and a bipolar plate.

The most significant advantages over alkaline technology are:

- possibility of working at high current densities and consequent considerable compactness;
- higher yield;
- greater purity of the hydrogen produced (up to 99.9998% as stated by Nel ASA), [63];
- absence of corrosive liquid electrolytes and consequent greater durability and reliability.

Anion exchange membrane electrolysers

This kind of electrolyser is based on an alkaline electrolyte, but it is also divided in two half-cells by a membrane (Figure 3.29) just as PEMEC. The difference with the

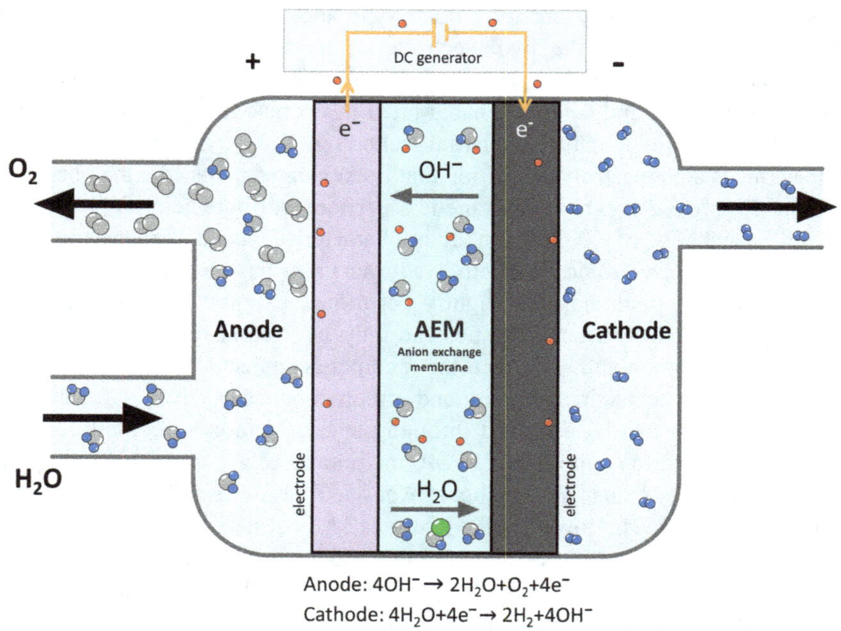

Figure 3.29 Anion exchange membrane electrolytic cell scheme

latter is that the exchanged ion is not a proton, but an anion: the hydroxyl (OH^-). Like in a traditional alkaline electrolyser, the water molecules move through the electrolyte, reach the cathode and are reduced to produce hydrogen. Whereas in a traditional alkaline electrolyser, the content of KOH is quite high, in AEMEC, it is as low as 1%. It circulates only at the anode and wets the membrane without crossing it. This is made possible by the difference of pressure between cathode (35 bar) and anode (1 bar). Thank to such a difference, the oxygen produced cannot reach the cathode and a highly pure (99.9%) hydrogen is obtained with only a very low moisture content [64].

The low KOH concentration is very important since it allows to use steel bipolar plates and transition metal catalysts, avoiding the use of expensive platinum catalysts.

The power consumption claimed by Enapter is 4.8 kWh/Nm^3_{H2} [65].

Solid oxide electrolysers

As previously mentioned, the specific consumption of electric energy can be reduced by increasing the operating temperature: in fact, while the total energy required by the reaction remains almost constant, the less valuable thermal component replaces the electrical component gradually the temperature rises. In the face of an increase in efficiency, there is an increase in construction costs and, although at an experimental level operating temperatures above 1000 °C have been reached, the temperature normally used is between 650 °C and 850 °C [66].

Figure 3.30 shows how a SOEC works: water, in the vapour state, is fed into the cathode and, under a voltage application, moves towards the electrolyte surface

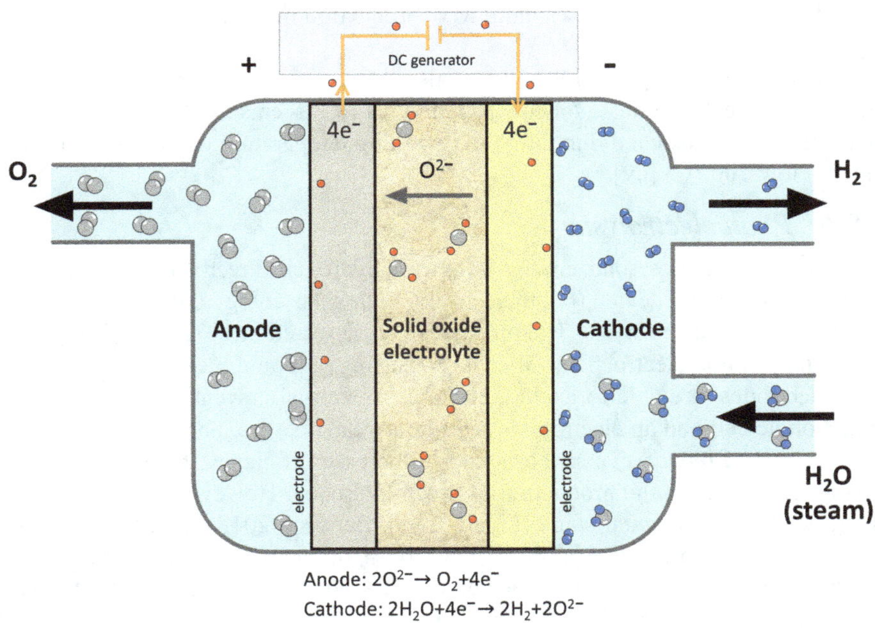

Figure 3.30 Solid oxide electrolytic cell scheme

where is reduced to pure hydrogen and oxygen ions. While the hydrogen is collected at the cathode, the oxygen ions cross the electrolyte and are oxidised to pure oxygen gas, which is collected at the anode.

Also in this case it is a technology derived from the fuel cell sector and the electrolytes developed for the latter are also suitable for use in electrolysis. As for solid oxide fuel cells, also for high temperature electrolysers the first configuration tested was the tubular one: this is the HOTELLY (High Operating Temperature ELectroLYsis) project developed between 1970 and 1987 by Dornier and Lurgi as well as other research activities conducted by Westinghouse. The tubular configuration offers the considerable advantage of realising the separation of the products, one inside and one outside the single cell. The HOTELLY project was stopped due to the high cost of the electrolyser and the lack of a short-term industrial application perspective after having conducted tests on electrolysers up to 100 cells operating at 1000 °C and up to 2500 operating hours [67].

The materials used must, of course, withstand high temperatures. The electrolyte is made with zirconium dioxide stabilised with diyttrium-trioxide (YSZ, yttria-stabilised zirconia). The cathode is a nickel cermet on a YSZ support, while the anode, also on a YSZ support, is made of lanthanum strontium manganite or is made with a mixture of oxides (lanthanum strontium cobalt ferrite).

The main problems are connected to the corrosion of the electrodes which can generate significant efficiency losses (up to 45%). In addition to the advantage of using a solid electrolyte, this type of electrolysers has the advantage of using heat rather than producing it as with other technologies. This allows for a significant increase in efficiency, even if it requires a high temperature heat source. During last year due to EU funding, different companies are investing to arrive at the commercial stage and in some cases multi MW production line, like Elcogen, Convion, SolydEra, SunFire [68–70].

Main topics are electrode materials (between new solution perovskite, nickelate-based materials, doped-LNO and LPNO seems to be promising) to increase performance and durability and working temperature decreasing (in same cases below 700 °C) [69].

3.5.3 Photoelectrolysis

Photoelectrolysis uses solar energy to extract hydrogen directly from water integrating into a single device (photoelectrode) both solar energy conversion to electricity and water electrolysis. The process was discovered in 1972 using TiO_2 [71].

Tandem photoelectrochemical cells, consisting of n- and p-type semiconductor photoelectrodes, are able to exclude the use of external current source. Joining a photovoltaic cell and an electrolysis cell into a single device, photoelectrolysis has the potential to reduce overall costs and to increase efficiency, being a very promising technique for the production of green hydrogen. However, this technology is still in the experimental stage (Figure 3.31). Between 2018 and 2020, colloidal semiconductor photo-active systems have been proposed, with interesting scalability pathway [72]. In general, scale up of this kind of technology will be relevant during the second half of 21st century [73].

Photoelectrodes are made of three layers: photovoltaic, catalytic, and protective or encapsulant. The first one is the semiconductor able to absorb sunlight (e.g. titanium dioxide or gallium/arsenide). Since the suitable semiconductors absorb only the most energetic portion of the solar spectrum, photo-sensitisers capable of widening the absorbed portion of the solar spectrum are added in order to increase light absorption and efficiency [74]. Photoelectrodes require different catalysts respectively capable to favour hydrogen and oxygen evolution reactions (HER and OER) and these catalysts constitute the second layers. Finally, the protective layer has the task to prevent the corrosion of the semiconductor immersed into the electrolyte. Of course it must be highly transparent in such a way to not obstacle sunlight in its path towards the photovoltaic layer.

This process requires that:

- there is an overlapping between energy band of the semiconductor and energy levels of the hydrogen and oxygen reduction reactions;
- the semiconductor shows a good stability under process conditions;
- the speed of charge transfer from the surface of the semiconductor is such to avoid corrosion and to reduce overvoltage energy losses.

To date, however, there are still some critical aspects to be resolved: low conversion efficiency, photocorrosion, product recombination and catalyst instability. The use of photoelectrodes equipped with cocatalysts has considerably improved these aspects. In particular, the kinetics of HER and OER are increased. For the photocathode, cocatalysts such as Mo sulphides, Ni sulphides and Co phosphides can be used, while for the photoanode, Ni-doped iron (III) oxide hydroxide, Co oxide and nickel oxide hydroxide [76]. This allows to avoid the use of catalysts based on Pt, Ru and Ir. However, to industrially manufacture high-efficiency and stable cells, it will also be necessary to be able to design new types of interface between the cocatalyst and the photoelectrode. Furthermore, the aspect

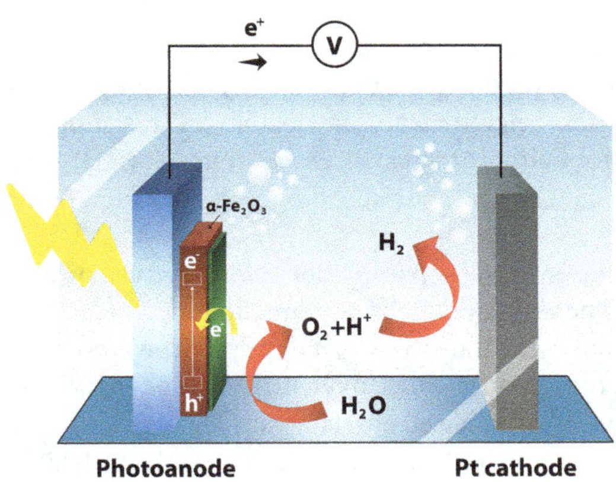

Figure 3.31 Schematic of a photoelectrochemical cell (adapted from [75])

of photocorrosion remains very critical, which requires new technologies for deposition of the cocatalyst on the photoelectrode.

A photoelectrode is basically a semiconductor which, immersed into an aqueous electrolyte, and exposed to sunlight, generates an electric voltage capable to make possible hydrogen and oxygen evolution reactions.

3.5.4 Plasma-driven electrolysis

Plasma-driven solution electrolysis is an electrochemical process in which an electric plasma involves a water electrolytic solution comprising a gas–vapour area [77].

Plasma-driven solution electrolysis in a electrolytic solution happen by a simultaneous events: a Faradaic electrolysis, a Joule heating of the electrolytic solution, evaporation at the electrode of the solution, the ionisation of the gas–vapour mixture in a small volume the proximity of the cathode, and, eventually, an electric arc discharges across the ionised gas–vapour mixture.

It is interesting to note that if in the aqueous solution are present alcoholic additives (methanol or ethanol), or acetic acid or glycerol [78], the energy yield of plasma-driven solution electrolysis is greater of many times the energy yield of common electrolysis technologies such as alkaline electrolysis, polymer electrolyte membrane electrolysis and solid oxide electrolysis. Moreover, no precious metals as catalysts are needed.

German-based company Graforce is starting commercialisation of a plasma electrolysis to produce also hydrogen starting from wastewater.

3.5.5 Seawater electrolysis

As already stated at the beginning of this chapter, drinking water is a commodity of primary importance and extracting hydrogen from drinking water is a regrettable option. One option is the use of seawater which is extremely abundant. Seawater contains various substances in the form of ions: mainly chlorine, but also other substances such as calcium and magnesium. These ions interfere with the electrolysis process of the water giving rise to side reactions. The main problems are the possible anodic production of chlorine which competes with that of oxygen, the need for efficient electrocatalysts capable of resisting in the presence of chloride, and the formation of substances, such as $Mg(OH)_2$ and $Ca(OH)_2$, which can precipitate on the electrode surface, clogging pores and partially reducing their activity.

A possible solution, at least partial, consists in the use of electrocatalysts based on transition metal hexacyanometalates, transition metal nitride and layered double hydroxide integrated with transition metal sulphide which are robust and highly selective towards the production of oxygen [79]. The same article reports also a comprehensive review of other electrocatalysts.

A recent article compared the direct seawater electrolysis with a fresh water electrolysis obtained by performing a preventive seawater desalination and purification. Although this requires increased capital and operating costs, the authors stated that they are marginal compared to those of water splitting and concluded that there is no convenience to directly electrolyse seawater [80]. Another recent article [81] also concluded the same, however, stated that it is necessary to regularly update the economic comparison. Actually we must not overlook that the comparison of the capital costs of reverse osmosis and electrolysis is greatly influenced by the different level of maturity of the two technologies. It is very likely that this cost difference will decrease as a result of economies of scale and improvements in electrolyser design. For example, AEMEC could be a promising technology for direct seawater splitting [81].

Finally, at the University of Shenzhen, a system has been developed that looks very promising for obtaining the direct electrolysis of sea water [82]. As in a conventional alkaline electrolyser, the electrodes are immersed in a concentrated electrolyte solution of potassium hydroxide from which hydrogen and oxygen evolve separately. The sea water remains separated from the electrolyte by special fluorinated membranes which allow the water to pass only if it is in the vapour state. When the electric current generates the splitting of the water molecules contained into the electrolyte, a pressure difference occurs between the sides of the membranes. This produces a spontaneous evaporation of the sea water and the vapour crosses the membrane and then liquefy again, replenishing the consumed water. According to the researchers, the cost of the membranes is very low and the efficiency of the system is in line with that of conventional alkaline electrolysers. This same system appears to be suitable also for the production of hydrogen from industrial wastewater.

3.6 Biological processes

There are several microorganisms capable of producing hydrogen both in the presence of light and in the dark, both in the presence of oxygen and in anaerobiosis, both by simply decomposing water and by decomposing organic substances.

3.6.1 Light-driven processes

This group of bioprocesses includes bio-photolysis and photo-fermentation.

The bio-photolysis of water is a process that carries out the splitting of water into gaseous hydrogen and oxygen in presence of light. This can be carried out by cyanobacteria and unicellular algae, which are photoautotrophic microorganisms, prokaryotic and eukaryotic, respectively. These microorganisms are equipped with pigments capable of capturing light radiations (chlorophyll and phycocyanin) and with special enzymes, i.e. biological catalysts, which allow to favour the bio-photolysis of water. Prokaryotes were the first living organisms to appear on Earth and are single-celled non-compartmentalised organisms. Eukaryotes, however, are uni- or multi-cellular organisms with a well-defined nucleus separated from the rest of the cell by the nuclear membrane.

The discovery that particular unicellular eukaryotic green algae, under certain conditions and for short intervals of time, produce hydrogen when exposed to sunlight dates back to the studies of H. Gaffron in 1939. Later it was discovered that also cyanobacteria, which are photosynthetic bacteria, have this feature.

The photosynthetic process is carried out by two photosystems that operate in sequence: in the first (PSII), the water is decomposed and oxygen and protons are released, while in the second (PSI) the protons are used to generate niconitamide adenine dinucleoid phosphate reduced to be used in the synthesis of carbohydrates.

Unicellular eukaryotic green algae can then convert protons coming from the water splitting into hydrogen thanks to the enzyme hydrogenase. The light radiation absorbed by the PSII generates electrons which are transferred to the ferredoxin using the light radiation absorbed by the PSI. The hydrogenase accepts the electrons carried by the reduced ferredoxin to generate hydrogen.

The enzyme linked to hydrogen production is hydrogenase, which can occur in two different types: uptake hydrogenase and bidirectional hydrogenase [83]. The first catalyses only the oxidation of hydrogen and is therefore not useful for its production:

$$H_2 \Rightarrow 2H^+ + 2e^- \tag{3.79}$$

The second, however, catalyses both the oxidation and the synthesis of hydrogen, bringing in this second case the ferredoxin (X), which is a protein capable of transporting electrons, from the reduced to the oxidised form:

$$2H^+ + 2X_r \Rightarrow H_2 + 2X_o \tag{3.80}$$

The light radiation absorbed by the PSII generates electrons which are transferred to the ferredoxin using the light radiation absorbed by the PSI. The hydrogenase accepts the electrons carried by the reduced ferredoxin to generate hydrogen.

Hydrogen production requires an anaerobic incubation time ranging from several minutes to a few hours for the synthesis and activation reactions of the hydrogenase to take place. Since hydrogenase is very sensitive to oxygen, the production of hydrogen requires a two-stage process to decouple the oxygen production coming from photosynthesis from hydrogen production.

This was demonstrated to be possible, by removing sulphur present in the cultivation medium of the green algae *Chlamydomonas reinhardtii*. This operation blocks the production of oxygen creating the condition of anaerobiosis necessary for the synthesis and activation of the hydrogenase [84].

Chlamydomonas reinhardtii is present mainly in soils rich in water and in lake basins. The preparation of cultures of this alga takes place in different incubation conditions (aerobiosis/anaerobiosis, bi-phasic cultures, presence/absence of light). In addition, different culture media have been tested, i.e. solutions, solid or liquid containing nutrients (amino acids, vitamins, ions, glucose, lipids and hormones).

Cyanobacteria are found in the marine environment and in different types of soils. They use air, water, minerals and light as nutritional sources. They are called nitrogen-fixing bacteria because they are equipped with heterocystic cells that have the task of fixing atmospheric nitrogen and converting it into ammonia. There are also non-heterocystic cyanobacteria which, however, are able to fix nitrogen, but at a much slower rate.

The reactions involved in the nitrogen fixation process are catalysed by the nitrogenase enzyme which acts only in prokaryotic cells by transforming inorganic nitrogen into organic nitrogen according to the following reaction:

$$N_2 + 8H^+ + 6e^- \Rightarrow 2NH_4^+ \tag{3.81}$$

Nitrogenase is an extremely oxygen-labile enzyme; however, the heterocystic cells contained in the cyanobacteria are not equipped with PSII so they do not produce oxygen and have a very thick wall that is impermeable to the oxygen of the surrounding environment [85]. PSI, however, is active and store energy by producing adenosine triphosphate (ATP) from adenosine diphosphate (ADP).

During nitrogen fixation, ATP is utilised as an energy source and is reconverted to ADP. In the absence of nitrogen, the nitrogenase enzyme cannot fix nitrogen. However, the microorganisms continue to store energy producing ATP if exposed to the light. Therefore, it is necessary to dispose energy and this is done by nitrogenase itself producing hydrogen and reconverting ATP to ADP.

$$2H^+ + 2e^- \Rightarrow 2H_2 \tag{3.82}$$

In this way, a complete splitting of the water is carried out. Some species of cyanobacteria also produce hydrogen through the enzyme hydrogenase, although they also own the NiFe hydrogenase which prefers oxidise hydrogen [86,87].

Photofermentation is carried out by some photo-heterotrophic bacteria (rhodobacteria) which have the ability to convert organic acids into hydrogen and carbon dioxide under anaerobic conditions and in the presence of light. Some may, however, have different types of metabolism: for example *Rhodopseudomonas Palustris*, a very common bacterium in nature, can have photo-autotrophic or photo-heterotrophic or chemo-autotrophic or chemo-heterotrophic behaviour.

These bacteria are mesophilic, i.e. they have an optimal growth temperature between 30 °C and 35 °C. They prefer organic acids as a carbon source and a neutral culture medium (pH = 7). Rates of hydrogen production vary according to the type of microorganism, the carbon source and the light intensity. On the basis of studies carried out, the maximum yield was obtained by using lactic acid as a carbon source.

Also in this case, the enzyme capable of catalysing the production of hydrogen is nitrogenase. The global generic reaction of the photo-fermentative processes is as follows:

$$CH_3COOH + 2H_2O \Rightarrow 4H_2 + 2CO_2 \tag{3.83}$$

The presence of ammonia can lead to inhibition of hydrogen production because the nitrogen fixation is obviously the preferred path for the bacteria.

A transparent reactor capable of creating an environment conducive to the growth of microorganisms such as algae and cyanobacteria is called a photobioreactor. In addition to the transparency of the reactor, the culture substrate must also be transparent. Therefore, if the substrate is not clear enough, a clarification pre-treatment is required. Moving away from the transparent surface, however, the brightness is reduced due to the absorption of light by microorganisms.

Essentially, photo-bioreactors are classified into vertical column, horizontal tubular and flat panel photo-bioreactors.

The vertical column photo-bioreactors consist of a transparent cylindrical column made of high-quality glass and surrounded by a water jacket that guarantees the maintenance of the temperature in the culture. Exposure is not optimal and, therefore, vertical reactors are usually used for mother culture intended to provide new microorganisms for inoculation.

Horizontal tubular photo-bioreactors are usually used for the production section. They consist of transparent plastic or glass tubes with a length ranging from 10 to 100 m and a diameter of 3–6 cm. The horizontal arrangement, actually slightly inclined to facilitate the collection of the gas, allows to collect about 88% of the radiation that would be obtained with a heliostat and about 95% of the radiation that occurs on an inclined plane of an angle equal to latitude.

The flat panel photo-bioreactors have a stainless steel structure and walls made of polycarbonate panels.

The light intensity has a positive effect on the production of hydrogen; however, it is necessary to keep in mind the so-called flash effect: a phenomenon of saturation by light radiation due to the inability of the cell to use more rapidly the energy that the photochemical apparatus is able to supply: therefore, many more photons are absorbed than can be processed. Experiments conducted with a stroboscopic light source have shown that intermittent lighting (hence the name of the effect) can generate the same amount of hydrogen as continuous lighting. In the presence of sunlight, it is possible to generate a similar effect by alternately subjecting the cells to light radiation by mixing.

Figure 3.32 shows a possible schematic of a plant for light-induced hydrogen bio-production.

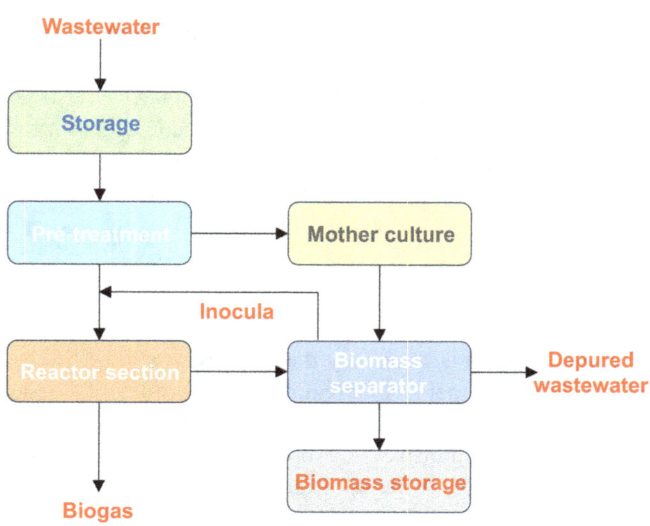

Figure 3.32 Scheme of a plant for biogas photosynthetic production

3.6.2 Dark fermentation

Dark fermentation is the fermentative conversion of an organic substrate to hydrogen (chemo-trophic hydrogen production). It is a complex process which involves different families of bacteria and can be considered like a stopped anaerobic digestion. In the latter process, hydrolytic microorganisms hydrolyse complex organic polymers to amino acids, sugars and fatty acids. These substances are then converted by acidogenic bacteria into a mixture of volatile acids. The third step is carried out by acetogenic bacteria which produce acetate and hydrogen. Finally, methanogenic bacteria convert acetate and hydrogen into methane and carbon dioxide. To produce hydrogen is important to avoid the last step which consumes hydrogen. This is possible by selecting only non-methanogenic bacteria. However, the process is not very efficient just in the third step and more than half COD (chemical oxygen demand) remains as a product of the acidogenesis. The production of chemo-trophic hydrogen can be carried out by mesophilic (25 °C–40 °C), thermophilic (40 °C–65 °C) or hyper-thermophilic (>80 °C) microorganisms.

The high temperature appreciated by hyper-thermophilic bacteria creates a highly selective environment towards other microorganisms. It follows that the activity of methanogenic bacteria is reduced and this favours the production of hydrogen over that of methane.

Among others, *Thermotoga neapolitana* and *Thermotoga maritima* [88] have aroused considerable and growing interest. They are very promising hyper-thermophilic bacteria for the production of hydrogen from various organic substrates such as agro-food waste and algal biomass [89]. Furthermore, *T. neapolitana* allows the recovery of by-products such as lactic acid [90] which have a commercial value and therefore allow to reduce production costs.

Finally, the coupling of the chemo-trophic production of hydrogen to processes such as the cultivation of micro-algae and photo-fermentation by non-sulphurous purple bacteria also appears feasible and very interesting.

3.7 Hydrogen from chemical hydrides

Hydrides, both binary and complex, capable of releasing hydrogen by reaction with water are commonly grouped in the category of chemical hydrides. The released hydrogen derives largely, and sometime completely, from water.

Ionic hydrides are obtained by directly reacting the metal with gaseous hydrogen: the reaction takes place with a reduction of the moles in the gas phase and is therefore favoured by high pressures. They generally react spontaneously, and sometimes violently, with water, decomposing into gaseous hydrogen and hydroxide and/or oxide, as happens for example with lithium:

$$LiH + H_2O \Rightarrow LiOH + H_2 \tag{3.84}$$

$$2LiH + H_2O \Rightarrow Li_2O + 2H_2 \tag{3.85}$$

Due to this characteristic, to protect them from humidity, they are sometimes transported as a suspension in light mineral oil.

Table 3.10 Binary chemical hydrides

Metal	Li	Na	K	Ca
Hydride	LiH	NaH	KH	CaH_2
Waste	Li_2O**	NaOH	KOH	$Ca(OH)_2$
Reaction	$2LiH+H_2O\rightarrow$	$NaH+H_2O\rightarrow$	$KH+H_2O\rightarrow$	$CaH_2+2H_2O\rightarrow$
	Li_2O+2H_2	$NaOH+H_2$	$KOH+H_2$	$Ca(OH)_2+2H_2$
$\Delta H°$ [kJ mol^{-1}]	−118	−82	−99	−222
Waste MW	60	40	56	148
Density*	11.76%	4.76%	3.45%	5.13%
	13.33%	5.00%	3.57%	5.41%
Recycle	$Li_2O+H_2O\rightarrow$	$NaOH\rightarrow$	$KOH\rightarrow$	$Ca(OH)_2\rightarrow$
	$2LiH+O_2$	$NaH+½O_2$	$KH+½O_2$	CaH_2+O_2
ΔH_{min} [kJ mol^{-1}]	690	368	385	794

*In the first line, the weight density is evaluated as the ratio between the mass of hydrogen produced and the mass of the hydride plus that of water, while in the second line it is evaluated as the ratio between the mass of hydrogen and the mass of waste produced.
**By increasing the amount of water, lithium also generates hydroxide as a residue, but the weight density of the hydrogen stored drops significantly.

The binary hydrides of some interest are listed in Table 3.10. The hydrides of beryllium, boron, magnesium, and aluminium are not listed because they are not ionic hydrides, but covalent, and therefore have a different behaviour. And there are no metal hydrides having a molecular weight higher than calcium since, as can be guessed from the table, the accumulation density of hydrogen would further decrease.

A first factor that determines the interest in a chemical hydride is the identification of a process capable of regenerating the hydride starting from the oxide or the hydroxide generated by the reaction with water so that, when fully operational, the entire process does not consume raw materials and does not produce waste (Figure 3.33). The only exchange between the system and the surrounding environment must be an exchange of energy that involves an energy supplied from the outside (E_s) to regenerate the hydride, an energy lost in the reaction that releases hydrogen (E_l). In particular cases, a part of the thermal energy developed by the reaction between hydride and water can be recovered. The reactions indicated in Table 3.10 for the recycling of the waste to hydride is the ideal reaction corresponding to the minimum energy demand.

A second fundamental factor is the energy density per unit of mass. This is interesting when the chemical hydride is considered for on board generation and compared with on board hydrogen storage. In this regard, it should be noted that there are three substances involved: the hydride, the water and the waste (oxide or hydroxide). It should also be noted that the use of hydrogen produces water again and, if this can be recovered on board (e.g. in the case of the use of hydrogen by PEMFCs), it must not be taken into account except in a minimal amount. Ultimately, the production of hydrogen generates a slight decrease in overall mass

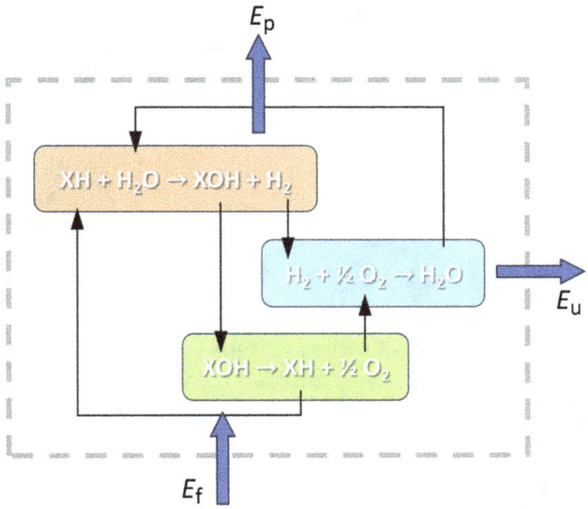

Figure 3.33 Water–hydride closed cycle

if the water is to be entirely stored, while it generates an increase in overall mass if the water can be recovered within the system. Therefore, the energy storage density can be considered equal to the ratio between produced hydrogen and sum of hydride and stoichiometric water stored in case of water not recyclable, but it increases to the ratio between produced hydrogen and produced waste if recycle of water is possible and the amount of water negligible with respect to the amount of hydride. For example, in case of sodium hydride:

$$\frac{m_{H_2}}{m_{NaH} + m_{H_2O}} < \frac{m_{H_2}}{m_{NaOH}}$$

A storage system based on binary chemical hydrides was developed and tested by Powerballs Technologies. These are spheres with a diameter of a few centimetres containing sodium hydride and covered with an impermeable layer made of polyethylene. Thanks to the waterproofing, they can be contained in water inside the tank (Figure 3.34). In the upper part of the tank, there is hydrogen in the gaseous state. When hydrogen is consumed, the pressure in the tank decreases. Once a minimum pressure threshold level is reached, a suitable mechanism cuts a sphere in half so that its contents are poured into water and the reaction between water and sodium hydride occurs. This reaction is very fast and the hydrogen that develops causes the pressure in the tank to rise again.

As highlighted above, chemical hydride systems must have a closed cycle in which there is no production of residual substances. In the mentioned case, sodium hydroxide and polyethylene are discharged at each refuelling. The latter is separated and recycled, while sodium hydroxide could have a market and therefore can be considered a by-product. However, both due to the existence of numerous production plants and because a massive use of sodium hydride would in any case lead to overproduction, it is also essential to provide for the recycling of sodium hydroxide.

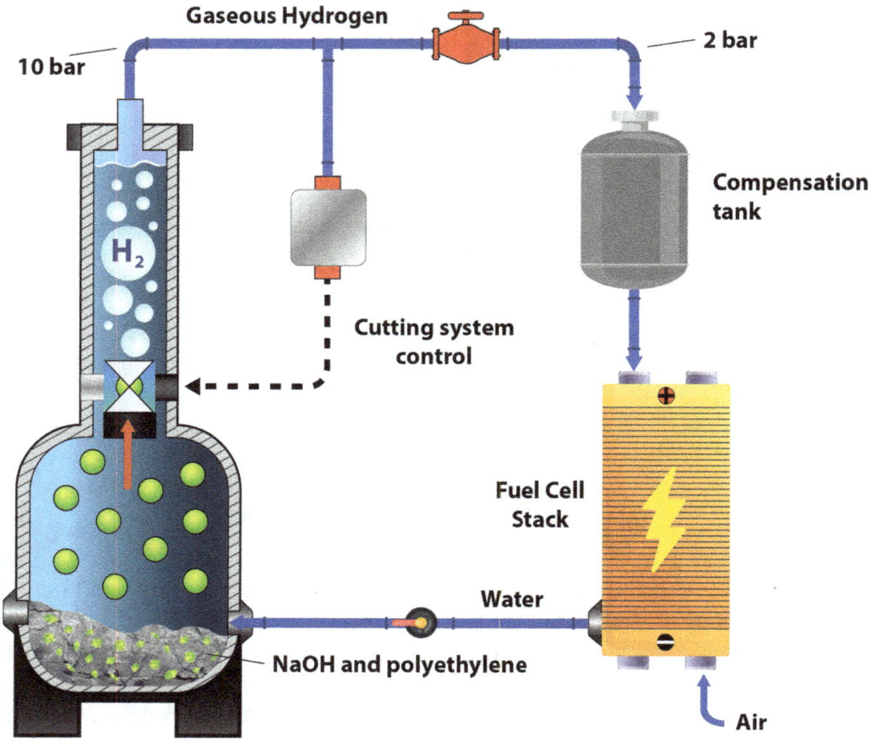

Figure 3.34 Powerballs Technologies system

Powerballs Technologies itself has identified a recycling process starting from the use of methane (Figure 3.35). In fact, at high temperature (>1000 °C), sodium hydroxide reacts with methane and decomposes according to the reaction:

$$NaOH + CH_4 \Rightarrow Na + CO + \frac{5}{2}H_2 \tag{3.86}$$

in such a way that the basic components for the production of the hydride are made available. Liquid sodium is completely separated, while only one fifth of the hydrogen is required by the hydride. The remaining fraction, which is equivalent to that made available by methane, is used together with carbon monoxide to feed the combustion chamber which provides the heat necessary for the decomposition reaction.

Sodium hydride essentially provides energy to the system, while hydrogen is supplied by water. Therefore, a possibility of recycling is to use sodium hydride as an energy storage from a renewable source. This is possible using an electrolysis process; however, the efficiency of the process is low.

In addition to binary hydrides, complex hydrides have aroused some interest. As already seen, some of these exhibit a behaviour partially similar to that of interstitial metal hydrides, while others are able to release hydrogen if reacted with water in the presence of suitable catalysts. Among the various complex hydrides, sodium tetrahydroborate has been studied above all, which had been tested, however with little success, by NASA. In the presence of a ruthenium catalyst, sodium tetrahydroborate reacts with water according to the reaction:

$$NaBH_4 + 2H_2O \Rightarrow NaBO_2 + 4H_2 \quad (3.87)$$

A scheme of the system is represented in Figure 3.36.

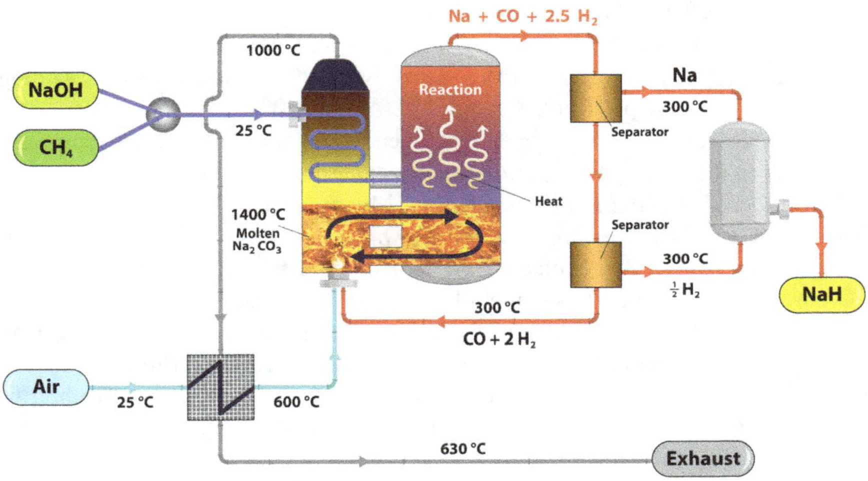

Figure 3.35 NaOH/NaH recycle plant scheme

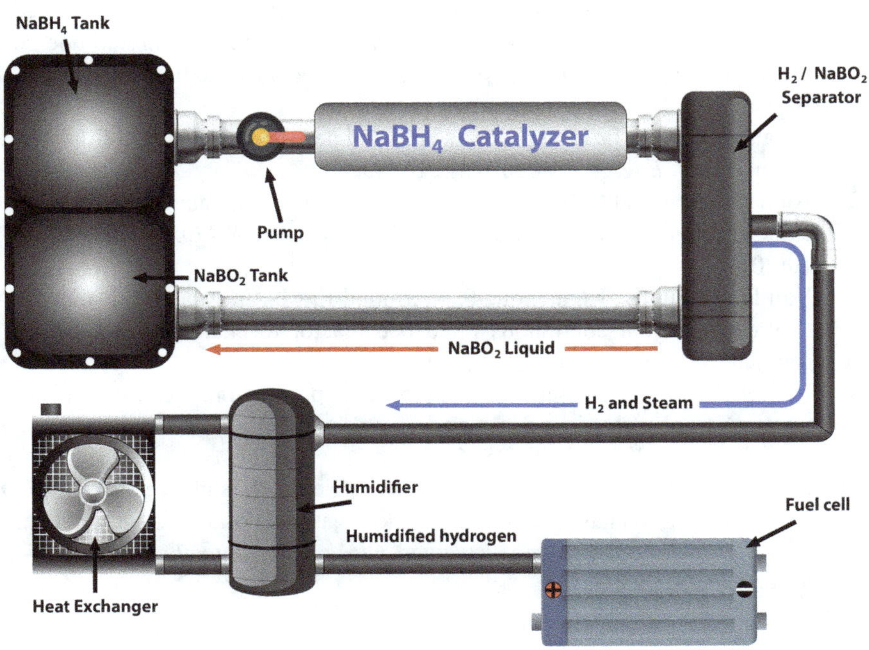

Figure 3.36 Millennium Cell system

References

[1] Prinzhofer A, Tahara Cissé CS, and Diallo AB. Discovery of a large accumulation of natural hydrogen in Bourakebougou (Mali). *Int J Hydrogen Energy*. 2018;43(42):19315–26.
[2] Marcaillou C, Muñoz M, Vidal O, Parra T, and Harfouche M. Mineralogical evidence for H_2 degassing during serpentinization at 300 °C/300 bar. *Earth Planet Sci Lett.*. 2011;303(3–4):281–90.
[3] Croft Production Systems. *Natural gas composition [Internet]*. 2023 [cited 2024 Aug 12]. Available from: https://www.croftsystems.net/oil-gas-blog/natural-gas-composition/. Accessed 12 August 2024
[4] Gaudernack B, and Lynum S. Hydrogen from natural gas without release of CO_2 to the atmosphere. *Int J Hydrogen Energy*. 1998;23(12):1087–93.
[5] Bartolazzi A. *Le Energie Rinnovabili*. Milan: Hoepli; 2005.
[6] U.S. EPA. *Domestic and global usage of gasification technology [Internet]*. [cited 2023 Oct 8]. Available from: https://archive.epa.gov/epawaste/hazard/wastemin/web/html/gasdom.html. Accessed 8 October 2023
[7] Buchholz T, Luzadis VA, and Volk TA. Sustainability criteria for bioenergy systems: results from an expert survey. *J Clean Prod*. 2009;17(Suppl 1):S86–98.
[8] Liu R, Xiao H, Guan S, Zhang J, and Yao D. Technology and method for applying biochar in building materials to evidently improve the carbon capture ability. *J Clean Prod*. 2020;273:123154.
[9] Gupta S, and Kua HW. Factors determining the potential of biochar as a carbon capturing and sequestering construction material: critical review. *J Mater Civ Eng*. 2017;29(9):04017086.
[10] NETL – National Energy Technology Laboratory, U.S. DOE. Gasifipedia [Internet]. 2023 [cited 2024 Aug 12]. Available from: https://netl.doe.gov/carbon-management/energy-systems/gasification/gasifipedia. Accessed 12 August 2024
[11] Breault RW. Gasification processes old and new: a basic review of the major technologies. *Energies*. 2010;3(2):216–40.
[12] Ahmad MI, Alauddin ZAZ, Soid SNM, *et al.* Performance and carbon efficiency analysis of biomass via stratified gasifier. *ARPN J Eng Appl Sci*. 2015;10(20):9533–7.
[13] Van De Steene L, Tagutchou JP, Mermoud F, Martin E, and Salvador S. A new experimental continuous fixed bed reactor to characterise wood char gasification. *Fuel*. 2010;89(11):3320–9.
[14] Calì G, Deiana P, Bassano C, *et al.* Syngas production, clean-up and wastewater management in a demo-scale fixed-bed updraft biomass gasification unit. *Energies*. 2020;13(10):2594.
[15] Rückel A, Oppelt A, Leuter P, Johne P, Fendt S, and Weuster-Botz D. Conversion of syngas from entrained flow gasification of biogenic residues with *Clostridium carboxidivorans* and *Clostridium autoethanogenum*. *Fermentation*. 2022;8(9):465.

[16] Weiland F, Hedman H, Marklund M, Wiinikka H, Öhrman O, and Gebart R. Pressurized oxygen blown entrained-flow gasification of wood powder. *Energy Fuels*. 2013;27(2):932–41.

[17] Göktepe B. *Entrained Flow Gasification of Biomass. Soot Formation and Flame Stability* (Doctoral thesis) [Internet]. Luleå, Sweden: Luleå University of Technology; 2015. Available from: https://www.diva-portal.org/smash/get/diva2:991197/FULLTEXT01.pdf.

[18] Higman C. Chapter 5 – Gasification process technology. *In: Advances in Clean Hydrocarbon Fuel Processing: Science and Technology*.Cambridge: Woodhead Publishing, ed. M. Rashid Khan; 2011. pp. 155–85.

[19] International Energy Agency (IEA). Technology Roadmap – Hydrogen and Fuel Cells. Paris, France; 2015.

[20] *World Energy Council*. World Energy Resources 2016. London, United Kingdom; 2016.

[21] Saveyn H, Eder P, Ramsay M, Thonier G, Warren K, and Hestin M. Towards a better exploitation of the technical potential of waste-to-energy. European Commission: JRC Science for Policy Report. Luxembourg: Publications Office of the European Union; 2016.

[22] Lui J, Chen WH, Tsang DCW, and You S. A critical review on the principles, applications, and challenges of waste-to-hydrogen technologies. *Renew Sustain Energy Rev*. 2020;134:110365.

[23] Rice University. *Making hydrogen from waste plastic could pay for itself [Internet]*. [cited 2023 Oct 9]. Available from: https://news.rice.edu/news/2023/making-hydrogen-waste-plastic-could-pay-itself. Accessed 9 October 2023

[24] Wyss KM, Silva KJ, Bets KV, *et al.* Synthesis of clean hydrogen gas from waste plastic at zero net cost. *Adv Mater*. 2023;35(48):2306763.

[25] Ding L, Yang M, Dong K, *et al.* Mobile power generation system based on biomass gasification. *Int J Coal Sci Technol*. 2022;9(1):34.

[26] Wiyono A, Gandidi IM, Berman ET, Mutaufiq, and Pambudi NA. Design, development and testing of integrated downdraft gasifier and multi IGCS system of MSW for remote areas. *Case Stud Therm Eng*. 2020;20:100612.

[27] Towler G, and Sinnott R. Specification and design of solids-handling equipment – chapter 18. In: *Chemical Engineering Design*. 2nd edn. 2013. eds. Gavin Towler and Ray Sinnott, Butterworth-Heinemann, Oxford (UK), pp. 937–1046.

[28] U.S Environmental Protection Agency, North Carolina State University (NCSU). Adjustable-throat Venturi scrubber with plunger. File: Adjthroatplunger.jpg – Wikimedia Commons [Internet]. 2006 [cited 2024 Aug 12]. Available from: https://commons.wikimedia.org/wiki/File:Adjthroatplunger.jpg. Accessed 12 August 2024

[29] RTI International. Warm gas desulfurization process technology [Internet]. 2021 [cited 2023 Oct 9]. Available from: https://www.rti.org/impact/warm-gas-desulfurization-process-technology. Accessed 9 October 2023

[30] Vamvuka D, Arvanitidis C, and Zachariadis D. Flue gas desulfurization at high temperatures: a review. *Environ Eng Sci*. 2004;21(4):525–47.

[31] Arachchige USPR. Amines' effect on CO_2 removal efficiency. *Int J Res.* 2019;6(3):775–779.

[32] Iliuta I, and Larachi F. CO_2 and H_2S absorption by MEA solution in packed-bed columns under inclined and heaving motion conditions – hydrodynamics and reactions performance for marine applications. *Int J Greenh Gas Control.* 2018;79:1–13.

[33] Burr B, and Lyddon L. *A Comparison of Physical Solvents for Acid Gas Removal.* Bryan, Texas, USA; 2008.

[34] Viparelli P, Eramo P, and Villa P. Aspetti tecnologici della produzione di idrogeno da idrocarburi e da biomasse. *La Termotec – Com Termotec Ital [Internet].* 2006;(7):30–47. Available from: https://www.cti2000.it/index.php?controller=pubblicazioni&action=show&id=28652.

[35] Moon DK, Lee DG, and Lee CH. H_2 pressure swing adsorption for high pressure syngas from an integrated gasification combined cycle with a carbon capture process. *Appl Energy.* 2016;183:760–74.

[36] Spazzafumo G. Cogeneration of power and hydrogen with integrated fuel processor counterpressure steam cycles. *Int J Hydrogen Energy.* 2004;29 (11):1147–50.

[37] Cicconardi SP, Perna A, Spazzafumo G, and Tunzio F. CPH systems for cogeneration of power and hydrogen from coal. *Int J Hydrogen Energy.* 2006;31(6):693–700.

[38] Cicconardi SP, Perna A, and Spazzafumo G. Combined power and hydrogen production from coal. Part B: comparison between the IGHP and CPH systems. *Int J Hydrogen Energy.* 2008;33(16):4397–404.

[39] Calabrò A, Deiana P, Fiorini P, Girardi G, and Stendardo S. Possible optimal configurations for the ZECOMIX high efficiency zero emission hydrogen and power plant. *Energy.* 2008;33(6):952–62.

[40] Galeno G, and Spazzafumo G. ZECOMIX: performance of alternative layouts. *Int J Hydrogen Energy.* 2010;35(18):9845–50.

[41] Giaconia A, Tarquini P, and Vignolini M. *Cicli termochimici di idrolisi per la produzione di idrogeno [Internet].* vol. 52. Energia, ambiente e innovazione; 2006. Available from: https://www.eai.enea.it/component/jdownloads/?task=download.send&id=967&catid=46&Itemid=1058, last access date: 2024-12-16

[42] Steinfeld A. Solar hydrogen production via a two-step water-splitting thermochemical cycle based on Zn/ZnO redox reactions. *Int J Hydrogen Energy.* 2002;27(6):611–9.

[43] Loutzenhiser PG, Meier A, and Steinfeld A. Review of the two-step H_2O/CO_2-splitting solar thermochemical cycle based on Zn/ZnO redox reactions. *Materials (Basel).* 2010;3(11):4922–38.

[44] Tarquini P. *Produzione di idrogeno con cicli termochimici alimentati da energia solare concentrata [Internet].* Available from: https://www.enea.it/it/seguici/documenti/parole-energia/TarquiniPresentazione.pdf. Accessed 12 August 2024

[45] Varsano F, Padella F, La Barbera A, and Alvani C. The carbonatation reaction of layered $Na(Mn_{1/3}Fe_{2/3})O_2$: a high temperature study. *Solid State Ion*. 2011;187(1):19–26.

[46] Vitart X, Carles P, and Anzieu P. A general survey of the potential and the main issues associated with the sulfur–iodine thermochemical cycle for hydrogen production using nuclear heat. *Prog Nucl Energy*. 2008;50 (2–6):402–10.

[47] Carles P, Vitart X, and Yvon P. CEA assessment of the sulphur–iodine cycle for hydrogen production. In: *Nuclear Production of Hydrogen – Fourth Information Exchange Meeting, 14–16 April 2009 Oakbrook, Illinois, USA*. 2010. pp. 167–77.

[48] Brown LC, Besenbruch GE, Lentsch RD, *et al. High efficiency generation of hydrogen fuels using nuclear power*. Final technical report for the period August 1, 1999 through September 30, 2002 [Internet]. General Atomics Report GA–A24285. 2003. Available from: http://www.osti.gov/servlets/purl/814014-tdQyiq/native/%0Ahttps://fusion.gat.com/pubs-ext/AnnSemiannETC/A24285.pdf. Accessed 16 December 2024

[49] Wang Q, and Macián-Juan R. Design and analysis of an iodine–sulfur thermochemical cycle-based hydrogen production system with an internal heat exchange network. *Int J Energy Res*. 2022;46(9):11849–66.

[50] Lewis M, and Masin J. An assessment of the efficiency of the hybrid copper–chloride thermochemical cycle. *In: AIChE Annual Meeting and Fall Showcase (Cincinnati, OH)* [Internet]. Cincinnati, OH, USA; [cited 2023 Oct 9]. Available from: https://aiche.confex.com/aiche/2005/techprogram/P20544.HTM. Accessed 9 October 2023

[51] Naterer GF, Suppiah S, Stolberg L, *et al.* Clean hydrogen production with the Cu–Cl cycle-progress of international consortium, II: simulations, thermochemical data and materials. *Int J Hydrogen Energy*. 2011;36(24):15486–501.

[52] International Energy Agency Hydrogen Implementing Agreement (IEA HIA). IEA/HIA task 25: high temperature processes for hydrogen production – chloride family of cycles [Internet]. 2010 [cited 2014 Dec 15]. Available from: ieahia.org/pdfs/Task25/Chloride-family-cycles.pdf. Accessed 15 December 2024

[53] Weidner JW, and Holland CE. *Production of Low Temperature Electrolytic Hydrogen*. United States Patent US 9,574,276 B2 [Internet]. United States; US 9,574,276 B2, 2017. Available from: https://patents.google.com/patent/US9574276B2/en. Accessed 16 December 2024

[54] Hancke R, Holm T, and Ulleberg Ø. The case for high-pressure PEM water electrolysis. *Energy Convers Manag*. 2022;261:115642.

[55] IRENA. *Green Hydrogen Cost Reduction. Scaling up Electrolysers to Meet the 1.5 °C Climate Goal [Internet]*. Abu Dhabi:International Renewable Energy Agency; 2020. Available from: https://www.irena.org/-/media/Files/IRENA/Agency/Publication/2020/Dec/IRENA_Green_hydrogen_cost_2020.pdf.

[56] Stolten D, and Krieg D. Alkaline electrolysis – introduction and overview. In: Stolten D (ed.). *Hydrogen and Fuel Cells: Fundamentals, Technologies and Applications*. Weinheim: Wiley-VCH; 2011.

[57] Lohmann-Richters FP, Renz S, Lehnert W, Müller M, and Carmo M. Review—challenges and opportunities for increased current density in alkaline electrolysis by increasing the operating temperature. *J Electrochem Soc*. 2021;168(11):114501.

[58] Hodges A, Hoang AL, Tsekouras G, *et al.* A high-performance capillary-fed electrolysis cell promises more cost-competitive renewable hydrogen. *Nat Commun*. 2022;13(1):1–11.

[59] Hysata. Our technology – Hysata [Internet]. 2023 [cited 2023 Oct 9]. Available from: https://hysata.com/our-technology/. Accessed 9 October 2023

[60] Lei Q, Wang B, Wang P, and Liu S. Hydrogen generation with acid/alkaline amphoteric water electrolysis. *J Energy Chem*. 2019;38:162–9.

[61] Sun X, Simonsen SC, Norby T, and Chatzitakis A. Composite membranes for high temperature PEM fuel cells and electrolysers: a critical review. *Membranes (Basel)*. 2019;9(7):83.

[62] Solvay. Aquivion® ion conducting polymers [Internet]. 2023 [cited 2023 Oct 9]. Available from: https://www.solvay.com/en/brands/aquivion-ion-conducting-polymers. Accessed 9 October 202

[63] NEL Hydrogen. Hydrogen generators [Internet]. 2019. Available from: https://nelhydrogen.com/wp-content/uploads/2020/03/High-Purity-Brochure-Rev-H-Single-Pages.pdf. Accessed 9 October 2023

[64] Enapter. *AEM Technology – Enapter Handbook* [Internet]. [cited 2023 Oct 9]. Available from: https://handbook.enapter.com/knowledge_base/aem_technology.html. Accessed 9 October 2023

[65] Enapter. Anion exchange membrane water electrolysis: how it works – Enapter [Internet]. 2020 [cited 2023 Oct 9]. Available from: https://www.enapter.com/it/newsroom/aem-water-electrolysis-how-it-works. Accessed 9 October 2023

[66] Wachsman ED, and Lee KT. Lowering the temperature of solid oxide fuel cells. *Science*. 2011;334(6058):935–9.

[67] Zahid M, Schefold J, and Brisse A. High-temperature water electrolysis using planar solid oxide fuel cell technology: a review. In: Stolten D (ed.). *Hydrogen and Fuel Cells – Fundamentals, Technologies and Applications*. New York: Wiley-VCH; 2011.

[68] Hauch A, Küngas R, Blennow P, *et al.* Recent advances in solid oxide cell technology for electrolysis. *Science*. 2020;370(6513).

[69] Nechache A, and Hody S. Alternative and innovative solid oxide electrolysis cell materials: a short review. *Renew Sustain Energy Rev*. 2021;149:111322.

[70] Lehtinen T, and Noponen M. Solid oxide electrolyser demonstrator development at Elcogen. *ECS Trans*. 2021;103(1):1939.

[71] Fujishima A, and Honda K. Electrochemical photolysis of water at a semiconductor electrode. *Nature*. 1972;238(5358):37–8.

[72] Agosti A, Natali M, Amirav L, and Bergamini G. Towards solar factories: prospects of solar-to-chemical energy conversion using colloidal semiconductor photosynthetic systems. *ChemSusChem*. 2020;13(18):4894–9.

[73] van de Krol R, and Parkinson BA. Perspectives on the photoelectrochemical storage of solar energy. *MRS Energy Sustain*. 2017;4(1):1–11.

[74] IEA. IEA agreement on the production and utilization of hydrogen. In: Elam CC (ed.). 1999 *Annual Report*. IEA/H2/AR-99. Paris: International Energy Agency; 1999.

[75] Zhang B, Li F, Yu F, *et al.* Electrochemical and photoelectrochemical water oxidation by supported cobalt-oxo cubanes. *ACS Catal.* 2014;4(3):804–9.

[76] Kumar M, Meena B, Subramanyam P, Suryakala D, and Subrahmanyam C. Recent trends in photoelectrochemical water splitting: the role of cocatalysts. *NPG Asia Mater.* 2022;14(1):88.

[77] Bespalko S, and Mizeraczyk J. Overview of the hydrogen production by plasma-driven solution electrolysis. *Energies*. 2022;15(20):7508.

[78] Saksono N, Batubara T, and Bismo S. Hydrogen production by plasma electrolysis reactor of KOH–ethanol solution. *IOP Conf Ser Mater Sci Eng.* 2017;162(1):012010.

[79] Mohammed-Ibrahim J, and Moussab H. Recent advances on hydrogen production through seawater electrolysis. *Mater Sci Energy Technol.* 2020; 3:780–807.

[80] Hausmann JN, Schlögl R, Menezes PW, and Driess M. Is direct seawater splitting economically meaningful? *Energy Environ Sci.* 2021;14(7):3679–85.

[81] Farràs P, Strasser P, and Cowan AJ. Water electrolysis: direct from the sea or not to be? *Joule*. 2021;5(8):1921–3.

[82] Xie H, Zhao Z, Liu T, *et al.* A membrane-based seawater electrolyser for hydrogen generation. *Nature*. 2022;612(7941):673–8.

[83] Bothe H, Schmitz O, Yates MG, and Newton WE. Nitrogen fixation and hydrogen metabolism in Cyanobacteria. *Microbiol Mol Biol Rev.* 2010;74 (4):529–51.

[84] Melis A, Zhang L, Forestier M, Ghirardi ML, and Seibert M. Sustained photobiological hydrogen gas production upon reversible inactivation of oxygen evolution in the green alga *Chlamydomonas reinhardtii*. *Plant Physiol.* 2000;122(1):127–35.

[85] López-Torrejón G, Jiménez-Vicente E, Buesa JM, Hernandez JA, Verma HK, and Rubio LM. Expression of a functional oxygen-labile nitrogenase component in the mitochondrial matrix of aerobically grown yeast. *Nat Commun.* 2016;7:1–6.

[86] Peters JW, Boyd ES, D'Adamo S, Mulder DW, Therien J, and Posewitz MC. Hydrogenases, nitrogenases, anoxia, and H_2 production in water-oxidizing phototrophs. In: Borowitzka MA, and Moheimani NR (eds). *Algae for Biofuels and Energy [Internet]*. Dordrecht: Springer Netherlands; 2013 [cited 2023 Oct 9]. pp. 37–75. Available from: https://link.springer.com/chapter/10.1007/978-94-007-5479-9_3.

[87] D'Adamo S, and Posewitz MC. Hydrogenase evolution and function in eukaryotic algae. In: Rögner M (ed). *Biohydrogen*. Berlin: De Gruyter; 2015. pp. 145–72.

[88] Nguyen TAD, Pyo KJ, Sun KM, Kwan OY, and Sim SJ. Optimization of hydrogen production by hyperthermophilic eubacteria, *Thermotoga maritima*

and *Thermotoga neapolitana* in batch fermentation. *Int J Hydrogen Energy*. 2008;33(5):1483–8.

[89] Pradhan N, Dipasquale L, D'Ippolito G, *et al.* Hydrogen production by the thermophilic bacterium *Thermotoga neapolitana*. *Int J Mol Sci*. 2015;16 (6):12578–600.

[90] Pradhan N, D'Ippolito G, Dipasquale L, *et al.* Kinetic modeling of hydrogen and L-lactic acid production by *Thermotoga neapolitana* via capnophilic lactic fermentation of starch. *Bioresour Technol*. 2021;332:125127.

Chapter 4
Storage and distribution

Apart from the geological deposits mentioned at the beginning of Chapter 3, molecular hydrogen is not available in nature and, like electricity, is an energy carrier. Therefore, once produced it needs to be stored and distributed. It also allows a practical storage in various conditions and with specific constrains. These operations can be carried out in very different ways and the choice is linked to aspects that can go beyond the particular application being studied and involve the entire hydrogen system. A specific application of hydrogen could, for example, favour one form of storage and distribution, but if all other applications connected to the first one involve the use of a different form, the latter would probably be the one chosen for the storage and distribution of hydrogen. It follows that, in addition to economic convenience and safety, the degree of penetration of hydrogen in the energy sector at different scale is a fundamental parameter.

Hydrogen has been and still is used in aerospace missions thanks to the fact that it is the fuel with the highest heating value per unit of weight. Therefore, for certain applications, distribution and on-board storage technologies have been successfully developed for decades.

Also for the ground storage of hydrogen, there are various technologies that are already adequate, based on both natural reservoirs (underground rock caverns, aquifers, porous rocks, salt mine, depleted hydrocarbon deposits) and artificial reservoirs. The underground storage of hydrogen is especially convenient for large quantities and/or long periods. Examples of underground storage of hydrogen or hydrogen-rich gases are found in Argentina (Diadema field, depleted hydrocarbon deposit), Germany (Kiel and Ketzin, town gas respectively in salt cavern and in aquifer), Czech Republic (Lobodice, town gas in aquifer) France (near Baynes, refinery gaseous products in aquifer), the United Kingdom (near Teeside, hydrogen in salt cavern) and USA (Clemens, Spindletop and Moss Bluff, hydrogen in salt caverns) [1–4]. Sambo *et al.* [4] also reported a list of other potential hydrogen underground storage.

However, the situation regarding on-board storage in the road transport sector is quite different since in this case the energy density on a volumetric basis and the refuelling time are of fundamental importance, in addition to the weight energy density that, although very high, must take into account the weight of the tank.

Precisely for this reason, the storage and distribution of hydrogen probably represent the bottleneck of the entire system. To overcome this problem, technologies other than the hydrogen storage in pure form have also been examined.

4.1 Compressed gas

The most basic form of accumulation is that in the gaseous state since molecular hydrogen is in this state in ambient conditions. Accumulation in gaseous form can take place in different ways: in pressurised tanks made of metal or composite material, in glass microspheres, in underground rock salt caves, in aquifers and porous rocks, in depleted hydrocarbon deposits.

Gaseous hydrogen is usually marketed in cylinders at a pressure of about 20 MPa, but tanks up to about 70 MPa are currently used for on-board storage. Regulatory framework appears more and more defined in many countries, like as EN 13807:2017 in EU for battery vehicles with multiple-element gas containers.

Approximating the tank of capacity v_i with a cylindrical tube of height h and internal diameter d_i, the latter is given by:

$$d_i = 2 \cdot \sqrt{\frac{v_i}{\pi \cdot h}} \tag{4.1}$$

and, obviously, it reduces as the pressure increases as a consequence of the reduction of the volume occupied by the hydrogen. Given σ the admissible tension for the material of which the tank is made, and p the pressure, the external diameter d_e can be approximated by the expression:

$$d_e = d_i \cdot \left(1 + \frac{p}{2 \cdot \sigma}\right) \text{ for } \frac{p}{2 \cdot \sigma} < 0.1 \tag{4.2}$$

from which the volume v_t of the tank is obtained:

$$v_t = \frac{\pi \cdot d_e^2 \cdot h}{4} \tag{4.3}$$

and, indicating with ρ_t the density of the material used, the mass m_t of the tank:

$$m_t = \frac{\pi \cdot \left(d_e^2 - d_i^2\right) \cdot h}{4} \cdot \rho_t \tag{4.4}$$

As can be seen from Figure 4.1, the effect of pressure has a significant beneficial impact on the energy density per unit of volume since the latter, even taking into account the tank, decreases since the reduction of the internal diameter prevails over the increase in thickness. Conversely, the effect of pressure on the energy density per unit of weight is rather negative since, despite the reduction of the internal diameter, the increase in thickness involves a progressive increase in the weight of the empty tank.

In the case of Figure 4.1 there is an accumulation weight density that fluctuates between 1.5% and 2.5%, far below the values considered acceptable for the accumulation of hydrogen on board vehicles (35 MPa and 70 MPa). Furthermore, the use of steel involves an increasing risk of embrittlement as pressure increases. To avoid this risk and improve the weight density, it is therefore necessary to choose materials other than steel, materials that have a better ratio between admissible

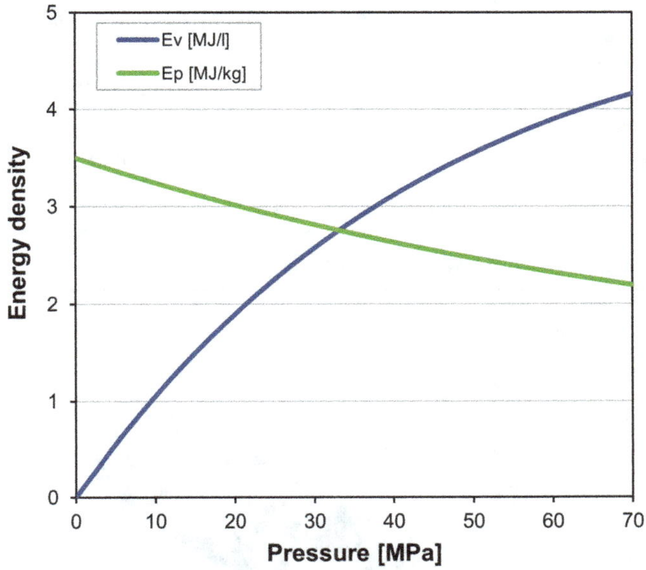

Figure 4.1 Energy densities of a fully loaded steel tank

tension and density and that, of course, are not permeable to hydrogen and do not react with it. Usually it is aluminium or high density polyethylene, externally reinforced with composite materials containing carbon fibres or Kevlar. Four classes of cylinders have been introduced to group these different solutions, named from T1 to T4 [5,6] where T indicates 'type':

T1: cylinder made of only steel (pressure up to 20 MPa);
T2: cylinder made of steel wrapped with carbon fibres (pressure up to 40 MPa);
T3: cylinder made of aluminium wrapped with carbon fibres (pressure up to 35 MPa);
T4: cylinder made of polymer wrapped with carbon fibres (pressure up to 70 MPa).

A not exhaustive list of regulation for hydrogen cylinders in different Countries and for different purposes is ISO 11119 series, EN 12245, EN 12257, UNECE Reg 134, EC79, ASME Section VIII Division I and III, ISO 19881, ISO 15869, EN 17533, EN 17339, JARI S 001 [7].

As example, Table 4.1 reports some data of tanks used for Toyota Mirai and Figure 4.2 shows their structure.

The minimum theoretical compression work between two thermodynamic states 1 and 2 corresponds to isothermal compression work and is given by:

$$l_c = R \cdot T_1 \cdot \ln\left(\frac{p_1}{p_2}\right) \tag{4.5}$$

where R is the gas constant and T the temperature.

Table 4.1 Carbon-based composite fibre cylinders (adapted from [8)])

	G2	G2L	G2XL-1
Nominal working pressure (bar)	700	700	700
Length (mm)	684–1467	1650–2060	2060
Diameter (mm)	299	486	702
Internal volume (l)	25.3–64.9	176.0–230.0	458.5
Tank mass (kg)	22.6–43.0	100.9–136.0	243.8
Hydrogen storage capacity @NWP (kg)	1.0–2.6	7.2–9.4	18.8

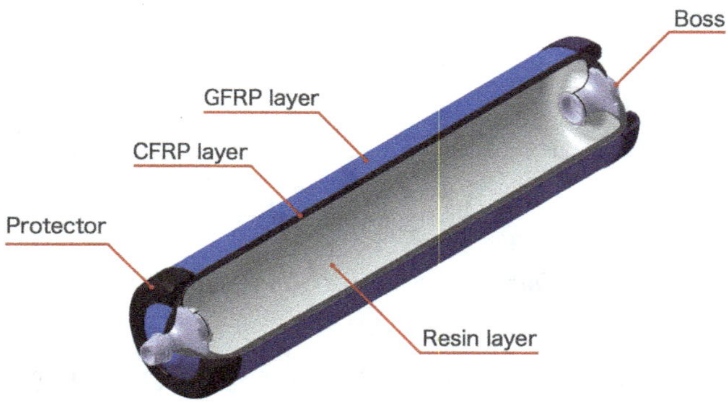

Figure 4.2 Construction details of a high-pressure hydrogen gas tank [copyright Toyota Motor Corporation]

In practice, a machine used for the exchange of work is not very suitable for exchanging heat and, therefore, the adiabatic compression work is closer to that of a real transformation:

$$l_c = \frac{k}{k-1} \cdot \frac{p_1 \cdot V_1}{m_1} \cdot \left[\left(\frac{p_2}{p_1} \right)^{\frac{k-1}{k}} - 1 \right] \tag{4.6}$$

where k is the ratio between specific heats at constant pressure and volume.

In Figure 4.3, the comparison between the specific works theoretically spent in the two transformations is shown and it is noted that adiabatic compression requires significantly more work already for compression ratios that are not too high. The greater work translates into an increase in the internal energy of the fluid, but the corresponding increase in temperature has only the negative effect of causing a lower density of the gas which would then spontaneously cool down leading to a reduction in pressure.

It follows that at the end of compression the gas must be brought back to room temperature before being stored and, therefore, it is preferable to resort to an intercooled compression which allows to reduce the work spent: again in Figure 4.3

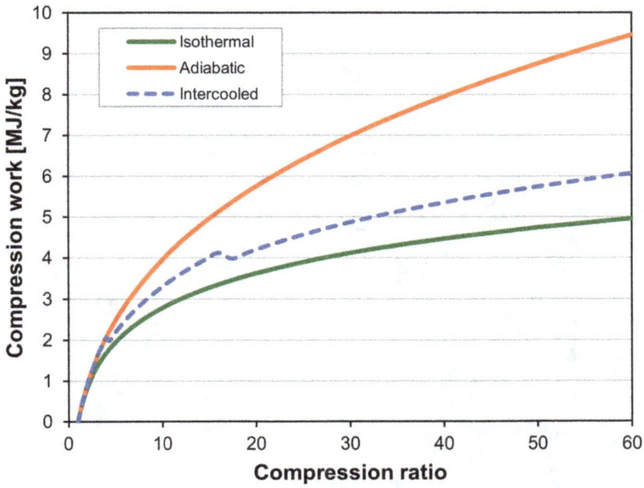

Figure 4.3 Compression work versus compression ratio for different kinds of thermodynamic transformation

you can see the effect of an ideal inter-cooling with a maximum ratio of each compression phase equal to four. Obviously, passing to real compression there is a further increase in the work spent, reaching, in some cases, 8% of the energy content hydrogen.

Current range of compression reaches up to 100 MPa in a 70 MPa hydrogen refuelling station for high mass refuelling a day; lower value is used for storage and utilisation in industrial sector, usually up to 45 MPa. Hydrogen is generally compressed by reciprocating piston compressors. To avoid contamination of gas with oil, self-lubricating seals are used, the wear of which is checked by monitoring the oil and gas vents. The cylinder is surrounded by an external cooling jacket in which a coolant circulates. Figure 4.4 shows an example of hydrogen compressor [9].

An alternative to avoid contact between hydrogen and lubricant is that of the diaphragm compressor in which a flexible membrane that delimits a chamber is moved alternately in order to suck and compress the hydrogen. Another alternative have taken place with the use of ionic liquids [10] characterised by being liquids that do not vaporise and that do not dissolve significant quantities of hydrogen. In the ionic liquid compressor it is not a piston that exerts the pressing action on the hydrogen, but a liquid mass that rises and falls alternately inside a cylinder. This allows a considerable reduction of mechanical elements in movement and, therefore, of vibrations and noise, while significantly increasing the interval between maintenance. The heat generated during compression is partially absorbed by the liquid, making the thermodynamic transformation closer to isothermal, resulting in improved efficiency.

Hydrogen compression can also be achieved through an electrochemical process. Thanks to the application of an external current and the presence of a

Figure 4.4 Hydrogen compressor Idro Meccanica TDE/H [courtesy of Idro Meccanica and SAFE SpA]

platinum-containing catalyst deposited on the anodic side of a proton exchange membrane, hydrogen is oxidised to protons that pass through the membrane to be reduced to hydrogen in the cathode compartment which is at higher pressure [11]. Since the membrane is selective, the compressor also allows for the possible purification of hydrogen.

Finally, hydrogen can be compressed also using metal hydrides as will be explained later in Section 4.3.1.

Compressors with a very wide suction pressure range can be used as boosters, i.e. fed with already compressed hydrogen coming from an accumulation. In a refuelling station for hydrogen car there is a main storage commonly at 20 MPa and a high-pressure storage (90 MPa) that actively transfer hydrogen to the car during the refuelling to obtain in the car a storage at 70 MPa, normally with a hydrogen cooler at −40 °C to prevent over heating of the car tank. A booster transfers hydrogen from low to high pressure tank to prepare station to the refuelling of a car, that normally happen without use of compressor and only by pressure difference between pressure in the station tanks and pressure in car storage.

The compressed gaseous hydrogen storage system on board must provide a high level of safety and, therefore, is equipped with valves suitable for this purpose. In particular, there are pressure regulation valves followed by safety valves which, in the event of a regulator malfunction, open due to the overpressure and make the hydrogen flow dispersing it outside the vehicle. Similarly, the tank itself is equipped with a safety valve that opens and disperses the hydrogen into the atmosphere if, for example due to the effect of heat, the pressure inside the tank exceeds a predetermined value. Finally, there are shut-off valves that allow isolation of the tank only or larger portions of the supply system [12].

Although the storage of gaseous hydrogen at high pressures is currently used on all vehicles available on the market, other technologies under development could solve the safety issues posed by the use of cylinders at high pressures, and could allow the optimisation of both volumetric and gravimetric densities.

Remaining in the field of compressed hydrogen, an alternative that in the past has aroused some interest is that of glass microspheres. The idea dates back to the 1970s [13].

These are hollow glass spheres with a diameter of a few tens of μm, characterised by a very thin wall (1 μm), but capable of withstanding very high pressures, at least up to 100 MPa. Obviously the spheres are closed and almost totally impermeable to hydrogen at room temperature, but they become more and more permeable as the temperature increases. They can be transported in the form of a fluid powder.

The charging phase takes place by introducing the microspheres into a tank which is then filled with hydrogen at the desired pressure. The tank is then heated to 300 °C–400 °C and kept under pressure for an adequate time so that the microspheres are filled with hydrogen under pressure. Once the charging phase is complete, the tank is cooled down to room temperature making the microspheres waterproof again. At this point it is possible to gradually reduce the pressure in the tank and transfer the microspheres to an accumulation tank.

Under the action of an initial pressure difference Δp, the internal pressure grows from instant t_0 to instant t_1 according to the law:

$$p_{i,1} = p_{i,0} \pm \Delta p \cdot \left(1 - e^{-\frac{t_1 - t_0}{\tau}}\right) \tag{4.7}$$

in which the + sign refers to the filling phase, while the − sign refers to the emptying phase, and τ is a time constant which depends on the external diameter and the wall thickness of the microspheres, the absolute temperature and the hydrogen permeability of the glass [14].

Figure 4.5 shows, as a function of the ratio between time and time constant, the trend of the ratio between the pressure variation inside the microspheres and the initial pressure difference: after a time interval equal to τ, the variation of pressure reached 63% of the total, while for 2τ and 3τ intervals it reached 86% and 95% respectively.

To obtain the release of the accumulated hydrogen, the microspheres must be heated. The higher the temperature, the higher the release rate: therefore, by controlling the temperature as a function of the hydrogen demand, it is possible to control the pressure in the tank and this means that hydrogen can be accumulated at very high pressures in a tank sized for much more modest pressures.

Since rupture of the storage tank or a microsphere does not cause propagation to the other microspheres in the bed, a safety factor of 1.5 is considered adequate for a bed with a microsphere density of 10,000 per mm^3 [15]. This

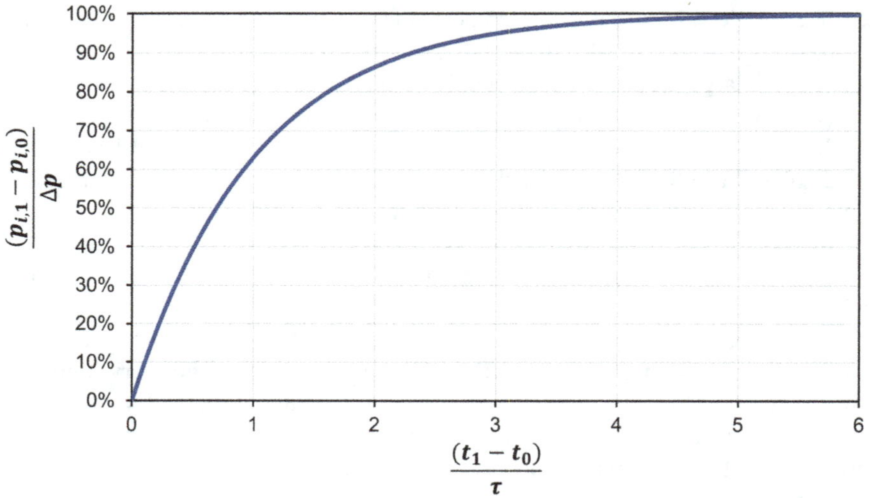

Figure 4.5 Pressure variation inside microspheres from the initial difference

density corresponds, for a bed of microspheres with a diameter of 50 μm, to a packing of 63%.

The densities shown in Figure 4.5 are related to the bed of microspheres only, but it is also necessary to take into account the weight and volume of the storage tank including the components used for heat exchange.

Although it would be less expensive and safer than traditional compressed or liquid hydrogen tanks [16], the interest in this storage system has decreased over time, apparently due to the slowness of the charging and release phases and the difficulty in making microspheres of regular shape and size to achieve theoretical energy densities.

4.2 Cryogenic liquid

The further reduction of the overall dimensions of the system requires ever greater technological efforts compared to the benefits obtained and also translates into an increase in the weight of the system per unit of stored energy. Consequently, it is possible to obtain positive results only by resorting to the phase passage, which allows to reduce both the overall dimensions and the weight of the system, at the expense of much heavier thermodynamic conditions. In fact, saturated liquid hydrogen can be obtained, at atmospheric pressure, at about 20 K and any pressure increase does not allow, however, to significantly raise this value. And, however, when calculating the energy density, we must not neglect the weight and volume of the tank and auxiliary equipment.

Gases are usually liquefied by the Linde process, a process that involves the sequence compression, cooling to room temperature, isoenthalpic (Joule-Thomson) expansion. Actually, this expansion involves a further cooling which, for suitable

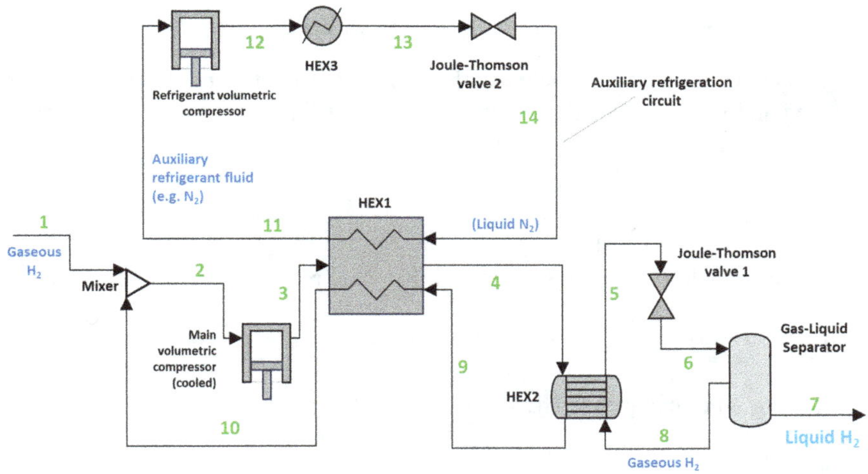

Figure 4.6 Linde liquefaction plant scheme

expansion ratios, allows to liquefy a fraction of the gas, whereas the fraction remaining in the gaseous phase is used to cool the compressed gas and, finally, recirculated to the compressor input.

This process cannot work with hydrogen since, at room temperature, it has a negative Joule-Thomson coefficient and, therefore, its isoenthalpic expansion would produce a heating (see Section 2.2). This behaviour is reversed, however, below 202 K at ambient pressure, called upper inversion temperature.

The Linde process with pre-cooling (Figure 4.6) is well suited for liquefying gases that have at ambient pressure an upper inversion temperature lower than ambient temperature, namely hydrogen, helium and neon. This is a cycle that uses an auxiliary fluid having an upper inversion temperature higher than the ambient temperature and a liquefaction temperature lower than the upper inversion temperature of the gas to be liquefied. Nitrogen is usually used as auxiliary refrigerant for the liquefaction of hydrogen. Its upper inversion temperature at ambient pressure is about 573 K [17] and its liquefaction temperature at ambient pressure is about 78 K [18]. Therefore, an elementary scheme of a plant that carries out this cycle is constituted by a closed auxiliary circuit that carries out the liquefaction of nitrogen by means of a simple Linde cycle and interfaces with the main open circuit by means of a heat exchanger that acts as a nitrogen evaporator (upper part of Figure 4.6). The main circuit carries out a similar process, but the cooling phase is carried out using liquid nitrogen and recirculated hydrogen gas as cold fluids (lower part of Figure 4.6): in this way the temperature of the hydrogen upstream of the lamination valve is lower than the inversion one, and it is therefore possible to obtain liquid hydrogen.

Consider the HEX1 heat exchanger which allows the main circuit [1–10] to transfer heat to the auxiliary refrigeration circuit [11–14]. The heat balance is given by:

$$\dot{m}_3 \cdot h_3 + \dot{m}_9 \cdot h_9 + \dot{m}_{14} \cdot h_{14} = \dot{m}_4 \cdot h_4 + \dot{m}_{10} \cdot h_{10} + \dot{m}_{11} \cdot h_{11} \tag{4.8}$$

Indicate with x_R the ratio between the flow of refrigerant and the global flow of hydrogen and with x_L the ratio between the flow of liquid hydrogen that is sent to the storage and the global flow of hydrogen:

$$x_R = \frac{\dot{m}_{11}}{\dot{m}_3}; x_L = \frac{\dot{m}_7}{\dot{m}_3}$$

Dividing each term of (4.8) by the global flow rate of hydrogen, we obtain:

$$h_3 + (1 - x_L) \cdot h_9 + x_R \cdot h_{14} = h_4 + (1 - x_L) \cdot h_{10} + x_R \cdot h_{11} \tag{4.9}$$

Moving on to exchanger HEX2, the heat balance is:

$$\dot{m}_4 \cdot h_4 + \dot{m}_8 \cdot h_8 = \dot{m}_5 \cdot h_5 + \dot{m}_9 \cdot h_9 \tag{4.10}$$

which becomes:

$$h_4 + (1 - x_L) \cdot h_8 = h_5 + (1 - x_L) \cdot h_9 \tag{4.11}$$

Being $h_6 = h_5$, the heat balance of the liquid/vapour separator is:

$$\dot{m}_5 \cdot h_5 = \dot{m}_7 \cdot h_7 + \dot{m}_8 \cdot h_8 \tag{4.12}$$

or:

$$h_5 = x_L \cdot h_7 + (1 - x_L) \cdot h_8 \tag{4.13}$$

And by joining (4.9), (4.11) and (4.13), we obtain the heat balance of the main circuit downstream the hydrogen compressor:

$$h_3 + x_R \cdot h_{14} = (1 - x_L) \cdot h_{10} + x_R \cdot h_{11} + x_L \cdot h_7 \tag{4.14}$$

From which we have:

$$x_L = \frac{h_{10} - h_3 + x_R \cdot (h_{11} - h_{14})}{h_{10} - h_7} \tag{4.15}$$

Figure 4.7 shows the entropic diagram of the process. Taking into account that $h_{10} \approx h_1 \approx h_2$, all the terms of (4.15) are connected to the hydrogen properties but x_R, h_{11} and h_{14}. However, optimal value of h_{11} is the highest compatible with T_3, so that the amount of liquefied hydrogen is limited only by the flow and the enthalpy of the refrigerant, usually nitrogen at about −200 °C.

Over time, this kind of limit has led to the study and development of more complex cycles capable of reducing specific energy consumption.

For large-scale plants (> 5 t_{LH2}/d) the current process used is the Claude cycle with pre-cooling which achieves a partial recovery of the work spent on compression by dividing the expansion between a Joule-Thomson valve and a turbine. Figure 4.8 shows a plant that adopts this type of process.

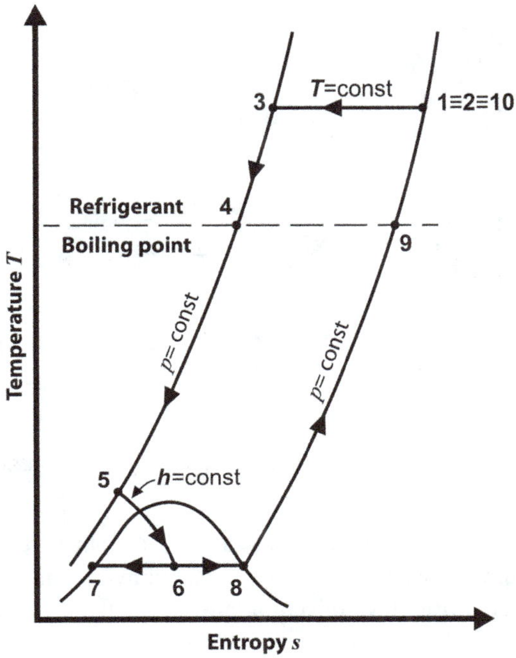

Figure 4.7 *Entropic diagram of the Linde liquefaction process*

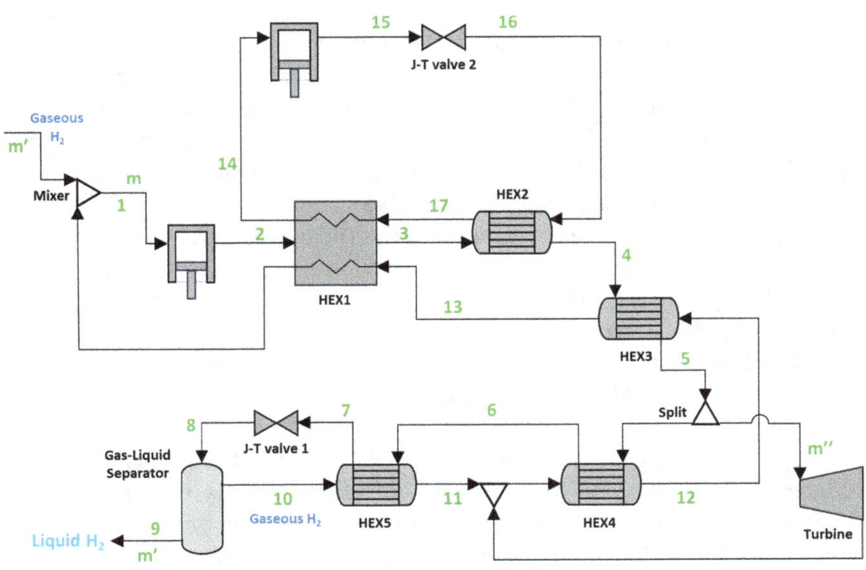

Figure 4.8 *Pre-cooled Claude process*

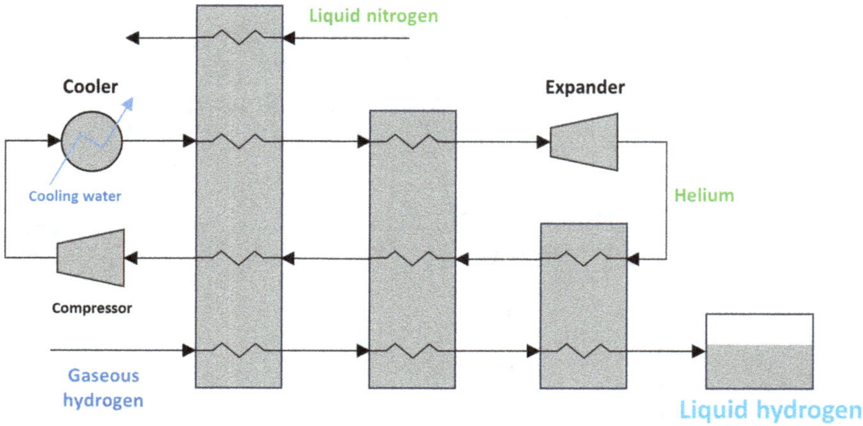

Figure 4.9 Pre-cooled Helium Brayton process

Hydrogen is compressed almost isothermally and then refrigerated by heat exchange with liquid nitrogen and with recirculated hydrogen. Liquid hydrogen is continuously extracted from the collection tank, while the cold gaseous fraction is sent back to the compressor together with the flow that has been diverted to the expander.

For smaller-scale plants, the pre-cooled Helium Brayton cycle (Figure 4.9) is used.

This cycle uses two different auxiliary fluids. Helium is compressed and cooled in three steps, respectively using water, liquid nitrogen and recirculated gaseous helium. Then it is expanded into a turbine in such a way as to reach a temperature lower than the hydrogen liquefaction temperature. Returning to the compressor helium absorbs heat from hydrogen, causing its liquefaction.

An interesting alternative to traditional liquefaction processes is the magnetic liquefaction. The magneto-caloric effect is a magneto-thermodynamic process in which a change in the magnetic field induces a reversible variation in temperature. This effect is obtained by applying and removing a magnetic field to a magneto-caloric alloy (e.g. gadolinium, silicon and germanium) placed in a thermally insulated environment.

Therefore, such an effect can be used to obtain the refrigeration of a fluid following a cycle of four steps: adiabatic magnetisation, isothermal magnetisation (heat removed by a refrigerant), adiabatic demagnetisation and isothermal demagnetisation (heat removed from the fluid to be refrigerated or liquefied). Whereas a compressed-gas refrigeration system has an efficiency around 38% of the theoretical Carnot efficiency, this system can reach around 50% of the theoretical Carnot efficiency [19].

In evaluating the work required for liquefaction, it is necessary to take into account, in addition to the irreversibility, the exothermic transformation from ortho-hydrogen to para-hydrogen, which requires the subtraction of a further 703 kJ per kg of converted ortho-hydrogen. In equilibrium conditions, the para-hydrogen content is 25% at room temperature, while liquid hydrogen is made up almost exclusively of para-hydrogen. Therefore, the conversion from normal-hydrogen to para-hydrogen will require a heat subtraction of 527 kJ/kg. Latent heat of vaporisation of liquid hydrogen is 446 kJ/kg, so boil-off (i.e. vaporisation) occurs during the storage period if the ortho-hydrogen is not converted before storage.

However, the kinetics of this transformation is very slow and, therefore, when the gaseous hydrogen is liquefied, a mixture is obtained that is still very rich in ortho-hydrogen. For a short-term accumulation, this slowness is an advantage since a smaller heat amount is subtracted, but for a long-term accumulation it is necessary to force the full conversion to para-hydrogen in order to avoid that the progressive release of heat cause the evaporation of a part of liquid hydrogen. Therefore, it is necessary to ensure that the conversion takes place during the liquefaction process and this is feasible through the use of suitable catalysts, which were usually based on hydrated iron trioxide. Sullivan *et al.* [20] found that chromic oxide supported by silica gel is less susceptible to poisoning and has a highly efficient catalytic surface for conversion at low temperatures.

Although the theoretical work required for the liquefaction of hydrogen starting from ambient conditions is 11.63 MJ/kg [12], the specific energy consumption for a Claude cycle is within the range of 39–46 MJ/kg$_{LH2}$ and within the range 44–49 MJ/kg$_{LH2}$ for the helium Brayton cycle [21].

The ultimate target set by US Department of Energy for the energy consumption of hydrogen liquefaction is 21.6 MJ/kg$_{LH2}$ [22]. Several promising processes towards such a target have been reported by Aziz [23].

Since the conversion from ortho- to para-hydrogen is exothermic and spontaneous, a certain amount of heat is released during storage time in relation to the initial concentration of ortho-hydrogen. Evaporation leads to an increase in pressure and, therefore, the tank must be equipped with a safety valve that allows the hydrogen to be purged, thus regulating the internal pressure. Therefore, the heat generated not only involves a loss of energy spent for liquefaction but can also cause a loss of hydrogen if the pressure exceeds the valve intervention threshold. It is also important that the valve does not allow air to enter from the outside to avoid the formation of solids which block the functioning of the valve itself.

However, the conversion is slow and the optimal concentration of accumulated para-hydrogen depends on the storage time. Gùrsu [24], calculated the amount of liquid hydrogen left inside the tank:

$$m_{L(t,I_0)} = e^{\left[1.57785 \cdot I_0 \cdot \left(\frac{1}{0.00114 \cdot I_0 \cdot t + 1} - 1\right)\right]} \tag{4.16}$$

where I_0 is the initial concentration of ortho-hydrogen present in the tank and t is the storage time of the hydrogen expressed in hours.

Boil-off depends also on heat transferred from surrounding environment, and therefore on the tank thermal insulation, shape and dimension. Thermal insulation is increased by using materials with low heat conductivity, and introducing a vacuum space between the outer and inner walls. Although liquid hydrogen is stored in a very well-insulated vessel and tank, perfect insulation does not exist.

Increasing the size of the storage results in a lower surface-to-volume ratio. Also the shape is important: a sphere has the lowest surface-to-volume ratio. Thanks to these expedients, boiling can be reduced from 2% to less than 0.1% per day [25,26].

When considering the mobility sector, there are other two causes of evaporation to be considered: the heat generated by the motion of the liquid inside the tank (called sloshing) and the flashing during the refuelling, as the hydrogen pass from a tank at higher pressure to a tank at lower pressure.

In addition, evaporation occurs during each transfer between two tanks due to the higher temperature of the pipes and the second tank. The entire transfer system must be closed, isolated and under vacuum both to minimise evaporation and to avoid the presence of la which could liquefy and/or solidify.

Pipes and containers for the transport and storage of liquefied hydrogen were developed for the aerospace sector in which liquid hydrogen is used as a propellant for launching missiles (like Space Shuttle or Ariane) and must have very strict thermal insulation requirements since the evaporation of liquid hydrogen would result in a source of risk. A fundamental characteristic of the materials used for liquid hydrogen handling is the resistance to thermal stress at low temperatures since the temperature goes from room temperature to liquid temperature with a difference greater than 250 °C. Compared to gaseous hydrogen, there is less risk of embrittlement since the solubility of hydrogen decreases with decreasing temperature [26]. Furthermore, there are no corrosion problems because liquid hydrogen is not corrosive.

The direct transport of liquid hydrogen in isolated pipelines is technically feasible, but becomes economically disadvantageous even for distances of the order of hundreds of meters when compared to road transport, notwithstanding this requires expensive super-insulated cryogenic vessels.

Tanks for liquid hydrogen are designed in such a way as to avoid transmission of heat from the external wall to the liquid: they usually consist of two walls of austenitic stainless steel between which multilayer insulation, spray foam, or hollow glass microspheres are alternately inserted to prevent the passage of heat by conduction or convection. Furthermore, to prevent direct heat radiation, protective panels with low heat emission based on plastic and aluminium are installed between the internal and external walls of the container. The largest storage system for liquid hydrogen, with a capacity of 8000 m^3 (about 565 t), is currently owned by NASA [27,28]. The system is composed of three different system: the two oldest were built in mid-1960s and a third one in 2018 for new private commercial companies and partners such as SpaceX and Blue Origin.

Most large liquid hydrogen containers have a spherical shape because the latter has the lowest surface area for heat transfer per unit of volume (Figure 4.10). Furthermore, as the diameter of the containers increases, the volume increases faster than the external surface so that larger containers, proportionally, cause less heat transfer losses. The evaporated hydrogen generates an increase in pressure inside the container and must be recycled in the liquefaction process or catalytically converted to water.

Cylindrical containers, however, are preferable due to their ease and cost-effectiveness of construction although about 1%–2% per day is lost by evaporation.

Figure 4.10 Cryogenic stationary storage for liquid hydrogen [copyright NASA]

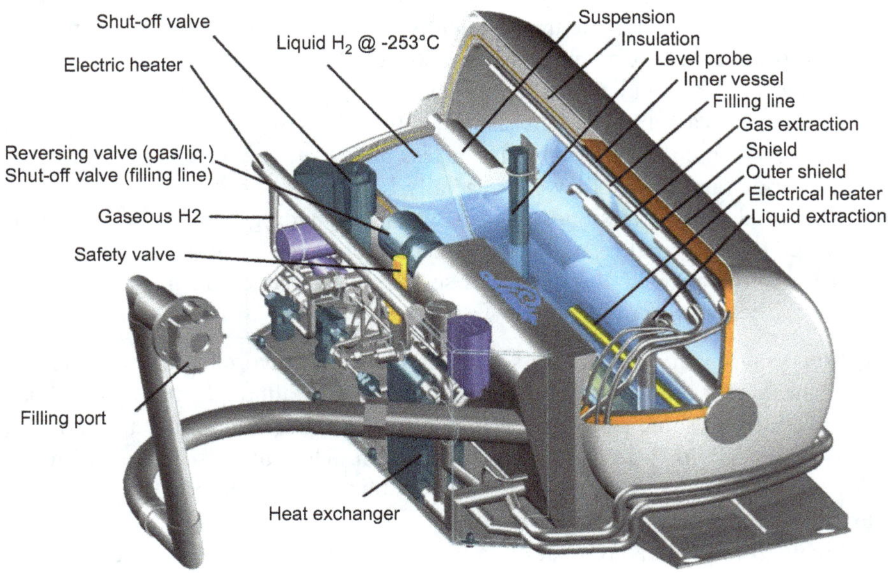

Figure 4.11 Cryogenic tank for on-board liquid hydrogen storage [29] [copyright Linde]

They are used for transportation and for on-board storage (Figure 4.11). In this last case the demand for hydrogen is higher than that of spontaneous evaporation and, therefore, the system is equipped with a heater fed by exhaust gases or consisting of an electrical resistance.

When liquid hydrogen is cooled to the triple point ($T = 13.8$ K, $p = 0.007$ MPa) it begins to form solid hydrogen. Slush hydrogen is a suspension of solid hydrogen in sub-cooled liquid. When the ratio between the two phases is unity, the gravimetric density increases by 15.5% compared to liquid hydrogen at its boiling temperature [30]. It has a fluid dynamic behaviour very similar to that in liquid form and also has the advantage of lower evaporation losses and, consequently, is safer than liquid hydrogen. It is mainly used in aerospace rockets [31].

4.3 Hydrides

The hydride is the anion of hydrogen (H^-) and it is the simplest negative ion being made up of only one proton and two electrons. The term is also used to describe the compounds of hydrogen with the other elements except for those of groups 16 and 17. On the basis of the prevailing bond, hydrides can be:

- ionic (or saline), i.e. binary compounds in which a strongly electropositive metal (usually alkaline or alkaline earth) reduces the hydrogen which therefore behaves like a halogen; they are solid compounds with crystalline structure with the exception of gallium hydride which is liquid; they generally react spontaneously, and sometimes violently, with water, releasing hydrogen gas and forming hydroxide or oxide (see Section 3.7);
- covalent (or molecular), i.e. binary compounds in which hydrogen shares the electron with a non-metal or a semimetal (elements of groups 13 and 14) with a covalent bond; they are non-polar or less polar with the exception of nitrogen trihydride (ammonia); they are generally liquid or gaseous;
- interstitials, i.e. compounds also in which the hydrogen nucleus (proton) is placed in the interstices of the metal matrix, while the electron becomes part of the electronic cloud that characterises the metal bond; the metal matrix is generally made up of an alloy.

Some non-binary salts are also commonly called hydrides. These salts can have a behaviour similar to interstitial hydrides, but, although more stable than ionic hydrides, in some cases they can also have a behaviour similar to the latter, that is to react with water to produce hydrogen. They are also called complex hydrides because an electropositive metal establishes an ionic bond with a 'complex' characterised by a covalent bond.

Finally, in the literature relating to the hydrogen storage, certain hydrocarbons are also called hydrides which can, under appropriate conditions of pressure and temperature and in the presence of suitable catalysts, release hydrogen generating another hydrocarbon characterised by a higher C/H ratio.

Taking into account the behaviour as a medium of hydrogen storage, hydrides can be classified as:

- metal hydrides, i.e. interstitial hydrides;
- complex hydrides, i.e. non-binary salts when they behave as interstitial hydrides;

- liquid hydrides, i.e. liquid covalent hydrides and liquid hydrocarbons that can undergo cycles of hydrogenation/dehydrogenation;
- chemical hydrides, i.e. ionic hydrides and non-binary salts when they behave as ionic hydrides.

4.3.1 Metal hydrides

Some metals and metal alloys are able to chemically absorb hydrogen forming interstitial hydrides which later can release the hydrogen reversibly [32]:

$$M + \frac{x}{2}H_2 \Leftrightarrow MH_x + \text{heat} \tag{4.17}$$

where M is the metal and MH_x is the metal hydride. As early as 1866, Thomas Graham had noticed that palladium could absorb significant amounts of hydrogen, but this discovery had virtually no application until the second half of the past century when the development of ternary-type hydrides based on metal alloys containing transition metals began. The first hydride of this type, $ZrNiH_3$, was developed in 1958 by George Libowitz.

The phenomenon of the formation of a hydride is usually described on a composition-pressure graph by means of isothermal curves that are obtained by varying the pressure of hydrogen at constant temperature (Figure 4.12). The metal

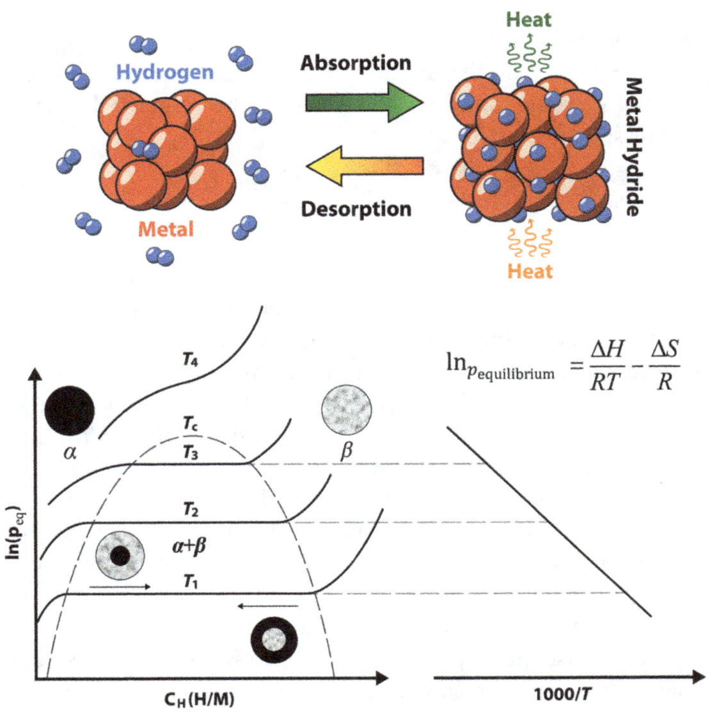

Figure 4.12 Hydriding and dehydriding process (based on [33])

matrix, in the presence of hydrogen at low pressure, initiates the dissociation of hydrogen molecules adsorbed by the metal surface. Atomic hydrogen penetrates inside the crystalline lattice dissolving in solid solution (phase α) until it reaches the saturation of the metal matrix: in this part of the curve a fairly modest increase in the concentration of hydrogen inside the metal matrix is obtained by gradually increasing the pressure. Once saturation is reached, a transition process starts from phase α to phase β (hydride) and the isotherm becomes practically parallel to the abscissa axis, that is also isobar, since the hydrogen is reversibly absorbed with negligible pressure variations. This part of the isotherm is called 'plateau' and is characterised by the coexistence of the two solid phases α and β. When the transition α-β is complete, the hydrogen concentration returns to increase modestly compared to the pressure.

For a metal hydride suitable to store hydrogen, the hydride phase is more stable than the metal and therefore the absorption of hydrogen is an exothermic process, while desorption process requires the supply of heat at a temperature the higher the more the hydride it is stable. As a consequence the storage of hydrogen in the form of metal hydrides can be considered a very safe system as hydrogen can be released only if the system is supplied with heat at an adequate temperature.

The free energy variation ΔG° connected to the hydride formation reaction is given by:

$$\Delta G^\circ = R \cdot T \cdot \ln p_{eq} = \Delta H^\circ - T \cdot \Delta S^\circ \tag{4.18}$$

in which R is the gas constant, T is the temperature, p_{eq} is the pressure corresponding to the plateau (equilibrium pressure) and ΔH° and ΔS° are, respectively, the enthalpy and entropy variations connected to the reaction. Extrapolating the equilibrium pressure we obtain:

$$\ln p_{eq} = \frac{1}{R} \cdot \left(\frac{\Delta H^\circ}{T} - \Delta S^\circ \right) \tag{4.19}$$

Putting the inverse of the temperature on the abscissa, this equation is represented by a straight line (Figure 4.12) whose slope is given by the enthalpy term which depends on the stability of the metal-hydrogen bond. The difference in entropy between a metal and its hydride is of the order of 10 J K^{-1}mol^{-1}, definitely negligible compared to the variation connected to the dissociation and adsorption of hydrogen (130.858 J K^{-1}mol^{-1}) and, therefore, the entropic term is practically independent of the material chosen to make the metal hydride [34].

Given the small size of the hydrogen atom, the number of hydrogen atoms in the crystals of many hydrides can be double or triple the number of atoms in the metal and this means that the energy density per unit of volume is very high, even higher than that of liquid hydrogen. However, due to the heaviness of the metal matrix, the weight energy density is modest (2%–7%).

The charging phase must take place at the lowest possible pressure in order to limit the energy required for compression. Since this phase is exothermic, the minimum pressure is also linked to the temperature of the medium available for the hydride refrigeration. It follows that, from the point of view of the charge, regardless of the system that will use the hydrogen, those hydrides are ideal that have an equilibrium pressure equal to the atmospheric one at a temperature just above the ambient temperature. The discharge phase, however, is endothermic and therefore it is important that the amount of heat required is as small as possible and at the lowest possible temperature. In this case the limit can be imposed by the pressure at which hydrogen is required by the system.

Substantially, therefore, the working pressure and temperature should fall respectively in the ranges 1–10 bar and 20 °C–100 °C, corresponding to an enthalpy variation between 15 and 24 kJ/mol.

If case of hydrides for on board storage, the energy density per unit of weight becomes a fundamental criterion in the choice. In this regard, it should be pointed out that both during charging and discharging phases the effective working range of a hydride is limited to the plateau region. In fact, it would be expensive to extend the charging phase beyond the end of the plateau to obtain a modest increase in accumulated hydrogen, while during the discharge phase it would be impossible to go below the pressure required by the system.

Other characteristics required of the hydride are a low cost, and therefore abundant and easily available materials, and a remarkable resistance to charge/discharge cycles.

Since the reaction rate is favoured by the extension of the surface, the hydrides are used in the state of fine powder, with particles of size in the range 10–100 μm. However, this fact is unfavourable from the point of view of thermal conductivity, which is fundamental for heat exchange both in the charge and discharge phases. Usually a gas-solid reaction begins where the nucleation process is easiest, i.e. at the grain boundary or at a crystal defect or impurity. It follows that both the preparation process and the past history of the hydride have an influence on the reaction kinetics since they affect the size of the particles, the number of defects and the presence of impurities.

Unlike what reported in Figure 4.12, the plateau of real systems is slightly inclined and does not coincide between the charge and discharge phases. In fact, metal hydrides undergo the phenomenon of hysteresis whereby the equilibrium pressure in the charge phase is greater than the equilibrium pressure in the discharge phase. There are several theories that attempt to justify this phenomenon, but none of them have yet convincingly explained it. As the charge/discharge cycles follow, the hysteresis can increase and cause a decrease in the storage capacity.

There are also other factors that can cause the degradation of the storage capacity of a metal alloy subjected to hydrogen charge/discharge cycles. These factors can be classified into intrinsic, which cause the deformation and the creation of defects in the crystal lattice, and extrinsic, which concern the hydride poisoning, i.e. presence of impurities in the supplied hydrogen. In order to operate as a hydrogen accumulator, the metal matrix must undergo a first charge at a pressure significantly higher than the equilibrium pressure so that this first operation proceeds very quickly. In case of poisoning an analogous process could sometimes help to restore full storage capacity.

There are low temperature hydrides and high temperature hydrides, which usually are less heavy and more expensive being based on magnesium, vanadium and their alloys, and also more negatively affected by the presence of pollutants (including H_2O and O_2 which could be found in electrolytic hydrogen). If the temperature at which the heat is available is sufficient, it is possible to use such light high temperatures hydrides, however, they are not able to use all the available heat. In this case it is possible to use mixed tanks in which the heat released is first used to dehydrogenate the hydride at high temperature and then the hydride at low temperature.

Ternary hydrides are classified into families identified by two letters (AB) which represent substances having, respectively, high and low affinity for hydrogen: A is the component that forms a stable hydride and which, generally, is a rare earth or an alkaline-earth metal, and B is the component which forms an unstable hydride and which, generally, is a transition metal. Nickel is widely used because it is an excellent catalyst for the dissociation of hydrogen. The letters can be followed by numbers which indicate the molar ratio between the two components and, in relation to the values that these subscripts can assume, there are several families of hydrides more or less numerous. For example Y_6Fe_{23} belongs to the rather unusual A_6B_{23} family, but the most studied, if we exclude the AB which has above all historical importance having been examined first, there are only two: AB_5 and AB_2.

The families of hydrides are also linked to the crystalline structure. The most compactly packed crystalline structures are the face-centred cubic and the compact hexagonal: in both cases the single atom is in contact with 12 other atoms (six on the same plane, three on the upper plane and three on the lower one). In the case of the body-centred cubic structure the single atom is in contact with eight other atoms (arranged at the vertices of the cube). As a result, the body-centred cubic structure is more suitable for the accumulation of hydrogen as the interstitial sites are larger and can therefore host a greater quantity of hydrogen atoms. Table 4.2 shows the hydrides family and their crystalline structures.

Most metal hydrides adsorb hydrogen up to a hydrogen/metal ratio (H/M) of 2. Hydrides with H/M>2 are ionic or covalent compounds and belong to the group of complex hydrides.

Table 4.2 Families of alloys capable to form hydrides

Alloy	**Example**	**Structure**
AB	TiFe	Cubic
A_2B	Mg_2Ni	Cubic
A_2B_7	Y_2Ni_7	Hexagonal
AB_2	ZrV_2	Hexagonal or cubic
AB_3	$CeNi_3$	Hexagonal
AB_5	$LaNi_5$	Hexagonal

The most studied hydrides have a cubic (AB, AB_2) or hexagonal (AB_2, AB_5) structure and usually reach capacities not exceeding 2% by weight: for example 1.37% for $LaNi_5$ and 1.89% for FeTi [35].

FeTi is the main compound of the AB family which is characterised by low-cost materials.

A_2B hydrides have a remarkable storage capacity, but high stability and the temperature and heat of decomposition increase with the stability of the hydride. The types of alloys most used are magnesium-based intermetallic compounds that have a good kinetics and do not require catalysts for the absorption of hydrogen, are light and less expensive of AB_5 alloys, but more sensitive to poisoning.

The AB_2 compounds have a good ability to accumulate hydrogen and allow charging and discharging phases of the order of minutes.

AB_5 compounds generally form hydrides with equilibrium pressures of a few MPa and temperatures below 100 °C. These are alloys that show a weak presence of pressure hysteresis, are resistant to the impurities present in gaseous hydrogen and their activation does not involve problems. The disadvantage is the high cost.

A high amount of stored hydrogen can be obtained by using light elements such as magnesium, which is abundant in nature and non-toxic. It forms a hydride that contains up to 7.6% by weight of hydrogen and the heat to be removed per mole of hydrogen is 76 kJ. The equilibrium pressure at room temperature is extremely low and remains such in the range of temperatures of application interest (at 100 °C it is only 0.36 mbar). The hydriding phase is therefore easily achievable, while the dehydriding phase requires very high temperatures: to have operating pressures close to the atmospheric one, it is necessary to reach about 280 °C–300 °C. In any case, the kinetics of the reactions is extremely slow. The use of nanostructured magnesium hydride, produced by high-energy grinding, is convenient compared to massive material: the presence of sub-micrometric grains reduces the diffusion path of hydrogen in the solid phase and the high density of defects and of grain boundaries provides preferential paths for gas escape. Good absorption and desorption kinetics at temperatures around 230 °C were obtained by grinding metallic magnesium or its hydride with or without the addition of catalysts. It appears that diniobium pentoxide is the best catalyst for magnesium [34].

Aluminium hydride contains 10% hydrogen by weight and its density is 148 kg m^{-3}. The heat to be removed per mole of hydrogen is 7.6 kJ. In theory it should decompose already at room temperature, but the experimental results obtained at DOW Chemical Company showed that the release of hydrogen below 150 °C is very low [36]. Researchers from Sandia National Laboratories and Brookhaven National Laboratories lowered this temperature by several tens of degrees by doping the hydride with lithium hydride [37]. An undoped hydride, but made using nanotechnology, released, at a temperature below 100 °C, a quantity of hydrogen close to the theoretical one.

Main hydrides and their properties are shown in Table 4.3.

Metal hydrides have the advantages of constant pressure adsorption and desorption, moderate temperature operation, and storage stability and safety (long-term

Table 4.3 Main metal hydrides and their characteristics (data from [38,39])

Hydride	Class	Hydrogen (% weight)	Plateau pressure (bar)	T (°C)	Heat of hydride formation (kJ/mol_{H2})
MgH_2	Elemental	7.6	1	279	74.5
Mg_2NiH_4	A_2B	3.8	1	291	64.1
$FeTiH_2$	AB	1.9	5	35	28.1
$TiMn_2H_3$	AB_2	2.0	14	25	30.1
$LaNi_5H_6$	AB_5	1.5	2–5	25	30.6

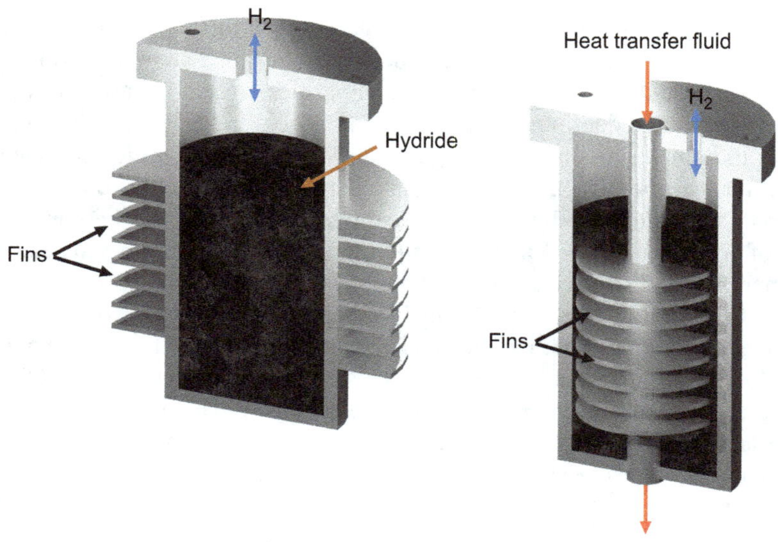

Figure 4.13 Examples of hydrides tanks and heat transfer (adapted from [40])

storage capability). Tanks with metal hydrides and alloys of different type are available on the market.

The metal hydride tanks (Figure 4.13) must be constructed bearing in mind the two functions they must perform: that of storage and that of heat exchanger. They are cylindrical in shape and are designed to withstand rather low pressures as hydrogen gas is present in very small quantities inside them. The volume is almost entirely occupied by the metal material devoted to hydrogen adsorption. In the case of rather small size, the heat exchange takes place through the external surface, suitably finned, whereas an internal circuit is provided with a heat-carrying fluid (generally water) for tanks of considerable size.

The response of the storage system is slow and therefore it is advisable that the tank can contain an amount of gaseous hydrogen sufficient to restart the process after a pause or to overcome a transient. Alternatively, a small gaseous hydrogen tank can be used which can be refuelled during the hydride discharge.

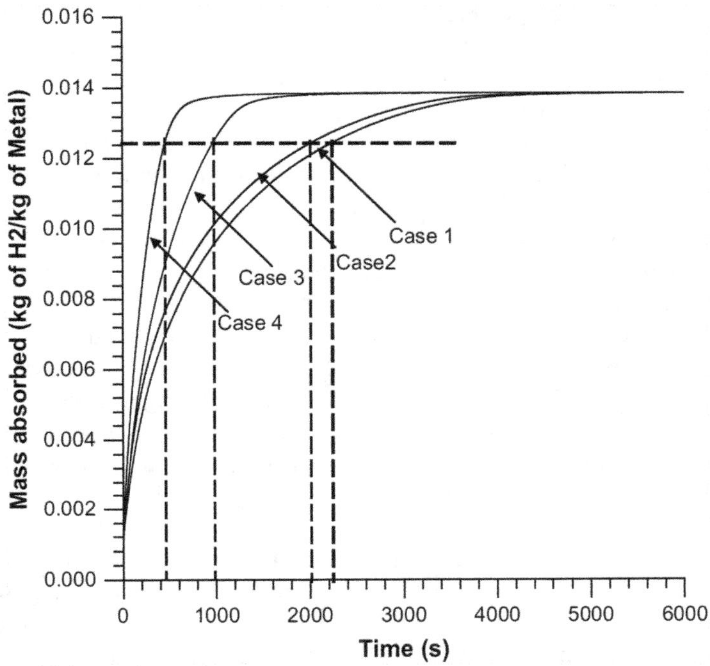

Figure 4.14 Time needed for metal hydride charging [40]

The hydrogenation of the tanks is a simple operation and can be considered feasible in 10–20 minutes; being satisfied with a partial refuelling (80%), the operation can be carried out in less than ten minutes [41]. Figure 4.14 shows a comparison between the speed of hydrogen absorption by four cylindrical metal hydride tanks. Case 2 and case 3 correspond to the configurations illustrated in Figure 4.13, while case 1 and case 3 refer to the same ones, but without fins.

One of latest and biggest hydrogen storage using hydrides was realised by GKN Hydrogen and utility Southern California Gas Co. in 2022 for U.S. Department of Energy's National Renewable Energy Laboratory in Flatirons Campus in Boulder, Colorado: two systems called HY2MEGA each able to store 250 kg of hydrogen was installed for a power-to-power solution [42].

Other applications of metal hydrides

Metal hydrides can find applications other than hydrogen storage:

- hydrogen compression;
- generation of mechanical energy;
- heat storage;
- electricity storage;
- separation of hydrogen and isotopes.

The relationship between pressure and temperature in the reversible hydriding/dehydriding process is given by (4.19). Therefore by using a heat source at an adequate temperature, it is possible to obtain a considerable increase in the pressure of the hydrogen after having accumulated it at low pressure. A review of the use of metal hydrides for hydrogen compression has been written by Lototskyy *et al.* [43]. Metal alloys such as, for example, $TiMn_2$ guarantee the possibility of reaching high pressures, while $TiCr_2$ provides characteristics of less hysteresis [44].

It has recently been proposed to use a metal hydride compressor coupled with an electrochemical compressor. The electrochemical compression stage is followed by the hydride stage which exploits the waste heat (at 80 °C) of the electrochemical compressor, reaching up to 80–100 MPa. The alloy identified as the best is $TiCr_{1.25}Fe_{0.75}$ which releases 13.9 kJ/kmol in the charging phase and requires 17.2 kJ/kmol in the discharging phase. At the time of writing this book, the article has not yet been published except in the form of a short abstract [45].

Thermal sorption compressors could be used also to reduce power consumption for hydrogen liquefaction [46].

This ability of hydrides to increase the pressure of hydrogen can also be used to generate work from low-temperature heat. In Figure 4.15, for example, an expander is placed between two tanks containing the same metal alloy. The first tank, containing hydride, is heated to a temperature T_1 by a heat source and releases hydrogen at the equilibrium pressure p_1. The hydrogen passes through an expander which generates mechanical work and returns hydrogen which is absorbed in the second tank by cooling to a temperature $T_2 < T_1$ corresponding to an equilibrium pressure $p_2 < p_1$. Once all the transferable hydrogen from the first to the second tank has passed through the expander, the cycle is reversed.

By exploiting the endothermicity of the dehydriding reaction, it is also possible to accumulate heat that can be returned by the reverse hydriding process. If

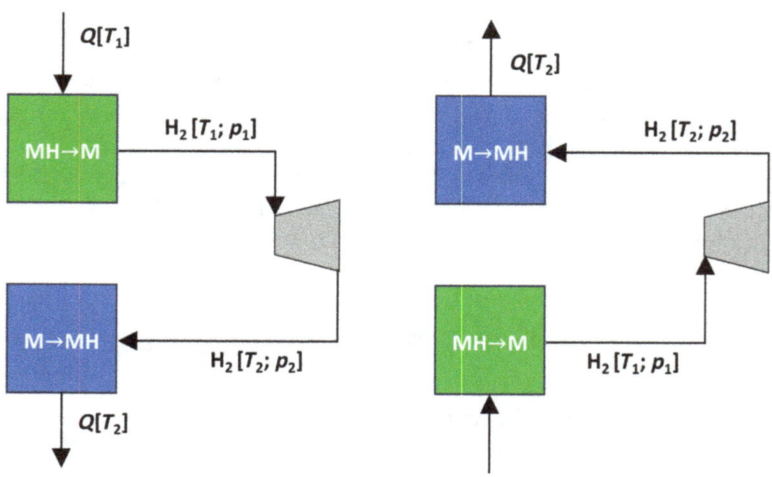

Figure 4.15 Block diagram of a 'hydride expander'

hydrogen is supplied at the pressure at which it was released, then, unless the hysteresis effect, the heat is released at the same temperature at which it was stored. By increasing the pressure it is possible to obtain heat at a higher temperature, while by decreasing the pressure it is possible to obtain heat at a lower temperature. By exploiting this capacity, it is possible, for example, to air-condition the rooms.

Hydrides have also been used in the accumulation of electrical energy by means of nickel-metal hydride batteries, based on the reversible reactions:

$$MH_x + OH^- \Rightarrow MH_{x-1} + H_2O + e^- \quad (4.20)$$

$$Ni(OH)_2 + OH^- \Rightarrow NiO(OH) + H_2O + e^- \quad (4.21)$$

$$MH_x + NiO(OH) \Rightarrow MH_{x-1} + Ni(OH)_2 \quad (4.22)$$

Finally, another interesting feature of metal hydrides is that the alloys that form them do not, in general, absorb substances other than hydrogen. Therefore it is possible to use the process to purify the hydrogen and also to separate deuterium from protium: after a few passages high purity deuterium can be obtained.

4.3.2 Complex metal hydrides

The existence of a two-phase regime in the hydrogen absorption and release process in the case of sodium tetrahydroaluminate was discovered in the mid-1970s [47]. Bogdanovich and Schwickardi, in 1996, obtained the pressure-concentration isotherms at 180 °C and 210 °C, highlighting that there are non-binary salts that behave as interstitial metal hydrides [48].

Within a complex hydride, hydrogen tends to arrange itself at the vertices of a tetrahedron at whose centre there is a trivalent element (especially boron and aluminium). The bond established is of the covalent type and the complex anion generated forms an ionic bond with a cation, for example $Li^+ \cdots (BH_4)^-$.

There are still many aspects to be clarified in the behaviour of complex hydrides, especially in relation to the possibility of hydrogen desorption. Unlike what happens in the case of interstitial hydrides, the release of hydrogen by tetrahydroborates and tetrahydroaluminates occurs with decomposition of the host salt (Figure 4.16). By

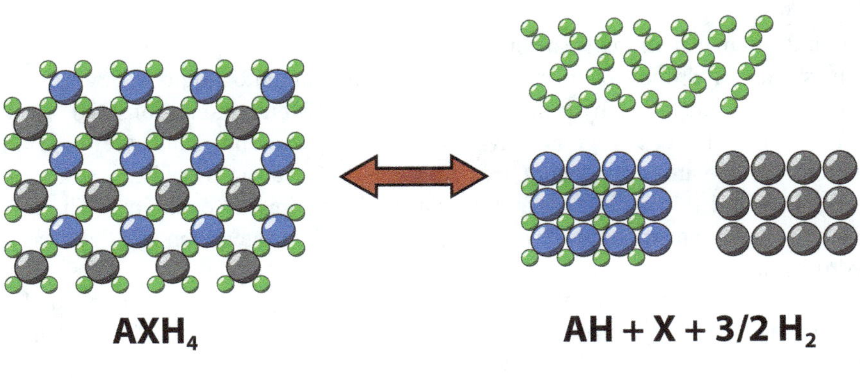

Figure 4.16 Complex hydride structure and hydriding-dehydriding process

Table 4.4 Weight density (%) of hydrogen in some complex hydrides

Mg_2NiH_4	3.62	Na_2LiAlH_6	7.04	$LiAlH_4$	10.62	$Ti(BH_4)_3$	13.09
Mg_2CoH_5	4.48	$NaAlH_4$	7.47	$NaBH_4$	10.66	$Mg(BH_4)_2$	14.94
Mg_2FeH_6	5.47	$Ca(AlH_4)_2$	7.90	Li_3AlH_6	11.23	$Ti(BH_4)_4$	15.04
$KAlH_4$	5.75	$LiNH_2$	8.78	$Ca(BH_4)_2$	11.56	$Al(BH_4)_3$	16.92
Na_3AlH_6	5.93	$Mg(AlH_4)_2$	9.34	$Fe(BH_4)_3$	12.05	$LiBH_4$	18.51

indicating the metal with A, and the trivalent element with X, the overall reaction takes the form:

$$A(XH_4) \Rightarrow AH + X + \frac{3}{2}H_2 \qquad (4.23)$$

and, in reality, it is the result of a more complex process not yet fully identified.

Table 4.4 shows the hydrogen content of some complex hydrides. It is not the storage density since not all hydrogen is released during dehydriding, see (4.23). For example, the hydrogen content of lithium tetrahydroborate is 18.2%, while the storage density is 13.6 since one atom of hydrogen remains bonded to lithium. However some complex hydrides can also react with water (see Section 3.7) in which case the hydrogen released can derive also from water.

Tetrahydroaluminates are much less stable than the corresponding tetrahydroborates and this difference is explained by the different electronegativity: 2.04 for boron and 1.61 for aluminium on the Pauling scale.

Most complex hydrides do not decompose appreciably at temperatures below the melting point [47,49,50]. The techniques for improving the absorption and desorption kinetics are essentially two: nanostructures, which increase the active surface, and the addition of catalysts.

There are complex hydrides which are liquid at room temperature and, therefore, are potentially more interesting as they are more easily handled. An example of these is aluminium tetrahydroborate, $Al(BH_4)_3$.

4.4 Liquid organic hydrogen carriers

A liquid organic hydrogen carrier (LOHC) is an organic liquid capable of reversibly storing hydrogen by turning it into another organic liquid. They are also incorrectly called liquid hydrides. Being liquid in ambient conditions (20 °C and 1 atm), they have the great advantage of ease handling and are generally suitable for the use of existing infrastructures. However, they have some disadvantages and this caused an interest rather discontinuous over time. The main issues are the large amount of energy required by dehydrogenation and the need for hydrogen purification [51].

The first experiences in the sector concerned the methylbenzene/methylcyclohexane couple [52]. Subsequently, in order to increase the accumulation density, the benzene/cyclohexane couple was taken into consideration which is, however,

Table 4.5 Characteristics of the main hydrocarbon/hydride couples

Hydrocarbon	Hydride	Reaction	Weight density %
Methylbenzene	Methylcyclohexane	$C_7H_8 + 3H_2 \leftrightarrow C_7H_{14}$	6.16
Ethylcarbazole	Dodecahydro-ethylcarbazole	$C_{14}H_{13}N + 6H_2 \leftrightarrow C_{14}H_{25}N$	6.50
Benzene	Cyclohexane	$C_6H_6 + 3H_2 \leftrightarrow C_6H_{12}$	7.19
Naphthalene	Decahydronaphthalene	$C_{10}H_8 + 5H_2 \leftrightarrow C_{10}H_{18}$	7.40
Biphenyl	Bicyclohexyl	$C_{12}H_{10} + 6H_2 \leftrightarrow C_{12}H_{22}$	7.46
Methylbenzene	n-heptane	$C_7H_8 + 4H_2 \leftrightarrow C_7H_{16}$	8.22

affected by a series of technological and safety problems [53]. More recently, the methylbenzene/n-heptane couple [54] and naphthalene/decahydronaphthalene [55] were studied. Table 4.5 shows the hydrogen storage weight density and the heat of reaction.

Although the methylbenzene/methylcyclohexane couple has the lower density, it could be more interesting thanks to a particular production process which avoids the intermediate production of hydrogen. Matsuoka *et al.* [56] investigated the electrochemical reduction of methylbenzene to methylcyclohexane using a proton exchange membrane and PtRu/C as a catalyst. Water fed to the anode generates protons which cross the membrane and react with the methylbenzene fed to the cathode. Only a residual amount of methylbenzene was found in the methylcyclohexane with no by-products.

The high temperature required by the dehydrogenation reaction (around 400 °C, for example, for methylcylohexane) means that applications with low temperature fuel cells do not allow to use waste heat and it is therefore necessary to burn a part of the hydrogen produced with consequent loss of efficiency.

In the case of mobile applications, the disadvantages, compared to some other hydrogen storage systems, mainly reside in the need for a rather heavy and bulky catalytic reactor and two tanks, one for the hydride and the other for the hydrocarbon. However, it is possible, for vehicles equipped with an internal combustion engine, to couple the dehydrogenation system to the engine in order to reduce weight and bulk [57] and one could think of the construction of a special variable geometry tank in which both hydride and hydrocarbon can be accommodated without mixing. A further disadvantage is constituted by the need to burn a certain quantity of hydrocarbon at the start in order to provide the heat necessary to activate the system. Alternatively, the vehicle can be equipped with a small gaseous hydrogen tank.

A good compromise to maintain the advantage of the existing distribution infrastructure avoiding the onboard issues connected to LOHCs is their distribution to the refuelling stations followed by local dehydrogenation to supply gaseous hydrogen to vehicles.

The only true liquid covalent hydride of some interest is ammonia which is liquid at room temperature and pressure of 0.8 MPa or at atmospheric pressure and temperature of −33 °C. At ambient temperature and pressure it is a toxic gas (lethal dose equal to 5%), and has a very pungent and penetrating odour, as well as having tear effects. It is the second largest chemical product in the world and, therefore, transport and distribution infrastructures are extremely widespread.

It can be produced starting from the constituent elements and can release hydrogen by catalytic reaction:

$$\frac{1}{2}N_2 + \frac{3}{2}H_2 \Leftrightarrow NH_3 \tag{4.24}$$

The weight density of hydrogen in ammonia is very high (17.65%) and for this reason its use was considered within the Euro-Quebec project (half of 1990s) as an alternative to liquefaction of hydrogen [58]. Later, it was shelved mainly due to toxicity, but, given the scarcity of efficient alternatives, a certain interest is periodically awakened. By 2030, different companies have announced world scale production of ammonia from renewable in Arabic Peninsula to be exported in Europe and Asia [59].

4.5 Adsorbent materials

Coming outside of the molecule we have to deal with the intermolecular forces. Although much weaker than the covalent and ionic bonds that hold the atoms together, they are responsible for important chemical-physical properties (e.g. solubility). The interactions between non-polar molecules are called van der Waals forces and are responsible for deviations from the ideal behaviour of gases.

The attractive interaction between neutral molecules E_A, called van der Waals attraction, is inversely proportional to the sixth power of the intermolecular distance r by means of a constant of proportionality k_A dependent on the nature of the interacting molecules:

$$E_A = -\frac{k_A}{r^6} \tag{4.25}$$

At the same time there is a repulsive interaction E_R, called van der Waals repulsion, which is inversely proportional to the twelfth power of the intermolecular distance by means of a proportionality constant k_R depending on the nature of the atoms that make up the interacting molecules:

$$E_R = \frac{k_R}{r^{12}} \tag{4.26}$$

To obtain the net energy of the intermolecular interaction, the repulsion and attraction energies are added. The function obtained, called Lennard-Jones potential, (Figure 4.17) intersects the abscissa axis only at one point, then reaches a minimum of energy for $r = 2^{1/6}r_0$ and finally goes up tending towards zero:

$$E = \frac{k_R}{r^{12}} - \frac{k_A}{r^6} = 4 \cdot E^0 \cdot \left[\left(\frac{r_0}{r}\right)^{12} - \left(\frac{r_0}{r}\right)^6 \right] \tag{4.27}$$

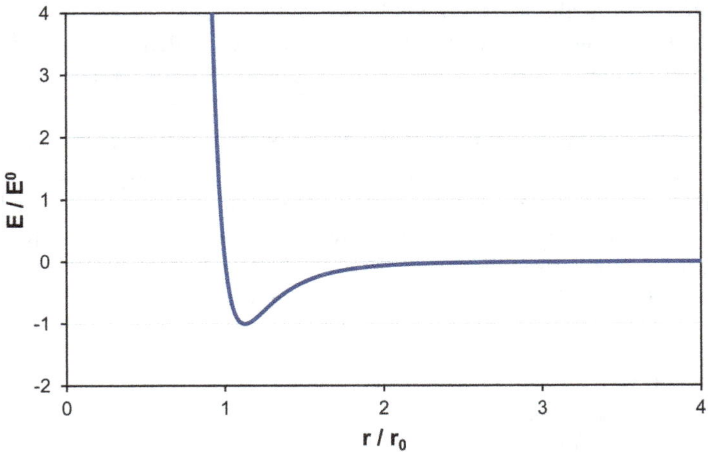

Figure 4.17 Lennard-Jones potential

where E^0 and r_0 are respectively the depth of the potential well and the distance at which the potential energy is zero.

One of the consequences of these interactions is the absorption of a gas on a solid surface: the molecules of the gas are attracted to the solid, but when they approach the repulsion force becomes bigger and bigger until the attraction force is balanced. It follows that a layer of gas molecules is formed at the equilibrium distance from the surface, that is, at the distance at which the potential energy of the molecule is minimal. This phenomenon is called physical adsorption or physisorption as there is no chemical bond. Instead, we speak of chemical adsorption or chemiadsorption when the atoms (or ions) of the adsorbed gas are fixed to the surface through real chemical bonds. The solid surface constitutes the adsorbent, while the adsorbed molecules are called adsorbate. The equilibrium distance corresponds to a distance of approximately one molecular radius of the adsorbate and, since the interaction is of low intensity, cryogenic temperatures are required to obtain significant physical adsorption. Adsorption is expressed as the amount of adsorbate with respect to the amount of adsorbent, generally as a percentage by weight, and depends, in addition to the temperature, on the nature of the two substances and on the pressure. Once a monolayer of adsorbed molecules is formed, other gaseous molecules interact with the surface of the monolayer itself and are in turn adsorbed.

In 1773 C. W. Scheele found that charcoal was capable of fixing gas on its surface and reported the news. Another 40 years passed before T. de Saussure, in 1814, started the first systematic studies on the phenomenon. The first to use the term adsorption was H. Kaiser in 1881. In 1909 Mc Bain described the adsorption of hydrogen on carbonaceous materials. In 1980 C. Carpetis and W. Peschka proposed the accumulation of hydrogen on activated carbon at cryogenic temperature. For hydrogen, the heat of adsorption is usually in the range of 3–10 kJ/mol [60].

This accumulation technology is completely reversible and has a very fast kinetics. The materials suitable for physical adsorption are those with a large specific surface area, such as, for example, activated or nanostructured carbon and carbon nanotubes. These materials have a relatively modest cost starting, for example, from biomass easily available at a local level to obtain activated bio-carbon [61]. Furthermore, a hydrogen storage system based on physical adsorption is constructively quite simple. Unfortunately, the temperatures required for adsorption are decidedly low, while the release of hydrogen presents no problems and can occur at room temperature.

In general, to evaluate the quality of a material used as an adsorbent, the following parameters are considered:

- the specific surface;
- the volume of micropores;
- the volumetric density.

However, the distribution and size of the pores are the aspects that best attest to the quality of an adsorbent material. For this purpose it is agreed to classify the pore sizes as follows:

- micropores (<2 nm);
- mesopores (from 2 to 50 nm);
- macropores (> 50 nm).

When hydrogen is in a state above the critical point, most of it is adsorbed by the micropores and the density of the adsorbed phase is greater than the non-adsorbed one. Therefore, to increase the absorption capacity, it is necessary to have materials made up mostly of micropores.

Traditional adsorbent materials are activated carbon, zeolites, dialuminium trioxide and silicon dioxide.

In the case of carbon, the absorption process is greatly favoured when the carbon has been thermally activated, i.e. treated with water vapour at 1000 °C. Activated carbon is characterised by a disordered structure consisting of atoms arranged in tiny layers to form a porous structure rich in micropores and with a considerable specific surface. According to R. Chahine and T.K. Bose [62] at a pressure of 35 bar and a temperature of 77 K, the density of accumulation by absorption of hydrogen is half that of hydrogen in the liquid phase (71 g/l at 20 K), and is also 3.4 times the density of hydrogen gas.

A comparison among different materials is shown in Table 4.6 adapted from [63]. The specific surface of the powders is measured with a standard method defined by Brunauer, Emmett and Teller, and generally the S_{BET} symbol is used. It is noted that the pore size is particularly important: in fact zeolites, dialuminium trioxide and silicon dioxide have specific surfaces that are comparable, but, while zeolites are characterised by micropores, dialuminium trioxide and silicon dioxide are characterised by mesopores and they achieve lower hydrogen adsorption values. Carbonaceous materials generally adsorb larger hydrogen amount since essentially have micropores.

Table 4.6 Total and micropore specific surfaces and adsorption capacities at 77 K and 1 bar of different materials (adapted from [63])

Material	Total			Micropore			H_2 % micro
	Volume (ml/g)	Specific surface (m^2/g)	H_2 $(ml/g)_{STP}$	Volume (ml/g)	Specific surface (m^2/g)	H_2 $(ml/g)_{STP}$	
Synthetic graphite	0.04	7	0	0.00	0	0	–
Al_2O_3 preshaped	0.80	233	7	0.00	0	0	0
Medium diameter CNF1	0.28	120	12	0.00	0	1	8
SiO_2 Becker AD 050	0.74	330	18	0.00	0	0	0
Hyperion CNF	2.75	238	22	0.00	0	0	0
Activated graphite 300	0.51	287	36	0.05	104	19	53
Zeolite ZSM5	0.28	431	80	0.16	388	78	98
Activated carbon Norit 990721	0.43	988	142	0.43	971	140	99
Activated carbon UOK A	0.65	1195	188	0.47	1085	178	95
Activated carbon Norit 990293	1.03	2029	238	0.92	1951	231	97

4.5.1 Carbon nanotubes

Fullerenes are macromolecules of microporous carbon which are made up of regular hexagonal rings at the vertices of which the carbon atoms are located. Graphite has a hexagonal crystal structure with carbon atoms arranged at the vertices of regular hexagons. This structure is organised in flat layers superimposed on each other, each of which is called graphene. Unlike graphite, the presence of pentagonal or heptagonal rings in fullerenes leads to non-planar shapes: spherical, ellipsoidal or tubular.

Theoretically a maximum of 60 hydrogen atoms can be adsorbed by a fullerene spherical surface so that a $C_{60}H_{60}$ composition can store 7.7% hydrogen by weight. Although the charging process is exothermic, an activation energy is required. Figure 4.18 shows the activation energy for both hydrogenation and dehydrogenation process. As shown in Table 4.7, even with a doping based on alkali metals it is difficult even to exceed half of the theoretical storage capacity.

Tubular-shaped fullerenes (Figure 4.19) are called carbon nanotubes and were first identified during a study of fullerenes synthesis using an electric arc [65]. Following this discovery, it was possible to produce nanotubes in large quantities by varying the operating conditions of the electric arc [66]. The nanotubes generated were of the multiple wall type, but single wall carbon nanotube were later produced [67] by adding to the graphite electrodes some metal catalysts (e.g. Co, Ni, Y). In 1996, using

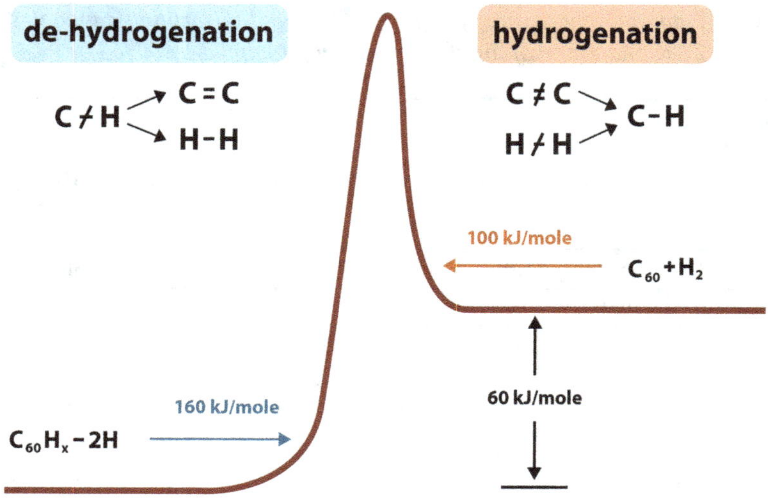

Figure 4.18 Activation energy for $C_{60}H_x \leftrightarrow C_{60}$ transformation (adapted from [64])

Table 4.7 Hydrogen storage capacity of Na, K, Li-doped fullerenes at 200 °C [64]

Starting material	Process parameters		Final product	wt% H_2
	Pressure, MPa	Time, h		
Na_3C_{60}	4	6	$Na_3C_{60}H_{19}$	2.30
Na_3C_{60}	10	6	$Na_3C_{60}H_{23}$	2.74
Na_3C_{60}	14	6	$Na_3C_{60}H_{26}$	3.19
Na_3C_{60}	14	24	$Na_3C_{60}H_{33}$	3.91
Na_6C_{60}	10	24	$Na_6C_{60}H_{36}$	3.99
K_3C_{60}	4	2	$K_3C_{60}H_5$	0.58
K_3C_{60}	10	2	$K_3C_{60}H_{10}$	1.10
K_3C_{60}	10	24	$K_3C_{60}H_{29}$	<3.0
Li_3C_{60}	10	6	$Li_3C_{60}H_{10}$	1.30
Li_3C_{60}	10	24	$Li_3C_{60}H_{24}$	<3.0
Li_6C_{60}	4	2	$Li_6C_{60}H_{7.5}$	0.95
Li_6C_{60}	10	6	$Li_6C_{60}H_{17}$	2.17
$Li_{20}C_{60}$	10	24	$Li_{20}C_{60}H_{27}$	<3.0

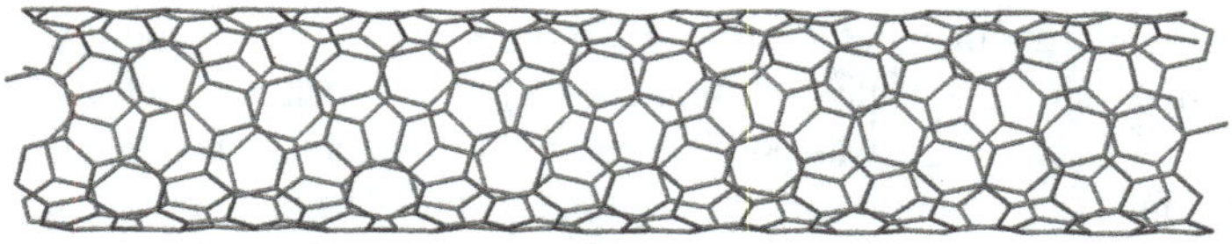

Figure 4.19 Structure of a carbon nanotube [68]

Table 4.8 Characteristics of a single-walled nanotube [71]

	Average diameter		1.2–1.4 nm
a	Distance between opposite carbon atoms		2.83 Å
b	Distance between corresponding carbon atoms		2.456 Å
c	Length of the bond between carbon atoms		1.42 Å
d	Distance between parallel bonds		2.45 Å
	Density	Armchair (10, 10)	1.33 kg/l
		Zig-zag (17, 0)	1.34 kg/l
		Chiral (12, 6)	1.40 kg/l

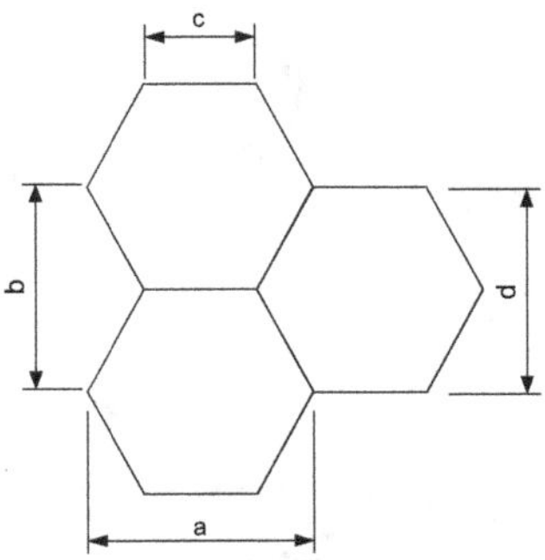

laser vaporisation of graphite, Richard Smalley obtained a large number of single-walled nanotubes with unusually uniform diameters and since then this has been the predominant method.

The first experiment to adsorb hydrogen on carbon nanotubes was carried out some years later [69].

Single-walled nanotubes consist of a graphene sheet, with a thickness equal to an atomic layer, rolled into a cylinder of a few nm in diameter and several μm in length. They are closed at both ends with a hemispherical shell and are light, flexible, thermally stable and chemically inert. Table 4.8 reports the characteristics of a single wall carbon nanotube. They are usually packed together to form arrays of 10–100 parallel tubes. They can be achieved with three different methods: arc discharge, laser ablation and chemical vapour deposition. The arc and laser processes operate on a carbon plasma that strikes a graphite bar maintained at about 1200 °C and doped with nickel and/or cobalt in an atomic concentration of 1%–2% with respect to carbon. Chemical deposition, which has the highest productivity [70], requires the use of transition metal catalysts and is based on the decomposition of light hydrocarbons or on the high pressure disproportionation of carbon monoxide.

Multi-walled nanotubes are formed from 2 to about 50 coaxial tubes in which the carbon atoms are distributed on longitudinally developing helices. The diameter of the tubes varies from a few nm for the internal ones to a few tens of nm for the external ones and the length is of the order of μm. They are usually obtained by catalytic pyrolysis of hydrocarbons, but they can also be obtained with an electric arc process.

A characteristic of nanotubes is chirality, a term that indicates the orientation of the graphite structure with respect to the axis of the nanotube. The chiral vector is defined as the vector orthogonal to the axis of the nanotube that joins an atom to itself: in practice, it represents the circumference of the nanotube. With reference to Figure 4.20, the chiral vector can be decomposed according to two main directions:

$$\vec{r} = a \cdot \vec{x} + b \cdot \vec{y}$$

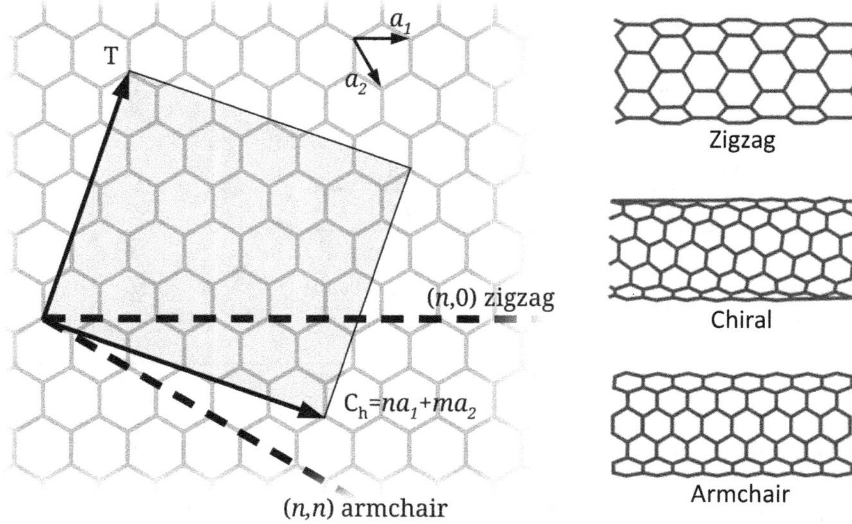

Figure 4.20 Chiral vector and nanotube configurations (adapted from [72])

The diameter of a nanotube is related to chirality through the relationship:

$$d = 0.0783 \cdot \sqrt{a^2 + a \cdot b + b^2}[\text{nm}] \tag{4.28}$$

The helicity of a nanotube is defined as the angle between 0° and 30° that the chiral vector forms with one of the main directions. Each configuration is defined by indicating in brackets the pair of values n and m, i.e. (n, m), and the two extreme configurations are called 'zig-zag' (null helicity) and 'armchair' (maximum helicity, i.e. 30°), while the intermediate configurations are generically defined chiral.

Chirality affects several properties of the nanotube: in particular, when the graphene sheet is cut along the (n, 0) direction, a structure with metallic electronic properties is obtained, while when the direction is (n, n), a structure with semiconducting electronic properties is obtained.

Another particular characteristic is the high Young's modulus, of the order of 1 TPa.

At a given temperature, hydrogen adsorption depends only on pressure: it has been estimated that, at 10 MPa and 77 K, carbon nanotubes with a diameter of 2.2 nm absorb 11.24% by weight of hydrogen, but experimentally only 8.25% is obtained [73].

4.5.2 Graphite nanofibres

Graphite nanofibres, as shown in Figure 4.21, can have different structures:

- platelets, formed by many overlapping sheets of graphene, with sides from 50 to 200 nm;

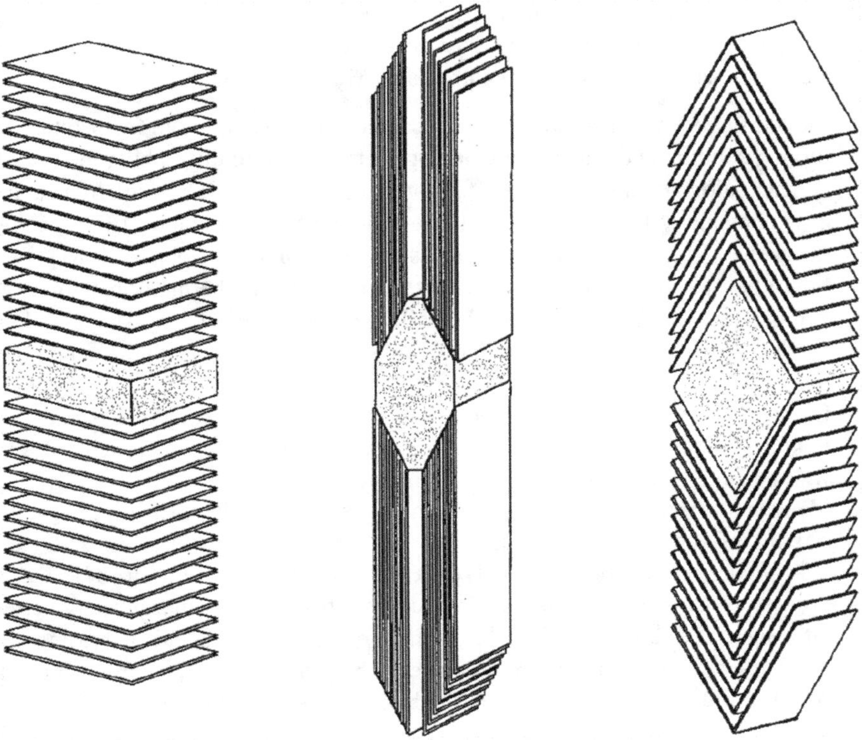

Figure 4.21 Shapes of graphite nanofibre [74]

- herringbone, formed by overlapping sheets of graphene folded at an angle;
- tubular, with graphene planes parallel to the length of the fibre.

The production takes place by chemical deposition from the vapour phase using catalysts based on Cu-Ni. During production, extraneous phases are also obtained such as carbon nanospheres, fullerenes, hydrocarbons which must be removed [75,76]. The high dispersion of experimental data could be caused by an incomplete elimination of these phases. The generated xamplers have dimensions from 5 to 200 nm in diameter and from 5 to 100 μm in length and must be subjected to an activation process (e.g. under vacuum at 1000 °C for 24 h).

Very high values of adsorption capacity (over 40% by weight) have been estimated [77], without anyone ever being able to repeat such a result. However, a recent article studied possible sorption mechanisms which could explain the feasibility of a very high hydrogen storage density (~20%–30% by weight) [78].

4.5.3 Metal organic framework and covalent organic frameworks

MOFs (metal organic frameworks) are materials made up of relatively simple organic units. They consist of metal ions bonded to organic ligands and contain micropores that can adsorb hydrogen gas. It seems that the trend of the hydrogen storage capacity has a slope equal to 0.019 mg/m^2 with respect to the surface area [79].

There are several methods of synthesis of MOFs:

- Slow diffusion or evaporations, which can be performed at room temperature by mixing solution and reagents and letting them thicken until reaching the critical concentration and the consequent formation of crystals [80]. It is an efficient and environmentally friendly method. For example, using a solution of DMF/chlorobenzene and $Zn(NO_3)_2$ and H_2BDC as reagents, MOF-5 is obtained [81].
- Solvothermal or hydrothermal process, carried out in a hermetic reactor at an autogenous pressure higher than the boiling point of the solvent. An example is IRMOF-8 with pore volumes of 0.603–0.693 S cm^3/g increasing the generation temperature from 120 °C to 150 °C [82].
- Electrochemical process, which generates the metal ion by anodic dissolution in mixtures containing organic binders and electrolytes. An example is Cu-MOF, with an average crystal size of 0.5–5 μm and a surface area of 1820 m^2/g [83].
- Mechanochemical process, which involves the mechanical grinding and mixing of the reactants. MOF-74 is an example obtained from ZnO and 2,5-dihydroxyterephthalic acid (H_4HDTA) with DMF as solution [84].
- Process assisted by microwaves, in which electromagnetic energy is converted into thermal energy, causing the heating of the material and increasing the number of collisions between molecules. The process is very energy efficient. An example are Zr-MOFs with high chemical and thermal stability [85].

Another promising category of adsorbent materials is that of covalent organic frameworks (COFs) which were synthesised in 2005 and showed high thermal stability (up to 600 °C) and high surface area (up to 1590 m^2/g) [86].

COFs are also able to capture carbon dioxide from air [87].

4.6 Comparison among the different storage systems

The systems examined have very different characteristics and this implies that they adapt to different extents to different uses. Apart, of course, from the cost, the fundamental characteristics for a comparison are the following:

- weight;
- size;
- safety;
- losses;

and for vehicles also:

- ease and speed of refuelling;
- supplying methods.

Moreover, the size and weight of the fuel that a vehicle must carry with it constitute a reduction in the payload capacity of the vehicle itself and, therefore, it is essential to consider the size and weight of the whole tank, rather than the same characteristics of the fuel. In any case, the fact that hydrogen has been chosen for aerospace missions is already a very significant concrete fact which testifies in its favour.

4.6.1 Weight and size

Figure 4.22 shows a comparison of different storage systems both on the basis of gravimetric and volumetric hydrogen density. Although compressed hydrogen is the currently preferred accumulation system, it can be seen that it is the least favourable in these respects. At opposite extremes are liquid hydrogen, with the greatest weight density, and metal hydrides, with the greatest volume density. Limited to these two aspects, the first would therefore be a candidate as suitable for the transport of hydrogen over long distances, while the more compact metal hydrides would be more suitable for stationary accumulation. LOHC and MOF (and more generally adsorption systems) have intermediate characteristics. Finally, light metal hydrides and chemical hydrides could satisfy both the requirement of lightness and that of compactness.

4.6.2 Safety

From the point of view of safety, the accumulation in the form of pure hydrogen, gaseous or liquid, is certainly more critical because an overpressure or a leak can create dangerous situations. An exception is the case of glass microspheres which must be heated to become permeable to hydrogen. Similarly to this case, if the hydrogen is accumulated in a combined form, a rupture of the tank is not enough to create a risk, but it is necessary to supply energy to cause it to escape. However, it is necessary to distinguish between the various storage methodologies in combined form because the compound that acts as a hydrogen carrier could itself be a fuel, for example methanol or methylcyclohexane. In this case there is no release of hydrogen since the required conditions are achievable only inside a reactor, but there can be combustion of the liquid and, therefore, the risk can be considered similar to that of traditional fuels.

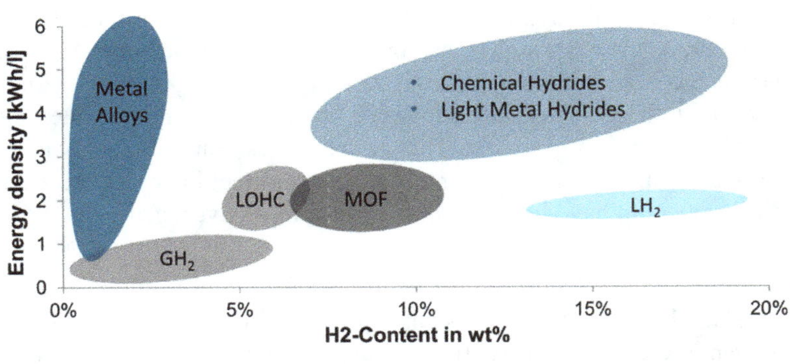

Figure 4.22 Volumetric vs gravimetric hydrogen density of main hydrogen storage systems [88]

As already stated in Section 4.2, liquid hydrogen is stored at a temperature that prevents uncontrolled combustion processes. However there could be problems because in the event of an accident that leads to the breakdown of the cryogenic tank, liquid hydrogen evaporates instantly but tends to spread horizontally and disperses more slowly due to the higher density with respect to air. Furthermore, the emission of gaseous hydrogen in closed environments due to evaporation is dangerous if the necessary precautions have not been taken and great care must be taken also in positioning the tank since, in the event of breakage, contact with hydrogen gas at very low temperature is very dangerous.

The storage of compressed hydrogen gas, although today is the dominant one, is the least safe since it is accumulated at room temperature and therefore the danger of fire in the event of an accident is real. However, it must be taken into account that in most cases flame develops mainly vertically and it is therefore easier to take the appropriate precautions.

4.6.3 Losses

Under normal conditions, the storage of gaseous hydrogen and any type of storage in combined form are practically free from direct losses. However, processes that require high temperature heat for dehydrogenation, for example the methylcyclohexane reaction, although not involving losses in the strict sense, may require that a significant percentage of the stored hydrogen be used to thermally feed the dehydrogenation process.

Storage using liquid hydrogen, however, is subject to possible losses of hydrogen due to evaporation. The insulation of the tank, however efficient, cannot be perfect. During the phases of hydrogen use, evaporation is not a problem, indeed it is intentional, because hydrogen is used in the gaseous state. But in periods when there is no use, the evaporation of liquid hydrogen generates a pressurisation of the tank. If the period of inactivity is prolonged, the pressure can reach the threshold value at which the safety valve is set and excess hydrogen must be released, possibly subjecting it to catalytic combustion. One way to avoid these losses is to cool the tank which, however, consumes energy. Another type of loss that results in energy consumption is that which can occur during tank refuelling because liquid hydrogen encounters surfaces that are initially at room temperature or, in any case, higher than that of hydrogen liquefaction. Also in this case there is the evaporation of a certain amount of hydrogen which is not lost, but recovered in the gaseous state. However, the energy spent on liquefying this amount of hydrogen has been lost. These energy losses are obviously not direct losses of hydrogen, but can also be interpreted as indirect losses, in the sense that the lost energy could have been used to produce more hydrogen.

4.6.4 Storage for stationary applications

The choice of a hydrogen storage system depends on many factors: the final use of the hydrogen, the energy density, the quantity and duration of storage, the maintenance and reliability of the system, the availability of power.

Unless the final use requires low temperatures, liquid hydrogen is usually not the chosen form. The same is true when an accumulation duration is expected for long periods, as the boil-off would lead to significant losses in the absence of evaporated hydrogen users.

The energy density can be related to both volume and weight and, in the case of hydrogen, these values are very different. It is a very important feature in both respects either in terms of transport or in terms of on-board storage, while in the case of stationary storage the volumetric density is above all important. This means that metal hydrides, with a high volumetric density but low weight density, could be preferred in stationary applications.

The amount of hydrogen to store affects aspects related to the economy of scale. For small quantities and if waste heat is available by the user, it can be interesting to accumulate it by means of metal hydrides which can be charged at low pressure, but can return hydrogen at higher pressure at the time of its use. However, as the quantity stored increases, the volume of the alloy used grows almost linearly and so does the cost, while the quantity of material required for a compressed gas tank undergoes a much smaller increase so that the compressed gas storage is generally simpler and cheaper than the others. By further increasing the amount of hydrogen to be accumulated, storage in liquid form can sometimes also be taken into consideration since by increasing the amount stored, losses due to evaporation are reduced. Finally, in the case of very large quantities, the most convenient method is the underground storage of hydrogen in the form of compressed gas.

In terms of maintenance, metal hydrides certainly involve fewer problems than storage as a compressed gas or cryogenic liquid because they do not require complex systems. Furthermore, they are also much safer because the risk of accidental hydrogen leaks is minimal and, if this occurs, it concerns very modest quantities since dehydrating requires heat.

The availability of energy is a very important factor because, once hydrogen is produced, the different storage technologies require different quantities and types of energy. The most energy-intensive process is undoubtedly liquefaction, while hydrading does not require energy, but the reverse process requires energy in the form of heat.

Therefore, for large capacities, of the order of 10^5–10^6 Nm^3, natural structures such as aquifers, rock or salt caves, porous substrates, depleted deposits could be used. In France and in the United Kingdom this type of storage has already been successfully tested up to pressures of 50 MPa and has been found to be cheaper by a couple of orders of magnitude compared to steel tanks. The European founded project Hystories [89] started in 2021, involve 15 European country to assess the potential of European sites to define underground site for hydrogen storage.

Going down to capacities of the order of 10^4 Nm^3, the use of spherical steel tanks with pressures around 10 MPa is preferable, while for lower capacities, cylindrical steel tanks with pressures around 20 MPa are more suitable. For single users that require no more than a few tens of Nm^3, single cylinders or cylinder packs of 50 l and pressure of 30 MPa up to 90 MPa are used.

The cryogenic liquid has increasing advantages as the quantity of stored hydrogen increases since the percentage losses due to evaporation are reduced as the volume of the tank increases, but the considerable energy expenditure of liquefaction must be taken into account. As a result, the solution is feasible especially if the hydrogen is required in liquid form by the end user. However, if hydrogen is required in gaseous form and is used on site, the negative impact of the expense for liquefaction can be reduced by using the hydrogen re-gasification phase to produce liquid air which, in turn, can be re-gasified by acting as a refrigerant fluid in the hydrogen liquefaction cycle.

The storage in metal hydrides, despite the various advantages listed, is currently confined to the storage of modest quantities of hydrogen due to the cost increase associated with the growth of the required material proportional to the amount of hydrogen to accumulate.

4.6.5 Storage for mobile applications

In the case of mobile applications, it is necessary to differentiate among the different types of vehicles:

- cars;
- heavy vehicles (trucks and buses);
- trains;
- boats;
- planes;
- aerospace vehicles.

Furthermore, for some of the vehicles mentioned, there may be differences related to the energy conversion device:

- internal combustion engines;
- gas turbines;
- fuel cells combined with electric motor.

Road and railways transport

For all means of transport, safety is a fundamental parameter. Especially when it comes to transporting people and, in particular, in an urban, densely populated and busy environment. In some respects, hydrogen is safer than other fuels currently in use (e.g. the very low density of hydrogen is a cause of very fast dispersion). However, during the transition to a system based also on hydrogen, numerous vehicles powered by traditional fuels will still be in circulation, flanked by battery-powered vehicles which also have problems with respect to combustion (at least in cases with lithium batteries). This fact makes it possible, in the event of an accident, to develop flames and heat that can reduce the safety of hydrogen storage systems such as of other kind of vehicles. Fire issue will be a theme also in the transition mobility towards carbon free mobility system.

The choice of the storage system largely depends on the storage capacity, that is, on the type of storage, on the size, on the range required for the vehicle, etc.

The weight and volume of the storage are important since the fuel consumption depends on the weight, while the carrying capacity of passengers and/or goods depends on the volume. Obviously, there is the possibility of creating larger vehicles to make room for a bulky accumulation, but it is not always possible: for example, city buses have limited maximum dimensions. Furthermore, the increase in the size of the vehicle leads to an increase in weight and therefore an increase in fuel consumption.

Especially for smaller vehicles, the geometric compatibility of the tank with the vehicle structure is equally important than the volume. The tank of a petrol or diesel car can be made of plastic, adapting to the unused cavities in the body of the car to allow maximum space to be allocated to passengers or other useful loads. A similar principle has been adopted in aircraft, where the kerosene tank is located in the wings. To have a low weight, a tank for the accumulation of hydrogen in the form of compressed gas should have a surface/volume ratio as low as possible, i.e. it should have a spherical shape. And also in the case of liquid hydrogen the spherical shape would be the optimal one, in order to reduce the heat exchange surface and therefore the need for thermal insulation. Obviously, the spherical shape is not compatible with the spaces available on board a vehicle and therefore a compromise must be reached that goes to the detriment of the lightness of the tank. This problem does not exist for other types of hydrogen accumulation.

In perspective, those systems that combine electric propulsion with low-temperature fuel cells and on-board production starting from chemical hydrides and water will become interesting. In this case, in fact, it is sufficient to take on board a modest quantity of water since it is continuously regenerated by the fuel cells.

Refuelling times vary greatly from one type of storage to another, but for a car they are preferably limited to a maximum of five minutes using compressed hydrogen. For heavy vehicles or trains, equipped with larger compressed hydrogen tanks, the times are lengthened at around 15–20 minutes. To obtain these performances usually hydrogen is cooled at −40 °C (as required by different standard for 70 MPa vehicle storage).

The existing trains are equipped with an accumulation in compressed gas cylinders located above the roof of the locomotive or in a specific carriage in the middle of passenger train (Alstom solution) [90], whereas locomotives for shunting in logistics centres have hydrogen tanks in the spaces adjacent to the fuel cells used to power the train (PESA solution) [91].

Boats

While for other mobile applications the weight of the storage plays an important role, it is not a problem for those boats equipped with ballast since the ballast can be reduced or eliminated in favour of the hydrogen storage system. This means that an accumulation in the form of metal hydrides can be particularly suitable.

For all ships that do not have ballast systems, the current hydrogen storage solutions are in compressed or liquid form. Compressed solution is adopted, for example, by Toyota for foiled hydrogen tender for Americas Cup in Barcellona in 2024.

An example of liquid hydrogen solution is that implemented in the Hydra project (see Section 6.1.5).

Airplanes

Hydrogen has an energy per unit of mass 2.8 times greater than kerosene. Therefore it has a great advantage in terms of lightness as well as in terms of consumption and load capacity. Due to the low density, the preferable form for accumulation is that of cryogenic liquid but some projects are using compressed hydrogen.

The idea of using liquid hydrogen as a propellant for the aviation sector dates back to the second half of the 1950s when it was tested on a B-57B Camberra [91]. In 1988, a Tupolev Tu-155 was tested and gave rise to the Russian-German project called Cryoplane with the goal to modify an Airbus A310 to operate with liquid hydrogen [92]. A Berlin-New York flight covers a distance of about 6400 km, which corresponds to a consumption of about 20,000 kg of hydrogen. In the liquid state, it occupies a volume about four times that necessary for a quantity of kerosene suitable to cover the same distance. Therefore, a problem arises in the allocation of the tank: in the case of the Airbus the tank was positioned in the upper part of the fuselage. Later an aircraft with a more limited range was also designed, a DO 328, and in that case the liquid hydrogen could be stored in two tanks attached to the wings.

A compressed hydrogen storage was used in two recent test flights. In January 2023, ZeroAvia conducted a first ten-minute flight test with a 19-seat Dornier 228 [92], equipped with a prototype hydrogen propulsion unit with fuel cells on the left wing of the aircraft, whereas the right wing was equipped with a standard Honeywell TPE-331 engine. A similar test was conducted in March 2023 by Universal Hydrogen [93]. The airplane was a Dash-8 modified to operate on hydrogen fuel cells equipped with two 15 kg hydrogen tanks, an 800 kW fuel cell stack and an electric motor under one wing. The other wing was equipped with a Pratt and Whitney engine. Universal Hydrogen expects to switch soon to liquid hydrogen.

Aerospace vehicles

The use of hydrogen in aerospace missions is certainly the one that has reached a notable maturity thanks to the fact that in the past the costs associated with these applications were not a problem. Hydrogen has been used both as a propellant for engines and as a source of electricity thanks to the use of fuel cells. Obviously the form of storage on board has always been that of cryogenic liquid.

Figure 4.23 shows the hydrogen and oxygen tanks location inside NASA's Space Launch System which should bring humans to the Moon and Mars in the next future.

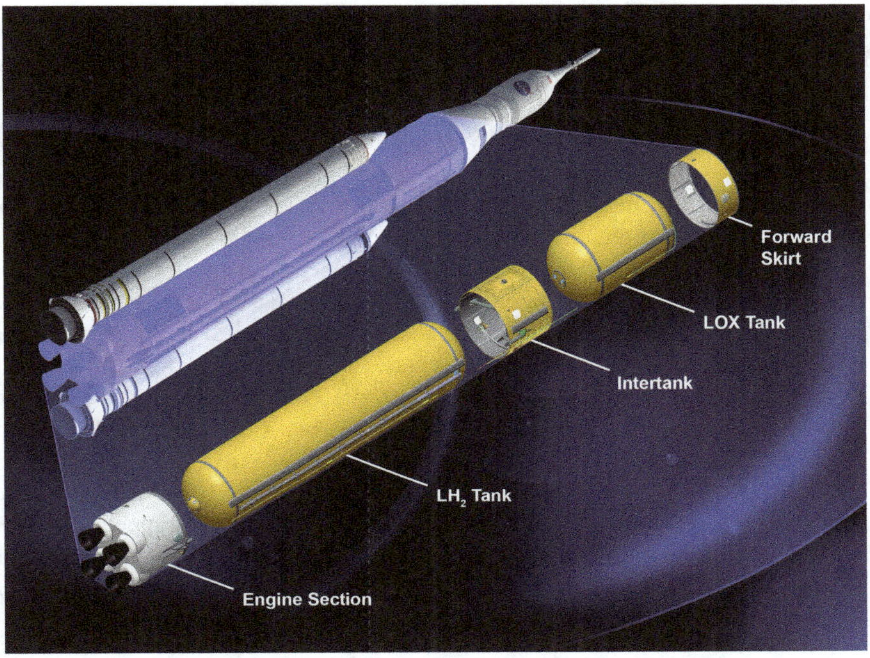

Figure 4.23 Picture of the liquid hydrogen and liquid oxygen tanks of the SLS [copyright NASA]

4.7 Distribution

Most of the hydrogen produced is currently used by the manufacturer itself and therefore does not need to be distributed. The transfer takes place, therefore, through suitable short extension pipes. With a view to expanding the market, the different alternatives must also be considered by adapting the normal modes of transporting other substances to the transport of hydrogen, although sometimes with some additional difficulties. The choice of the hydrogen transport system is mainly influenced by the quantity and distance.

Transport by hydrogen pipeline is currently suitable for large quantities and not too long distances, since the operating costs are the lowest, while the present capital cost is very high. For small quantities and large distances, however, the transport of hydrogen by road is preferable: taking into account the high costs of liquefaction compared to those of simple compression and, however, the greater capacity of cryogenic tankers, the convenience shifts from hydrogen compressed to liquid hydrogen as distance and quantity increase. The air transport of liquid hydrogen has also been hypothesised to cover long distances in short times and thus with reduce evaporation losses [94].

In addition, the production and storage method and the final use of hydrogen can assume a certain importance.

4.7.1 Truck transport

Given the smallness of the current market, the transport of hydrogen takes place almost exclusively by road in four different ways:

- single cylinders which generally contain 50 l at 20 MPa, i.e. 10 Nm^3 of hydrogen;
- cylinder packs consisting of several cylinders (up to 15, i.e. 150 Nm^3 of hydrogen) packed together (Figure 4.24);
- tank wagons, semi-trailers or containers equipped with special safety systems that transport a pack of cylinders under pressure at 30 MPa each having a capacity of 300–400 Nm^3 (up to a total of 4000 Nm^3 of hydrogen); for the construction of the cylinders, suitable steels are used in order to avoid the risks due to embrittlement and, in general, they are CrMo-quenched and tempered steels. In many cases composites cylinders (carbon fibre based) con transport around 1000 kg of hydrogen at 50 MPa [95].
- cryogenic tankers, capable of transporting up to about 30,000 Nm^3 of liquid hydrogen and consisting of an internal stainless steel tank, wrapped in a layer of insulating material contained, in turn, in a carbon steel casing; thanks to the insulation and the size, the daily evaporation losses are limited to 0.3%–0.6%.

The most common storage method is that of hydrogen gas under pressure. However, as already stated, due to the low energy density, liquid hydrogen can become convenient as the quantity or distance increases.

Currently, the rail transport of hydrogen in this form is not yet carried out. However, some companies are planning this development. For example, DB Cargo in Germany is planning liquid hydrogen transportation from port to end users [96].

Figure 4.24 Hydrogen road transportation [95]

4.7.2 Ship transport

The transport of liquid hydrogen by ship has also been studied, for example as part of the Euro-Quebec project. Similarly to what is already happening for natural gas, this is a feasible option if there is a need for a long distance transportation. The evaporated hydrogen would be used as a fuel to power the engines or other on-board utilities.

Canada has developed numerous ship designs for transoceanic hydrogen transport. One of these involves the use of five small barges transported in a larger ship, which can be separated at the end of the voyage. Each of them would carry 21,200 kg of hydrogen without any loss during 50 days of travel. Other projects instead envisage the use of several spherical containers or a single tanker with a capacity of 7000 tons. None of these ships have yet been built but those dedicated to the transport of liquefied natural gas are already able to transport 125,000 m^3 of gas (equivalent to 9000 tons of hydrogen).

Liquid hydrogen was also tested for continental hydrogen transportation by Kawasaki Heavy Industries Japanese Company: in February 2022 the ship Suiso Frontier completed the first liquid hydrogen transportation with a ship from Australia to Japan. The ship has a 1250 m^3 liquefied hydrogen tank. Furthermore the company obtained the approval in principle from Nippon Kaiji Kyokai (ClassNK) for a 160,000 m^3 liquefied hydrogen carrier [97].

C-Job Naval Architects in collaboration with LH2 Europe has designed a class of tankers equipped with three liquid hydrogen storage tanks with a total capacity of 37,500 cubic meters (Figure 4.25). The hydrogen evaporated during navigation will be used to power fuel cells that will provide electricity for propulsion and on-board services.

Figure 4.25 Artist impression of the liquid hydrogen carrier design by LH2 Europe and C-Job Naval Architects

4.7.3 Pipeline transport

While in the case of transoceanic transport the cheapest solution is liquid hydrogen, in the case of land transport of large quantities of hydrogen the most convenient method is that of gas pipelines which, after the investments necessary for their construction, require very low operating costs. Furthermore, the choice of distributing hydrogen in this way allows a constant supply of hydrogen and the elimination of the risks associated with road transport. However, the higher costs are justifiable only in the presence of large single users or smaller users concentrated in a restricted area.

The pipeline transport of hydrogen, even over large distances, is a long-tested process. None of the existing plants has ever suffered accidents or had any noteworthy safety problems. Although the data has no statistical significance given the small sample, is still a fact of considerable importance. The first hydrogen pipeline of significant length, more than 100 km, was built in Germany in the late 1930s. Subsequently, others were also built in other countries accounting for more than 4500 km in 2016, most of which in EU and USA [98]. Figure 4.26 shows the companies which own most of the hydrogen pipelines in the world and the distribution of those pipelines by geographical areas. A 1000 km long pipeline has been hypothesised from south to north of Oman, and this should be the longest hydrogen pipeline if will be constructed [99]. Most of the existing pipelines have diameters of 25–30 cm and carry hydrogen at an average pressure of 1–2 MPa.

Being a gas, hydrogen can be managed, with appropriate precautions, similarly to what happens for natural gas. Compared to the methane pipeline, the hydrogen pipeline has partly different characteristics related to the chemical-physical differences of the two gases. Due to the small size of the hydrogen molecule there are greater sealing problems and it is therefore necessary to intervene on the gaskets. Furthermore, at the same pressure, the flow of hydrogen is three times faster than

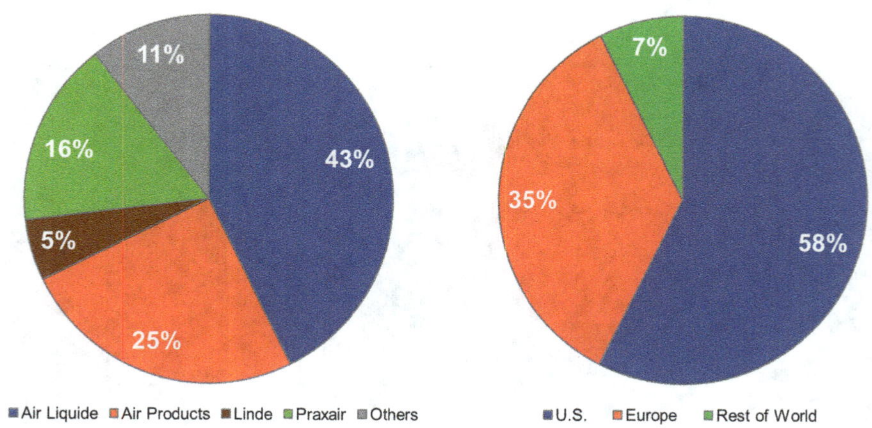

Figure 4.26 Hydrogen pipelines, distribution by companies and by geographical areas

that of natural gas, but the amount of energy transported is about three times less. This is a factor to be taken into account when sizing the pipeline, valves, fittings and compression stations [16]. The most delicate aspect concerns the embrittlement of the steel. This phenomenon is known and is particularly evident in special high-strength steels. Plastic based pipelines are under study [100]. It depends on the fact that the molecule dissociates and the atomic hydrogen diffuses into the steel. The embrittlement involves a decrease in the ability to deform plastically and leads to a sudden fracture.

Therefore, it is necessary to provide systems for identifying any leaks as well as considering precautions to avoid ignition triggers (explosion-proof materials and systems) given the chemical-physical characteristics of easy ignition of combustion.

It is therefore not possible to use the current network of pipelines for the transport of hydrogen. Due to the materials used, the greater care required in the production and commissioning phase and the need for more extensive monitoring of the network, the costs of the hydrogen pipeline are higher than those of a gas pipeline.

In order to use the existing network for natural gas, it has also been proposed to transport mixtures containing up to 20% of hydrogen and then carry out the separation in a dedicated hydrogen destination station. Recently, in Italy, SNAM has successfully experimented the transport of mixtures up to 10% of hydrogen by methane pipeline for their use in a pasta production company. Figure 4.27 shows a comparison between hydrogen and natural gas in term of power consumption and flow reduction vs pipeline length.

An innovative experimental technique for the transport of liquid hydrogen consists of a gas pipeline containing a superconducting material. Liquid hydrogen would act as a refrigerant for the superconductor and would allow the transport of electricity across long distances without the large energy losses of conventional power lines. Obviously, it is necessary to carry out an energy balance to evaluate the costs of pumping and continuous cooling of hydrogen during transport and an economic balance that also takes into account the costs necessary for thermal insulation [102,103].

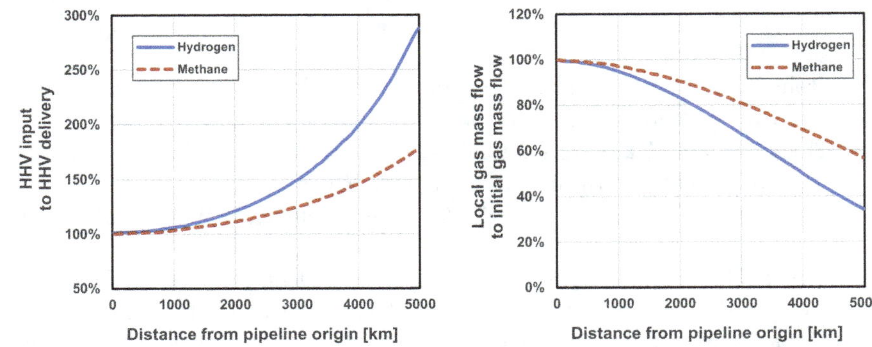

Figure 4.27 Energy consumption and flow reduction versus distance from pipeline origin [101]

References

[1] Nejat Veziroglu T, Sherif SA, and Barbir F. Hydrogen energy solutions – chapter 7. In: Agardy FJ, and Nemerow NL (eds). *Environmental Solutions: Environmental Problems and the All-Inclusive Global, Scientific, Political, Legal, Economic, Medical, and Engineering Bases to Solve Them.* New York: Academic Press; 2005. pp. 143–80.

[2] Kruck O, Crotogino F, Prelicz R, and Rudolph T. *Overview on all known underground storage technologies for hydrogen [Internet].* (H2020 Project HyUnder – Grant Agreement N. 303417). Deliverable No. 3.1. 2013. Available from: http://hyunder.eu/wp-content/uploads/2016/01/D3.1_Overview-of-all-known-underground-storage-technologies.pdf. Accessed 16 December 2024

[3] Liebscher A, Wackerl J, and Streibel M. Geologic storage of hydrogen – fundamentals, processing, and projects – chapter 26. In: Stolten D, and Emonts B (eds), *Hydrogen Science and Engineering: Materials, Processes, Systems and Technology.* Weinheim: Wiley-VCH; 2016. pp. 629–58.

[4] Sambo C, Dudun A, Samuel SA, Esenenjor P, Muhammed NS, and Haq B. A review on worldwide underground hydrogen storage operating and potential fields. *Int J Hydrogen Energy.* 2022;47(54):22840–80.

[5] Worthington Industries. *High pressure cylinders manufacturer [Internet].* [cited 2023 Oct 9]. Available from: https://www.worthingtonindustries.eu/gas-cylinders/high-pressure/.

[6] Li M, Bai Y, Zhang C, *et al.* Review on the research of hydrogen storage system fast refueling in fuel cell vehicle. *Int J Hydrogen Energy.* 2019;44(21):10677–93.

[7] Faber Cylinders. *Hydrogen – Faber Italy [Internet].* [cited 2023 Oct 9]. Available from: https://faber-italy.com/product/hydrogen/.

[8] Toyota. TOYOTA fuelcells | Technology [Internet]. 2024 [cited 2024 Aug 27]. Available from: https://www.toyota.co.jp/fuelcells/en/technology.html.

[9] Idromeccanica.it. Compressori idraulici [Internet]. [cited 2023 Oct 9]. Available from: https://idromeccanica.it/compressori-idraulici/#tde-h.

[10] Predel T, Schlücker E, Wasserscheid P, Gerhard D, and Arlt W. Ionic liquids as operating fluids in high pressure applications. *Chem Eng Technol.* 2007;30(11):1475–80.

[11] Rohland B, Eberle K, Ströbel R, Scholta J, and Garche J. Electrochemical hydrogen compressor. *Electrochim Acta.* 1998;43(24):3841–6.

[12] Hobein B, and Krüger R. Physical hydrogen storage technologies – a current overview – chapter 18. In: Stolten D (ed). *Hydrogen and Fuel Cells: Fundamentals, Technologies and Applications.* Weinheim: Wiley-VCH; 2010. pp. 377–94.

[13] Teitel RJ. Microcavity hydrogen storage – final progress report [Internet]. Report BNL 51439. Prepared for US Department of Energy and Environment, Energy Storage and Conversion Division, Under Contract No. DE-AC02-76CH00016. Upton, New York; 1981. Available from: https://www.osti.gov/servlets/purl/5719826. Accessed 16 December 2024

[14] Rambach GD, and Hendricks C. Hydrogen transport and storage in engineered glass microspheres. *In: Proceedings of the 1996 US DOE Hydrogen Program Review, NREL/CP-430-21938, 1–2 May, Miami, FL*. Miami, FL; 1996. pp. 765–772.

[15] Rambach GD. Hydrogen transport and storage in engineered glass microspheres. *In: Proceedings of the US DOE Hydrogen Program Review Meeting, Sandia National Laboratory, UCR-JC-11708, 18–21 April 1994*. 1994.

[16] Morgan D, and Sissine F. Hydrogen: technology and policy. *In: Congressional Research Service Report for Congress for Congress "Hydrogen: Technology and Policy" 95-540 SPR, 28 April 1995*. 1995.

[17] Dortmund Data Bank (DDBST GmbH). Dortmund Data Bank (DDB) – Pure component properties – Joule–Thompson coefficient – nitrogen [Internet]. 2023 [cited 2023 Nov 2]. Available from: http://ddbonline.ddbst.com/DDBSearch/onlineddboverview.exe?submit=Details&systemcomplist=1056.

[18] Engineering ToolBox. Nitrogen – thermophysical properties [Internet]. [cited 2023 Oct 9]. Available from: https://www.engineeringtoolbox.com/nitrogen-d_1421.html.

[19] Institute of Slush Hydrogen. The world's first hydrogen liquefaction by magnetic refrigeration [Internet]. 2018 [cited 2023 Oct 9]. Available from: https://slush-ish-mag-english.com/technology2.html.

[20] Sullivan NS, Zhou D, and Edwards CM. Precise and efficient *in situ* ortho — para-hydrogen converter. *Cryogenics (Guildf)*. 1990;30(8):734–5.

[21] Teng J, Wang K, Zhu S, *et al.* Comparative study on thermodynamic performance of hydrogen liquefaction processes with various ortho-para hydrogen conversion methods. *Energy*. 2023;271:127016.

[22] US DOE Hydrogen and Fuel Cell Technologies Office. DOE technical targets for hydrogen delivery [Internet]. [cited 2023 Oct 9]. Available from: https://www.energy.gov/eere/fuelcells/doe-technical-targets-hydrogen-delivery.

[23] Aziz M. Liquid hydrogen: a review on liquefaction, storage, transportation, and safety. *Energies*. 2021;14(18):5917.

[24] Gursu S, Lordgooei M, Sherif SA, and Veziroğlu TN. An optimization study of liquid hydrogen boil-off losses. *Int J Hydrogen Energy*. 1992;17(3):227–36.

[25] Chang HM, Ryu KN, and Baik JH. Thermodynamic design of hydrogen liquefaction systems with helium or neon Brayton refrigerator. *Cryogenics (Guildf)*. 2018;91:68–76.

[26] Godula-Jopek A, Jehle W, and Wellnitz J. (eds). Hydrogen storage technologies: new materials, transport and infrastructure. Weinheim: Wiley-VCH; 2012.

[27] Fesmire J, Swanger A, Jacobson J, and Notardonato W. Energy efficient large-scale storage of liquid hydrogen. *IOP Conf Ser Mater Sci Eng*. 2022; 1240(1):012088.

[28] NASA. Innovative liquid hydrogen storage to support space launch system [Internet]. 2023 [cited 2023 Oct 9]. Available from: https://www.nasa.gov/missions/artemis/innovative-liquid-hydrogen-storage-to-support-space-launch-system/.

[29] Rao AG, Yin F, and Werij HGC. Energy transition in aviation: the role of cryogenic fuels. *Aerospace*. 2020;7(12):181.

[30] Ohira K. Slush hydrogen production, storage, and transportation – chapter 3. In: Gupta RB, Basile A, and Veziroğlu TN (eds). *Compendium of Hydrogen Energy – Vol 2: Hydrogen Storage, Distribution and Infrastructure*. Sawston, UK: Woodhead Publishing Series in Energy; 2016. pp. 53–90.

[31] Jin T, Li YJ, Liang ZB, Lan YQ, Lei G, and Gao X. Numerical prediction of flow characteristics of slush hydrogen in a horizontal pipe. *Int J Hydrogen Energy*. 2017;42(6):3778–89.

[32] Sakintuna B, Lamari-Darkrim F, Hirscher M. Metal hydride materials for solid hydrogen storage: a review. *Int J Hydrogen Energy*. 2007;32(9):1121–40.

[33] Principi G, Agresti F, Lo Russo S, and Palade P. Materials for hydrogen storage in solid state: a survey. *In: World Hydrogen Technologies Convention 2007*. Montecatini Terme, Italy; 2007.

[34] Akiba E. Metal hydrides – chapter 19. In: Stolten D (ed). *Hydrogen and Fuel Cells: Fundamentals, Technologies and Applications*. Weinheim: Wiley-VCH; 2010. pp. 395–414.

[35] Maddalena A, Petris M, Palade P, *et al.* Study of Mg-based materials to be used in a functional solid state hydrogen reservoir for vehicular applications. *Int J Hydrogen Energy*. 2006;31(14):2097–103.

[36] Baranowski B, and Tkacz M. The equilibrium between solid aluminium hydride and gaseous hydrogen. *Zeitschrift fur Phys Chemie Neue Folge*. 1983;135:27–38.

[37] Sandrock G, Reilly J, Graetz J, Zhou WM, Johnson J, and Wegrzyn J. Accelerated thermal decomposition of AlH_3 for hydrogen-fueled vehicles. *Appl Phys A Mater Sci Process*. 2005;80(4):687–90.

[38] Abe JO, Popoola API, Ajenifuja E, and Popoola OM. Hydrogen energy, economy and storage: review and recommendation. *Int J Hydrogen Energy*. 2019;44(29):15072–86.

[39] Nagar R, Srivastava S, Hudson SL, *et al.* Recent developments in state-of-the-art hydrogen energy technologies – review of hydrogen storage materials. *Sol Compass*. 2023;5(January):100033.

[40] Askri F, Ben Salah M, Jemni A, and Ben Nasrallah S. Optimization of hydrogen storage in metal-hydride tanks. *Int J Hydrogen Energy*. 2009;34 (2):897–905.

[41] DeLuchi MA. Hydrogen vehicles: an evaluation of fuel storage, performance, safety, environmental impacts, and cost. *Int J Hydrogen Energy*. 1989;14(2):81–130.

[42] EnergyTech.com. GKN Hydrogen, SoCalGas work with National Renewable Lab to deploy Green H2 storage [Internet]. 2022 [cited 2023 Oct 9]. Available from: https://www.energytech.com/energy-storage/article/21245739/gkn-hydrogen-socalgas-work-with-national-renewable-lab-to-deploy-green-h2-storage.

[43] Lototskyy MV, Yartys VA, Pollet BG, and Bowman RC. Metal hydride hydrogen compressors: a review. *Int J Hydrogen Energy*. 2014;39(11):5818–51.

[44] Charbonnier V, Enoki H, Asano K, Kim H, and Sakaki K. Tuning the hydrogenation properties of $Ti_{1+y}Cr_{2-x}Mn_x$ laves phase compounds for high pressure metal-hydride compressors. *Int J Hydrogen Energy*. 2021;46 (73):36369–80.

[45] Sakaki K, Charbonnier V, Shinzato K, Kim H, and Asano K. Development of hydrogen storage materials for hydrogen compression and purification. *In: Book of Abstracts HYdrogen POwer THeoretical & Engineering Solutions International Symposium – HYPOTHESIS XIX*. Hiroshima, Japan; 2024. pp. 75–6.

[46] Rusanov AV, Solovey VV, and Lototskyy MV. Thermodynamic features of metal hydride thermal sorption compressors and perspectives of their application in hydrogen liquefaction systems. *J Phys Energy*. 2020;2(2):021007.

[47] Dymova TN, Eliseeva NG, Bakum SI, and Dergachev YM. Direct synthesis of alkali metal alumohydrides in melts. *Dokl Akad Nauk SSSR*. 1974;215 (6):1369–72.

[48] Bogdanović B, and Schwickardi M. Ti-doped alkali metal aluminium hydrides as potential novel reversible hydrogen storage materials. *J Alloys Compd*. 1997;253–254:1–9.

[49] Dymova TN, Dergachëv YM, Sokolov VA, and Grechanaya NA. Pressure of $NaAlH_4$ and Na_3AlH_6 dissociation. *Dokl Akad Nauk SSSR*. 1975;224:591.

[50] Züttel A. Materials for hydrogen storage. *Mater Today*. 2003;6(9):24–33.

[51] Niermann M, Beckendorff A, Kaltschmitt M, and Bonhoff K. Liquid organic hydrogen carrier (LOHC) – assessment based on chemical and economic properties. *Int J Hydrogen Energy*. 2019;44(13):6631–54.

[52] Taube M, Rippin DWT, Cresswell DL, and Knecht W. A system of hydrogen-powered vehicles with liquid organic hydrides. *Int J Hydrogen Energy*. 1983;8(3):213–25.

[53] Cacciola G, Giordano N, and Restuccia G. Cyclohexane as a liquid phase carrier in hydrogen storage and transport. *Int J Hydrogen Energy*. 1984;9 (5):411–9.

[54] Newson E, Haueter T, Hottinger P, Von Roth F, Scherer GWH, and Schucan TH. Seasonal storage of hydrogen in stationary systems with liquid organic hydrides. *Int J Hydrogen Energy*. 1998;23(10):905–9.

[55] Hodoshima S, Takaiwa S, Shono A, Satoh K, and Saito Y. Hydrogen storage by decalin/naphthalene pair and hydrogen supply to fuel cells by use of superheated liquid-film-type catalysis. *Appl Catal A Gen*. 2005;283(1–2): 235–42.

[56] Matsuoka K, Miyoshi K, and Sato Y. Electrochemical reduction of toluene to methylcyclohexane for use as an energy carrier. *J Power Sources*. 2017; 343:156–60.

[57] Klvana D, Touzani A, Chaouki J, and Bélanger G. Dehydrogenation of methylcyclohexane in a reactor coupled to a hydrogen engine. *Int J Hydrogen Energy*. 1991;16(1):55–60.

[58] Drolet B, Gretz J, Kluyskens D, Sandmann F, and Wurster R. The euro-québec hydro-hydrogen pilot project [EQHHPP]: demonstration phase. *Int J Hydrogen Energy*. 1996;21(4):305–16.

[59] Hydrogen Europe. Beyond Europe [Internet]. [cited 2023 Oct 10]. Available from: https://hydrogen.revolve.media/2021/beyond-europe/.

[60] Andersson J, and Grönkvist S. Large-scale storage of hydrogen. *Int J Hydrogen Energy*. 2019;44(23):11901–19.

[61] Pedicini R, Maisano S, Chiodo V, Conte G, Policicchio A, and Agostino RG. Posidonia Oceanica and wood chips activated carbon as interesting materials for hydrogen storage. *Int J Hydrogen Energy*. 2020;45(27):14038–47.

[62] Chahine R, and Bose TK. Characterization and optimization of adsorbents for hydrogen storage. *In: Hydrogen Energy Progress XI: Proceedings of the 11th World Hydrogen Energy Conference*. Miami, FL: International Association for Hydrogen Energy; 1996. pp. 1259–1263.

[63] Nijkamp MG, Raaymakers JEMJ, van Dillen AJ, and De Jong KP. Hydrogen storage using physisorption-materials demands. *Appl Phys A Mater Sci Process*. 2001;72(5):619–23.

[64] Loutfy RO, and Wexler EM. Feasibility of fullerene hydride as a high capacity hydrogen storage material. In: *Proceedings of the 2001 US DOE Hydrogen Program Review: 17–19 April 2001, Baltimore, Maryland*, NREL/CP-610-30535 [Internet]. Golden, CO: National Renewable Energy Laboratory; 2001. pp. 467–77. Available from: https://www.nrel.gov/docs/fy01osti/30535.pdf. Accessed 16 December 2024

[65] Iijima S. Helical microtubules of graphitic carbon. *Nature*. 1991;354(6348):56–8.

[66] Ebbesen TW, and Ajayan PM. Large-scale synthesis of carbon nanotubes. *Nature*. 1992;358(6383):220–2.

[67] Iijima S, and Ichihashi T. Single-shell carbon nanotubes of 1-nm diameter. *Nature*. 1993;363(6430):603–5.

[68] Edumol Molecular Visualizations, Flicrk.com. Nanotube | Carbon nanotube | Edumol Molecular Visualizations | Flickr [Internet]. 2014 [cited 2024 Aug 31]. Available from: https://www.flickr.com/photos/edumendo/15112442038.

[69] Dillon AC, Jones KM, Bekkedahl TA, Kiang CH, Bethune DS, and Heben MJ. Storage of hydrogen in single-walled carbon nanotubes. *Nature*. 1997;386(6623):377–9.

[70] Dai H, Rinzler AG, Nikolaev P, Thess A, Colbert DT, and Smalley RE. Single-wall nanotubes produced by metal-catalyzed disproportionation of carbon monoxide. *Chem Phys Lett*. 1996;260(3–4):471–5.

[71] The Nanotube Site. Physical properties of carbon nanotubes [Internet]. https://nanoten.com/NTSite/. 2020 [cited 2023 Jul 22]. Available from: https://web.archive.org/web/20041209090650/http://www.pa.msu.edu/cmp/csc/ntproperties/.

[72] Wikimedia Commons – Kebes and Andel. File:CNTnames.svg – Wikimedia Commons [Internet]. 2018 [cited 2024 Aug 31]. Available from: https://commons.wikimedia.org/wiki/File:CNTnames.svg.

[73] Darkrim F, and Levesque D. High adsorptive property of opened carbon nanotubes at 77 K. *J Phys Chem B*. 2000;104(29):6773–6.

[74] Bessel CA, Laubernds K, Rodriguez NM, and Baker RTK. Graphite nanofibers as an electrode for fuel cell applications. *J Phys Chem B*. 2001;105 (6):1121–2.

[75] Chiang IW, Brinson BE, Smalley RE, Margrave JL, and Hauge RH. Purification and characterization of single-wall carbon nanotubes. *J Phys Chem B [Internet]*. 2001 [cited 2023 Oct 10];105(6):1157–61. Available from: https://pubs.acs.org/doi/abs/10.1021/jp003453z.

[76] Chiang IW, Brinson BE, Huang AY, *et al.* Purification and characterization of single-wall carbon nanotubes (SWNTs) obtained from the gas-phase decomposition of CO (HiPco Process). *J Phys Chem B*. 2001 [cited 2023 Oct 10];105(35):8297–301.

[77] Park C, Tan CD, Hidalgo R, Baker RTK, and Rodriguez NM. Hydrogen storage in graphite nanofibers. In: *Proceedings of the 1998 US DOE Hydrogen Program Review 28–30 April 1998, Alexandria, Virginia, vol II* NREL/CP-570-25315. Alexandria, VA: National Renewable Energy Laboratory; 1998. pp. 525–38.

[78] Nechaev YS, Denisov EA, Cheretaeva AO, *et al.* On the problem of "super" storage of hydrogen in graphite nanofibers. *J Carbon Res*. 2022;8(2):23.

[79] Goldsmith J, Wong-Foy AG, Cafarella MJ, and Siegel DJ. Theoretical limits of hydrogen storage in metal-organic frameworks: opportunities and trade-offs. *Chem Mater*. 2013;25(16):3373–82.

[80] Raptopoulou CP. Metal-organic frameworks: synthetic methods and potential applications. *Materials (Basel)*. 2021;14(2):310.

[81] Li H, Eddaoudi M, O'Keeffe M, and Yaghi OM. Design and synthesis of an exceptionally stable and highly porous metal-organic framework. *Nature*. 1999;402(6759):276–9.

[82] Orefuwa SA, Yang H, and Goudy AJ. Rapid solvothermal synthesis of an isoreticular metal–organic framework with permanent porosity for hydrogen storage. *Microporous Mesoporous Mater*. 2012;153:88–93.

[83] Varsha MV and Nageswaran G. Review—direct electrochemical synthesis of metal organic frameworks. *J Electrochem Soc*. 2020;167(15):155527.

[84] Beamish-Cook J, Shankland K, Murray CA, and Vaqueiro P. Insights into the mechanochemical synthesis of MOF-74. *Cryst Growth Des*. 2021;21(5): 3047–55.

[85] Ren J, Segakweng T, Langmi HW, *et al.* Microwave-assisted modulated synthesis of zirconium-based metal-organic framework (Zr-MOF) for hydrogen storage applications. *Int J Mater Res*. 2014 [cited 2023 Oct 10];105 (5):516–9.

[86] Côté AP, Benin AI, Ockwig NW, O'Keeffe M, Matzger AJ, and Yaghi OM. Porous, crystalline, covalent organic frameworks. *Science*. 2005;310 (5751):1166–70.

[87] Lyu H, Li H, Hanikel N, Wang K, and Yaghi OM. Covalent organic frameworks for carbon dioxide capture from air. *J Am Chem Soc*. 2022 [cited 2023 Oct 10];144(28):12989–95.

[88] Reuß M, Grube T, Robinius M, Preuster P, Wasserscheid P, and Stolten D. Seasonal storage and alternative carriers: a flexible hydrogen supply chain model. *Appl Energy*. 2017;200:290–302.

[89] Hystories Project. Hydrogen storage in European subsurface – Hydrogen technologies could be one of the pillars of future European energy and transport systems [Internet]. [cited 2023 Oct 10]. Available from: https://hystories.eu/.

[90] Ferrovie.it. Alstom, pronta la prima Power Car per i treni a idrogeno della Val Camonica – Comunicato stampa Alstom [Internet]. 2023 [cited 2023 Oct 10]. Available from: https://www.ferrovie.it/portale/articoli/13778.

[91] PESA.pl. SM42-6Dn hydrogen [Internet]. [cited 2023 Oct 10]. Available from: https://pesa.pl/en/produkty/hydrogen-ready/sm42-6dn-hydrogen/.

[92] ACE (Aerospace Central Europe). The third hydrogen aviation cycle [Internet]. 2021 [cited 2023 Oct 10]. Available from: https://aero-space.eu/2021/05/24/the-third-hydrogen-aviation-cycle/.

[93] EnTraT (Energy Transition Team). Universal Hydrogen first test flight [Internet]. 2023 [cited 2023 Oct 10]. Available from: https://www.entrat.eu/universal-hydrogen-first-test-flight.html.

[94] Amos WA. Costs of storing and transporting hydrogen [Internet]. NREL/TP-570-25106. Golden, CO (United States); 1999 [cited 2023 Oct 10]. Available from: http://www.osti.gov/servlets/purl/6574/.

[95] Calvera Hydrogen. Calvera Hydrogen develops the largest ever hydrogen transport tube trailer model for Shell Hydrogen [Internet]. 2018 [cited 2023 Oct 10]. Available from: https://www.calvera.es/calvera-hydrogen-develops-the-largest-ever-hydrogen-transport-tube-trailer-model-for-shell-hydrogen/.

[96] DB Cargo. Hydrogen by rail [Internet]. 2022 [cited 2023 Oct 10]. Available from: https://www.dbcargo.com/rail-de-en/logistics-news/hydrogen-by-rail-8897790#:~:text=The hydrogen is left bound,to carry it by rail.&text=When it reaches the place,out and used in combustion

[97] Kawasaki Heavy Industries L. Kawasaki obtains AIP for large, 160,000 m^3 liquefied hydrogen carrier [Internet]. 2022 [cited 2023 Oct 10]. Available from: https://global.kawasaki.com/en/corp/newsroom/news/detail/?f=20220422_3378&wovn=it.

[98] Hydrogen Tools. Hydrogen pipelines [Internet]. [cited 2023 Oct 10]. Available from: https://h2tools.org/hyarc/hydrogen-data/hydrogen-pipelines.

[99] Zawya.com. Study moots 1,000 km hydrogen pipeline to serve North Oman [Internet]. 2023 [cited 2023 Oct 10]. Available from: https://www.zawya.com/en/projects/industry/study-moots-1-000-km-hydrogen-pipeline-to-serve-north-oman-cldgrh20.

[100] PVC4PIPES. TEPPFA position paper: PVC pipes fit for the hydrogen economy [Internet]. 2021 [cited 2023 Oct 10]. Available from: https://pvc4pipes.com/teppfa-position-paper-pvc-pipes-fit-for-hydrogen-economy/.

[101] Bossel U, Eliasson B, and Taylor G. The future of the hydrogen economy: bright or bleak? *Cogener Compet Power J*. 2003;18(3):29–70.

[102] Trevisani L, Fabbri M, and Negrini F. Long distance renewable-energy-sources power transmission using hydrogen-cooled MgB_2 superconducting line. *Cryogenics (Guildf)*. 2007;47(2):113–20.

[103] Prats Campmany F. *Economical study of electric power transmission with superconducting lines for HVDC systems [Internet]*. Barcelona: Escola Tècnica Superior d'Enginyeria Industrial de Barcelona; 2021. Available from: https://upcommons.upc.edu/handle/2117/350954. Accessed 16 December 2024

Chapter 5
Hydrogen energy utilisation

Hydrogen can be used to produce either thermal energy, or electrical energy or mechanical energy. However, its actual use will be largely influenced by the balance between the availability of hydrogen and its cost, and demand for electricity and fuel. If in the coverage of the energy needs, electricity were available at a shortage compared to the demand and hydrogen available in excess, then it would be reasonable to produce electricity for stationary use starting from hydrogen. In the opposite case, such a use would be completely irrational and hydrogen would instead be used as an electric energy storage and as a fuel both for the transport sector and for the production of high-temperature heat or as a raw material for the chemical industry.

The costs of converting the energy system is very high and some of the available technologies are not yet mature. However, the progressive reduction of available fossil resources and, with a faster time evolution, the progressive increase of fossil fuel costs due to CO_2 emission, will make the latter more expensive and the alternatives more competitive. Therefore, it can be assumed a transition scenario still whit predominance of fossil sources finally followed by the target scenario in which these sources disappear completely from the energy mix.

In the transition scenario, the decarbonisation of the system is however possible. There are two possible ways: sequestration of carbon dioxide downstream of combustion or sequestration of carbon in its various forms upstream of final uses. This second way therefore implies a predominant role for hydrogen. Whereas the first is applicable only to medium-large stationary users, the second route is mandatory for the transport sector, for the residential sector and for industries up to a certain size.

5.1 Combustion

This section deals with systems that involve direct combustion of hydrogen with production of thermal energy and possible subsequent conversion into mechanical and electrical energy.

The main characteristics of hydrogen that influence its combustion, differentiating it from that of other common fuels, are:

- very wide combustion interval of the air/hydrogen (oxygen/hydrogen) mixture, with the possibility of using particularly lean mixtures;

Table 5.1 Main differences among fuels

	Hydrogen	Methane	Gasoline
Density (kg/m^3)	0.090	0.717	750
Lower heating value (MJ/kg)	119.93	50	44.03
Volume of the cylinder occupied at stoichiometric ratio (%)	29.53	9.48	2
Diffusion rate in air (cm^2/s)	0.61	0.16	0.05
Flame propagation speed (m/s)	2.65–3.25	0.37–0.45	0.37–0.43
Self-ignition temperature (°C)	585	540	228–471
Minimum ignition energy (mJ)	0.02	0.29	0.24

- high diffusion rate in air (oxygen), and therefore more uniform mixture;
- high flame propagation speed;
- high self-ignition temperature;
- low ignition energy.

These differences could have, although not always, significant consequences from a thermodynamic point of view. Table 5.1 reports a comparison among hydrogen, methane and gasoline.

5.1.1 Internal combustion engines

The history of the hydrogen engine begins in 1807 with the patent of the Swiss François Isaac de Rivaz. In 1813, the patent was followed by the construction of a vehicle equipped with an atmospheric engine with manual electric ignition (Figure 5.1): the vehicle travelled only a few hundred meters. Although the engine were atmospheric, the inventor suggested the possibility of ignition by compression [1].

In 1854, the Italians Niccolò Barsanti and Felice Matteucci had a working prototype of a single-cylinder hydrogen engine built by the Benini Foundry (Figure 5.2). Three years later, encouraged by the tests carried out on the first engine, they had a twin-cylinder built which was used to operate machine tools in the workshop of the Maria Antonia railway station in Florence. After allowing a provisional protection certificate dated 1854 to lapse, Barsanti and Matteucci obtained the patent in 1857 [3].

In 1860, the Frenchman Joseph Étienne Lenoir built a vehicle, called Hippomobile, equipped with a single-cylinder hydrogen engine (Figure 5.3). It was the first internal combustion engine vehicle to be placed on the market. It had a 4% efficiency and a speed of 6 km/h.

Subsequently, gasoline took over as a fuel and the only significant activity in the field of hydrogen-fuelled internal combustion engines is to be ascribed to the German Rudolf Arnold Erren during the period 1920–1940. Only in the 1970s, thanks to both the greater familiarity with hydrogen due to aerospace missions and the growing concerns about the availability of fossil resources, the number of prototype hydrogen vehicles has steadily increased.

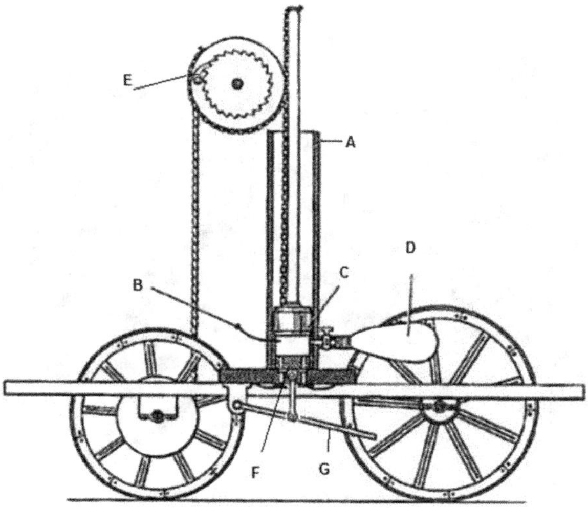

Figure 5.1 First vehicle equipped with a hydrogen engine; A = cylinder, B = spark ignition, C = piston, D = balloon containing hydrogen fuel, E = ratchet, F = opposed piston with air in and exhaust out valves, G = handle for working opposed piston [2]

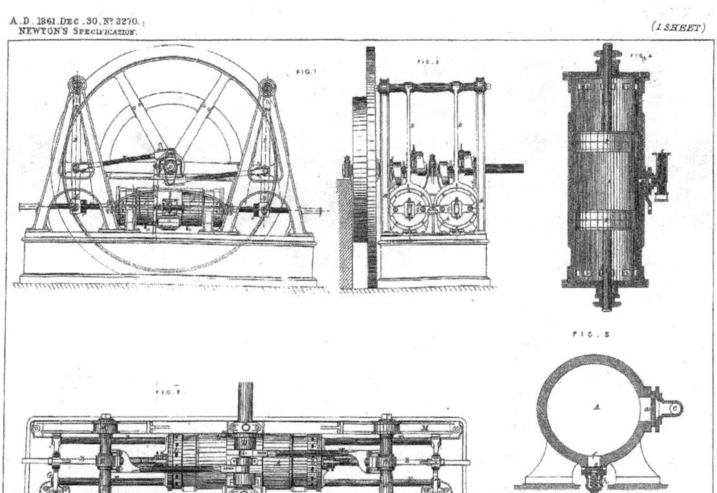

Figure 5.2 The engine of Barsanti and Matteucci (courtesy of Fondazione Barsanti and Matteucci, Italy [3])

Figure 5.3 *Hippomobile*

Figure 5.4 *Hydrogen-fuelled Toyota Corolla (copyright Toyota Motor Corporation)*

Although most of the existing prototypes are equipped with fuel cells, the reciprocating internal combustion engine could still play an important role in the transition phase of the transport sector from fossil fuels to hydrogen. In fact, compared to fuel cells, the internal combustion engine has a greater maturity, a lower cost, a greater versatility towards fuels and their purity and, above all, the possibility of operating even with different fuel mixtures. More recently Toyota developed a GR Yaris and a GR-Corolla (Figure 5.4) equipped with a 1.6-litre

Figure 5.5 MAN H4576 hydrogen combustion engine (copyright MAN Engines)

hydrogen-fuelled internal combustion engine and asked Yamaha to develop another hydrogen-fuelled internal combustion engine, a 5.0-litre V8 [4]. Also MAN developed a hydrogen-fuelled internal combustion engine (Figure 5.5).

While in the case of the use of hydrogen in a gas turbine there are no significant variations from the thermodynamic point of view compared to the use of other fuels, in the case of an internal combustion reciprocating engine there are considerable differences both in the performances relative to the limit cycle and those relating to the indicated cycle. However, there are also major problems relating to safety.

In this regard, the risk mainly consists in the pre-ignition of the mixture and the backfire. Both of these problems also exist with other fuels, but in the case of hydrogen, the probability is higher because the ignition energy is lower and the flame propagation rate is higher.

Pre-ignition consists in the ignition of the mixture before the piston reaches the predetermined position. The characteristics of the hydrogen responsible for the pre-ignition are the low ignition energy (0.02 mJ for hydrogen against, for example, 0.24 mJ for gasoline) and the wide flammability range in air (4%–75% for hydrogen vs., e.g. 1%–7.6% for gasoline). The lower limit represents the minimum value at which the flame is capable of self-sustaining; however, in the 4%–9% range, the flame is able to propagate, but not to achieve complete combustion. Due to this difference in behaviour, the value 4% is defined as the limit of incoherent flames, and the value 9% as the limit of coherent flames. Although the presence inside the combustion chamber of surfaces or particles at high temperature with very low thermal content does not determine the pre-ignition of traditional fuels, it is sufficient to cause the pre-ignition of the air/hydrogen mixture.

The air-hydrogen mixture has a very high flame propagation speed (2.65–3.25 m/s against, for example, 0.37–0.43 m/s for the air-gasoline mixture). Moreover, it also increases with the temperature. For such a reason, the pre-ignition of the mixture can favour the development of uncontrolled processes characterised by sudden increases in pressure inside the cylinders, with the consequent production of shock waves and detonation.

The detonation causes a decrease in the thermal efficiency of the engine and in the power developed. This is due to the fact that the detonation occurs when the compression phase is still in progress and therefore the movement of the piston encounters a resistance higher than normal. Similar losses, albeit to a lesser extent, are still related to pre-ignition even in the absence of detonation. Prolonged operation in conditions of detonation can cause serious damage to the engine, even breakage, while occasional operation has practically no consequences.

The sources of thermal energy that can trigger the pre-ignition and cause any backfires are the high-temperature areas inside the cylinder, that is: the spark plug, oil, walls, valves, residual hot gases and any deposits. Some of these sources can be easily neutralised by adequate refrigeration. Consequently, the main cause of pre-ignition is, generally, the particulate matter due to the greater heat capacity compared to gas. In the case of hydrogen, the particulate matter is caused by the combustion of the lubricating oil and is therefore present in minimal quantities. Furthermore, the use of lean mixtures reduces these problems since it lowers the flame propagation speed, increases the energy needed to ignite the mixture and reduces the combustion and the hot spot temperatures [5].

The backfire problem is caused by the presence of an air-hydrogen mixture in the adduction manifold with the inlet valve open: the possibility of pre-ignition and the high flame propagation speed, higher than that of delivery of the mixture, allows the flame itself to go up the adduction manifold.

An engine with carburettor feed or with continuous feed in the adduction manifold does not have a regular operation since, due to their functional characteristics; they would be easily subjected to pre-ignition and backfire with the aforementioned consequences. The first techniques tested to avoid pre-ignition were the injection of water into the intake manifold and the recirculation of the cooled flue gases.

The exhaust gases contain a high percentage of inert substances and therefore their partial recirculation lowers the flame propagation speed and increases the ignition energy. Recirculation must of course be limited since, by reducing the space available for the mixture, it involves volumetric losses. Values of this recirculation can be pushed up to a maximum of 20% of the mixture entering the cylinder. In addition to regular engine operation, a reduction in NO_x emissions is achieved, due to the reduction of the maximum cycle temperature due to the smaller quantity of fresh mixture.

The injection of water into the adduction manifold or into the carburettor allows to mix water, hydrogen and air and produces similar results. It is not clear whether the injection of water does not lead to an increase in volumetric losses [6] or whether the quantities of water required would be such as to reduce the volumetric efficiency [7].

Figure 5.6 BMW Hydrogen 7 [8]

Remaining into the field of premix engines, one of the prototypes developed was the BMW Hydrogen 7 (Figure 5.6), whose engine was derived from the standard gasoline engine and, in reality, was not a real hydrogen engine, but an engine capable of running with both gasoline and hydrogen. The hydrogen, stored as a liquid in a 7.8 kg tank, was injected into the intake manifold at a pressure slightly above atmospheric. The injection took place near the intake valve so that the manifold was crossed only by air and that an accidental pre-ignition did not give rise to backfire.

An alternative, again in order to avoid the risk of backfire, could be that of timed injection in the intake manifold [9]. The intake phase in the single cylinder is thus divided into three sub-phases of which only the central one is involved in the injection of hydrogen. In this way, in the first and third sub-phases, there is, in the intake manifold, only the passage of air which firstly cools the cylinder, reducing the risk of pre-ignition and subsequently washes the manifold itself, blocking the propagation of any backfire. However, this solution is effective at low rpm when, due to the simultaneous opening of the intake and exhaust valves, the air/hydrogen mixture comes into contact with the high-temperature exhaust gases. However, the pre-ignition in the compression phase occurs at high loads and a high number of revolutions, making the solution adopted ineffective [10].

Finally, again in the case of premix engines, it is possible to resort to the injection of liquid hydrogen into the intake manifold with the result of obtaining a colder air/fuel mixture and, therefore, less prone to pre-ignition.

In the case of direct injection, the fundamental parameter is the timing for injection, that is related to the pressure [11]. If the injection takes place shortly after the intake valve is closed (e.g. at 1 MPa), there is no danger of backfire, but the risk of pre-ignition remains, while if the injection is very delayed, i.e. carried out in a stroke of very advanced compression, near the top dead centre (for example at 8 MPa), the pre-ignition is also prevented. However, the high supply pressure leads to a reduced emptying of the storage tank and, consequently, to a reduction in the vehicle range. The case of direct injection of liquid hydrogen is not affected by this latter effect, which also represents a further obstacle to pre-ignition. An example of a vehicle with a low-pressure direct injection engine was the MAN Omnibus employed for some years at Munich airport.

Obviously, these different options also have a considerable influence on the time available for the air/fuel mixture to homogenise. However, this aspect is made less critical by the high diffusion rate of hydrogen.

Regarding the ignition system of the mixture, the two methods studied and tested are the traditional spark produced by a spark plug and the hot surface. In the case of low and medium pressure fuelling, pre-ignition can be obtained with richer mixtures compared to an excess of air of 1.2 and therefore spark ignition is more suitable, possibly with refrigeration of the spark plug points and injection of cold fuel. If, however, the power supply is at high pressure, problems of local instability of the flame can occur and for this reason also the ignition with a hot surface deserves to be considered since, in the experiments carried out, it showed a very good behaviour in terms of stability of the flame combustion. Unfortunately it has negative aspects in terms of energy consumption and prevention of pre-ignition.

Given the instability problems, experiments were carried out to establish the positions of the spark plug that allow regular combustion in direct injection engines [5] and the conclusion was that areas with too rich a mixture should be avoided. Furthermore, these experiments have highlighted the opportunity to reduce the interval between injection and ignition as much as possible and, therefore, to ensure homogeneous mixing, a high vorticity combustion chamber is required.

Combustion can be improved if a secondary flow of hydrogen is premixed in the adduction manifold: in this case, in fact, a slow combustion begins before injection into the main flow cylinder. Overall, there is a reduced ignition delay and a lower oscillation of the combustion pressure in the cylinder. This premixing must be more contained if the ignition system is of the hot surface type.

Finally, it should be noted that most of the prototypes made were obtained by modifying spark ignition engines: the specific design of a hydrogen engine with a combustion chamber and a cooling system designed taking into account the characteristics of hydrogen could solve the problem more effectively [12].

The performance of a thermal engine is decisive to define its potential field of use and economic feasibility. In this case, it is appropriate to evaluate this performance in relative terms, since there are already reference models which are the gasoline, diesel and natural gas engines.

While not taking into account the properties of the burnt gas, of the heat transmission through the cylinder walls, of the valves opening times and of the pressure drops crossing the valves, it is possible to make a qualitative comparison between the efficiency of a hydrogen engine and one with gasoline using approximate mathematical expressions. The thermal efficiency of the cycle ideally followed by an internal combustion engine is a function of the quantities of heat exchanged during the two non-adiabatic transformations. As a first approximation, considering the specific heat values at constant pressure and constant volume to be constant, this efficiency can be expressed through the relationship:

$$\eta = 1 - \rho^{(1-k)} \tag{5.1}$$

From this relationship, it can be seen that the thermal efficiency of an engine increases both with the compression ratio ρ and with the ratio between the specific heats of the gas k.

Despite the low ignition energy of the air-hydrogen mixture and the wide mixing range, the self-ignition temperature of the mixtures is so high (585 °C against 350 °C for diesel fuel) that ignition by compression is practically impossible. This fact, negative from the point of view of the difficulty of creating a diesel cycle, can be usefully exploited: in fact, in the event that the pre-ignition is prevented with suitable systems, the high self-ignition temperature makes it possible to significantly increase the compression ratio compared to that of a gasoline engine. As shown in (5.1), this possibility favours thermal efficiency, which constitutes a first factor in favour of the hydrogen engine.

Another very important aspect of the combustion of the air/hydrogen mixture is the width of the flammability range that allows the use of very lean mixtures. This is advantageous from the point of view of efficiency since the burnt gases are very diluted and, having a lower temperature, the ratio between the specific heats is higher.

Obviously, the real thermal efficiency of an internal combustion engine is lower than the ideal one due to irreversibility which is graphically highlighted in the deformations that the indicated cycle presents compared to the ideal one (Figure 5.7). However, it is interesting to note that these irreversibilities are sometimes related to the fuel used.

In a non-supercharged engine, the intake phase is achieved by means of the vacuum that the piston creates in the cylinder when it moves from top dead centre to bottom dead centre. The lower pressure also means that the aspirated mixture is more rarefied. The heating of the air entering the cylinder by the walls is added to this effect, with a consequent further decrease in density. Therefore, there is a lower filling of the cylinder or a loss of volumetric efficiency. This depression is necessary to overcome the distributed pressure drops, which are of modest importance, and the pressure drops through the throttle and the inlet valves. The presence and use of the throttle valve can reduce the inlet pressure of the mixture up to 0.03 MPa: the work spent by a gasoline engine to empty and fill the cylinder can reach 1/3 of the work done by the flue gases [9].

The amplitude of the flammability range of the air/hydrogen mixture is such as to allow a qualitative adjustment of the engine. This means that, except in extreme

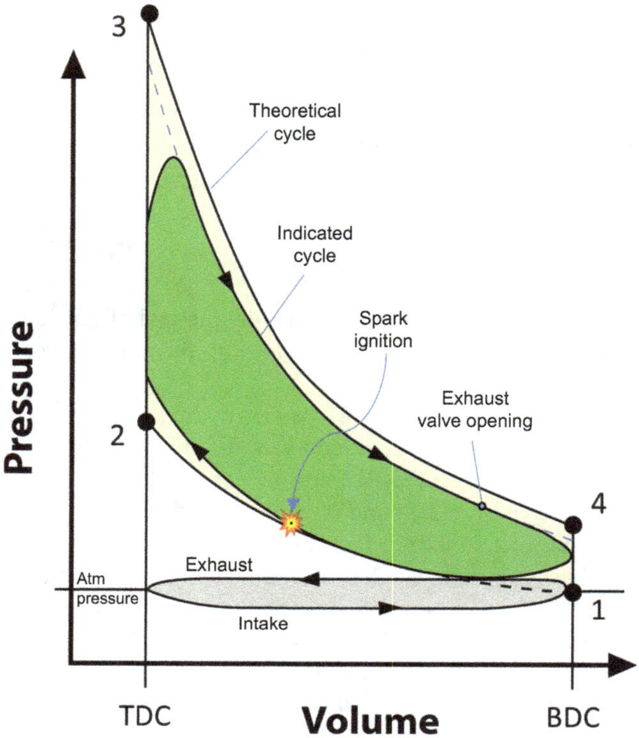

Figure 5.7 Beau de Rochas, limit (yellow) and indicated (green) cycles

cases, the engine power can be controlled varying the fuel flow without throttling the flow of air into the manifold. It follows the possibility of eliminating the throttle valve, or at least reducing its use to almost zero.

By defining the equivalence ratio R_{eq} as:

$$R_{eq} = \frac{\left(\frac{\dot{m}_{air}}{\dot{m}_{fuel}}\right)_{actual}}{\left(\frac{\dot{m}_{air}}{\dot{m}_{fuel}}\right)_{ideal}} = \frac{\dot{m}_{air,actual}}{\dot{m}_{air,ideal}} \tag{5.2}$$

the need to throttle the flow of combustion air occurs, for a hydrogen engine, only when an equivalence ratio of the order of 0.26–0.33 is reached. For higher values, the throttle is always open and the value of the equivalence ratio increases to provide higher power. A gasoline engine reaches the maximum thermal efficiency for $R_{eq} = 0.9$ while the hydrogen engine can improve further up to $R_{eq} = 0.4$. Since in current traffic conditions, and especially in cities, the engines frequently operate at low load, it is possible to achieve significant fuel economy by using a hydrogen-powered engine.

In the compression phase, as in the gasoline engine, a departure from the ideal transformation occurs due to the heat exchange between the mixture and the cylinder walls. This loss is more pronounced in the hydrogen engine as the higher compression ratio leads to a greater heating of the mixture and, therefore, greater losses.

The ignition of the mixture is not instantaneous and, therefore, the actual combustion is not isochoric. The gap between the indicated cycle and the ideal cycle is more significant in this phase. The parameters that influence the duration of combustion are the dosage and homogeneity of the mixture and the speed of propagation of the flame. A too rich mixture causes slow ignition and incomplete combustion due to lack of oxygen. A too lean mixture, however, causes slow combustion propagation. The better propensity to form flammable mixtures, the higher diffusion coefficient and the higher flame propagation speed are further points in favour of hydrogen. Combustion is much closer to the isochoric and, consequently, the advance required for ignition is much more modest.

As combustion progresses, the dissociation of the species produced by the combustion itself also intervenes increasing the amount of unburned products. The main products of gasoline combustion are water vapour and carbon dioxide: since the latter has a greater propensity to dissociate than water vapour, the losses due to this phenomenon are more sensitive in a gasoline engine than a hydrogen engine.

Finally, the expansion and exhaust phases do not present significant differences between the gasoline engine and the hydrogen engine.

The overall efficiency of the vehicle is however linked, in addition to the thermal efficiency of the engine, to the weight of the vehicle itself and this entails the need to also consider elements external to the engine, first of all the storage system. This may diminish the advantage just attributed to hydrogen. In general, it can be assumed that an increase of 1% in weight corresponds to a reduction in overall efficiency of 0.8% [7].

The consumption of the BMW Hydrogen 7 [1], measured by the FTP-75 test, was 3.7 kg of hydrogen (2.1 on the highway) per 100 km, corresponding to about 14 km/l of gasoline (8 on the highway); however, it was not an engine specifically designed for hydrogen.

Experimental data obtained with modified engines showed efficiency increases, in relative terms, from 15% to 50% [7] and, therefore, the improvement with engines designed and built specifically for hydrogen could be even higher.

The premix engine with timed injection of gaseous hydrogen into the manifold turns out to be a good compromise between the characteristics of the different types of hydrogen engines examined previously: in particular it is quite simple, safe, efficient and with low NO_x emissions when the engine is fuelled with lean mixtures. Experiments conducted on an engine of this kind have shown that the optimal values of the actual thermal efficiency measured on the bench, referred to the lower calorific value, when changing the equivalence ratio (0.3–1.3) and the motor rotation speed (1000–2000 rpm), are reached for very lean mixtures with equivalence ratio values between 0.4 and 0.6 [9]. As speed increases, efficiency improves by up to 45%.

The comparison in terms of power is much less obvious. The power W of an internal combustion engine is a function of the work per cycle L, the rotation speed

n and the number of strokes per cycle n_c:

$$W = \frac{n}{n_c} \cdot L \tag{5.3}$$

and the work is given by the product of the displacement V by the average effective pressure p_{me}:

$$L = V \cdot p_{me} \tag{5.4}$$

ultimately, for a four-stroke engine, we get:

$$W = \frac{n}{2} \cdot V \cdot p_{me} \tag{5.5}$$

But the effective average pressure is related to the volumetric efficiency λ_a, to the heating value of the mixture $H_{c,m}$ and to the effective thermal efficiency η_e:

$$p_{me} = \eta_e \cdot \lambda_a \cdot H_{c,m} \tag{5.6}$$

Finally, in the case of premixing, the calorific value of the mixture is linked to the stoichiometric air/fuel ratio L_{st}, to the filling coefficient λ, to the density of the mixture ρ_m and to the lower heating value of the fuel LHV:

$$H_{c,m} = \frac{\rho_m \cdot LHV}{1 + \lambda \cdot L_{st}} \tag{5.7}$$

while in the case of direct injection, indicating the air density with ρ_a, we have:

$$H_{c,m} = \frac{\rho_a \cdot LHV}{\lambda \cdot L_{st}} \tag{5.8}$$

Assuming $\lambda = 1$ and constant all the other quantities listed, with the exception of the heating value of the mixture, a theoretical comparison can be made between the powers that can be obtained in the various cases (Table 5.2).

As can be seen from Table 5.2 [1], a hydrogen-fuelled engine gives a power significantly lower than that of a gasoline engine: in fact the stoichiometric air-hydrogen mixture has, for the same volume, a heating value equal to 80%–85% of that of the air-gasoline mixture [5,6]. This is due, on the one hand, to the low

Table 5.2 Comparison between the power of gasoline engine and hydrogen engine

Fuel	**Gasoline**	**Hydrogen**		
			Gas	**Liquid**
Feeding	Premixing	Premixing	Direct injection	Premixing
Temperature [K]	293	293	293	210
Heating value [MJ/m^3]	3.59	2.97	4.21	4.14
Relative power	1	0.83	1.17	1.15

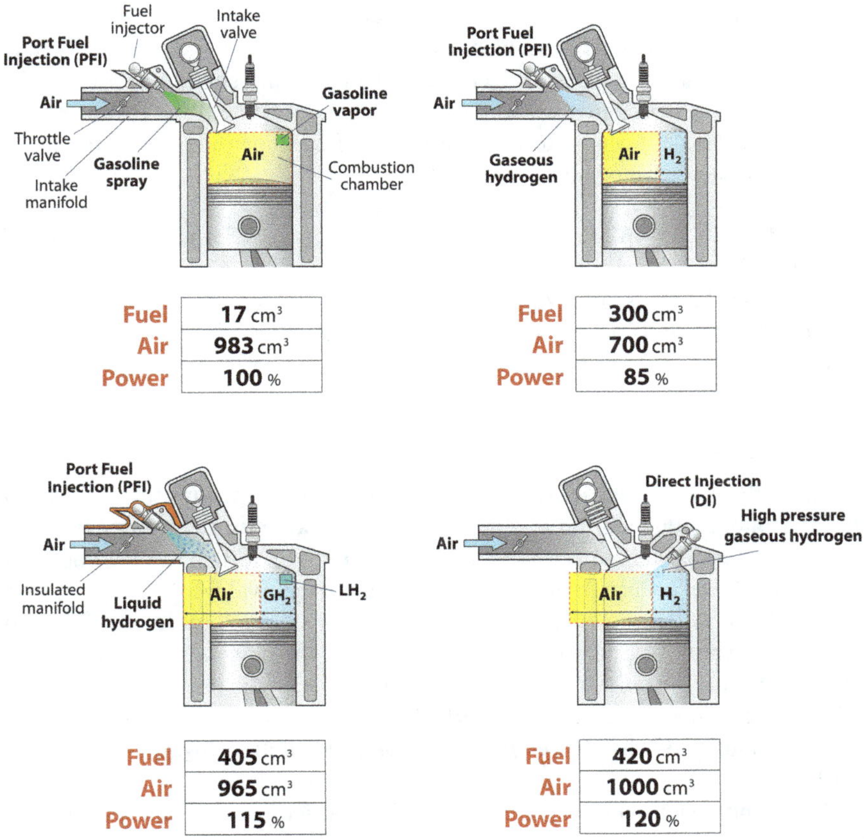

Figure 5.8 Power comparison among different options. Based on [5].

density of hydrogen, and on the other hand, to the lower ratio between lower and upper heating value. In fact, due to its low density, hydrogen occupies a portion of the cylinder larger than that occupied by gasoline vapours and requires less air (Figure 5.8). Using liquid hydrogen or direct injection, this filling defect is eliminated and to overturn the comparison in favour of hydrogen is possible.

With engines operating at speeds of 4000–5000 rpm, the time available for injection is very short. However direct injection can also be carried out with liquid hydrogen which, thanks to the smaller volume to be injected, facilitates the operation. In this case, the hydrogen must be injected about 5° before the top dead centre, when the pressure is high and it is therefore necessary to have a pump suitable for treating cryogenic liquids. The positioning of this pump is a critical point: if it is external to the tank and sucks hydrogen in the boiling phase, it is subject to cavitation, while if it is internal to the tank, the heat flow due to its operation accelerates the boil-off. Furthermore, injectors capable of operating at very high speeds in a wide range of temperatures are required.

In reality, another unfavourable phenomenon to hydrogen gas must also be taken into account. Taking into consideration one mole of a liquid hydrocarbon, the combustion reaction produces an increase in moles in the gaseous phase equal to half the moles of hydrogen present in the hydrocarbon, while if the hydrocarbon is gaseous the increase is equal to half the moles of hydrogen present in the hydrocarbon decreased by one unit:

$$C_nH_m + \left(n + \frac{m}{2}\right)O_2 \Rightarrow nCO_2 + mH_2O \tag{5.9}$$

The same happens for hydrogen and, therefore, if the latter is gaseous, the effect of combustion is that of a contraction in the number of moles in the gaseous phase:

$$H_2 + \frac{1}{2}O_2 \Rightarrow H_2O \quad \Delta H_0 = -242.0\,\text{kJ/mol} \tag{5.10}$$

Since the volume is defined by the position of the piston, this contraction translates into a reduction in pressure, i.e. in work and, ultimately, in power.

For the purposes of power, the choice of the ignition instant is very important, in fact the flame propagation speed varies as the equivalence ratio varies and therefore there is the risk of not obtaining a complete combustion if the ignition instant is not varied. Up to a unit equivalence ratio, the power/equivalence ratio is almost linear: the deviation may be due to the decrease in the ratio between specific heats as the temperature increases [9]. This trend confirms the ability of the hydrogen engine to vary power by means of a qualitative regulation without a throttle valve.

The compression ratio increases the combustion temperature, and therefore the possibility of backfire and the temperature of the exhaust gases. However, the ratio between the section of the cylinder and the volume of the combustion chamber increases and this allows a better expulsion of the gases and a more rapid cooling of the residual gases. Overall, the increase in the compression ratio has a beneficial effect.

The temperature of the exhaust gases is also affected by the rotation speed, since the shorter time available reduces the heat exchange.

As for the emissions, the combustion products do not contain carbon compounds except in minimal quantities due to the combustion of lubricating oil [12]. The only pollutants present are nitrogen oxides due to the simultaneous presence of oxygen and nitrogen contained into the air. Since the formation of these oxides is closely linked to the maximum temperature of the process, it is evident that the use of lean mixtures, by lowering the combustion temperature, disfavours their formation: in fact it has been experimentally demonstrated that for excess air values higher than two there is no formation of NO_x [1].

It is certainly difficult to evaluate, in a statistically correct way, the duration of a hydrogen engine since hydrogen-powered vehicles subject to regular use are not a significant sample. We must therefore rely on the properties of hydrogen to evaluate which of them, and to what extent, are favourable and which are against its use in internal combustion engines.

A gaseous fuel can be distributed to the cylinders more evenly than a vaporised liquid. This fact, combined with the high diffusion coefficient of hydrogen, leads to greater uniformity of the mixture and a more uniform distribution of pressures inside the cylinder. The greater uniformity of pressure results in smoother operation and therefore less wear and fewer mechanical problems on the components. The ease of formation of the mixture also translates into the reduction of its fluctuations during acceleration and deceleration transients. These advantages are partially lost in the case of liquid hydrogen injection.

The higher flame speed of the air-hydrogen mixture could, on the other hand, originate phenomena of combustion instability which, in the long run, could damage the engine.

Hydrogen does not cause carbon deposits on engine components and the combustion of the lubricating oil, in a properly maintained engine, is negligible [7]. In addition to reducing wear and pre-ignition problems, the lack of carbon deposits is favourable in terms of constant consumption since the spark plugs do not get dirty.

Unlike gasoline, hydrogen has no lubricating power and this promotes wear on components: it is therefore necessary to make certain components, such as injection valves and their seats, in wear-resistant material and with a very precise finish. The cost of these components will therefore be much higher than the similar components of traditional engines. However, engine oil is not diluted by hydrogen and no acids are formed that can attack it. Its oxidation is also delayed due to the lower combustion temperature. The properties of the oil, therefore, are preserved intact longer.

One problem that does not seem to cause much concern is hydrogen embrittlement [13].

This brief examination of the characteristics of hydrogen which may have to do with the durability aspect of an engine certainly does not allow us to draw absolute conclusions, but suggests that hydrogen engines, if well designed, may have lower oil consumption and lower maintenance costs compared to a gasoline engine [7].

An interesting case of a hydrogen engine is the Wankel engine, used on the Mazda RX-8 Hydrogen. This engine is of the rotary type, that is, it has a piston with a rotary rather than an alternating motion. It follows that there is no need for any crank mechanism to transfer motion to the crankshaft. The operation is as illustrated in Figure 5.9: the piston, of triangular section with curvilinear sides, rotates inside a suitably shaped cylinder so that, during the movement of the piston,

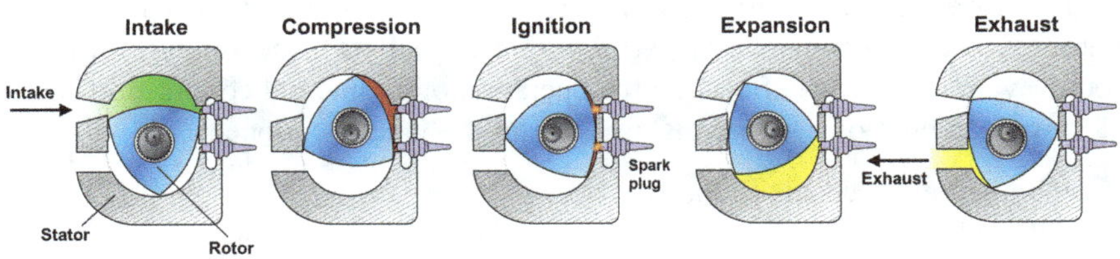

Figure 5.9 Working principle of a Wankel engine

isolated chambers are created along its perimeter. Each of these chambers has an evolution that follows all phases of engine operation from intake to exhaust, operations that take place through lights without the need for valves. The absence of crank mechanisms and valves results in less noise, less mechanical losses and a lower risk of mechanical breakage. The critical aspects, however, are the tightness of the blades that separate the chambers and the shape of the combustion chamber which is very elongated and entails difficulties in achieving complete combustion.

Wankel engine lends itself well, at least in principle, to the use of hydrogen. The absence of valves could suggest a high risk of backfire, but the air/hydrogen mixture enters an area where only the intake phase takes place and therefore the temperatures are the most modest of the entire cycle: the only hot surface is that of the piston, but it can be pre-cooled by initially sending only air. However, the great homogeneity of the mixture combined with the remarkable flame speed ensures that the combustion is complete despite the shape of the combustion chamber. The critical point is that of the seal between chambers since the reduced size of the hydrogen molecule makes it more likely to leak towards the lower pressure chambers.

Furthermore, the Wankel engine is more compact and lighter than a reciprocating engine [14]: and, therefore, allows a greater quantity of hydrogen to be stored on board for the same overall volume and weight.

Another solution of considerable interest is the one adopted on the Ford H2RV, which is a hybrid that uses a 2.3 litre hydrogen turbocharged engine [15]. Hybridisation entrusts the coverage of power peaks to the storage battery, while the heat engine operates at a constant speed with lean mixtures, in conditions of high performance and zero nitrogen oxide emissions and in such a way as to avoid pre-ignition. Consumption was 1.39 kg of hydrogen per 100 km, about 25% higher than the Ford Focus FCV (equipped with fuel cells) and about 25% lower than the gasoline version.

5.1.2 Blend-fuelled internal combustion engines

A solution that could truly create a bridge to a hydrogen-based transport system is based on blend fuelled internal combustion engines.

These types of engines combine a decidedly contained demand for hydrogen with a more than proportional percentage reduction in emissions with respect to the percentage of fuel replaced by hydrogen. This is due to a better progress of combustion and, consequently, less unburned materials in the exhaust gases and a lower production of nitrogen oxides connected to the possibility of using leaner mixtures.

The lower ignition energy and the greater speed of the flame front of the air/hydrogen mixture compared to gasoline, diesel and methane mixtures with air produce various positive effects on the operation of an internal combustion engine if a certain amount of hydrogen is mixed with the fuels mentioned:

- unburned products are reduced;
- leaner blends can be used;
- the advance is reduced by bringing the combustion closer to an isochoric transformation.

It follows that, at the expense of a decrease in power, there is an improvement in efficiency and a reduction in unburned and carbon dioxide emissions that go beyond the simple replacement of carbon atoms with hydrogen atoms thanks to both the reduction in consumption of fuel and the improvement of combustion. Furthermore, the leaner mixtures lead to a lowering of the combustion temperature and, consequently, to a lower production of nitrogen oxides.

Since the 1980s, prototypes of reciprocating internal combustion engines have been successfully tested which use hydrogen combined with traditional liquid and gaseous fuels. In general, car manufacturers do not show interest in this technology since blends do not constitute a definitive solution to the environmental problem and, albeit in a much smaller quantity since hydrogen is the secondary fuel, require the presence on board of hydrogen, with the need therefore for a distribution infrastructure. However, in the first decade of this century, these systems were the object of growing interest in Italy (in 2007 a prototype of this kind was presented, the FIAT Panda Aria) and abroad (e.g. in Sweden and the USA) and were therefore carried out numerous test campaigns to demonstrate its potential in terms of reducing emissions and consumption compared to methane.

As an example of a successfully completed project, we can cite the Malmö-Hythane Project [16], with which a 5%–7% reduction in energy consumption and 10%–20% in the emission of climate-altering gases were obtained, together with a reduction of NO_x and CO emissions always greater than 10%. The conversion to Hythane8 (8% by volume of hydrogen) of the buses tested did not involve modifications to the vehicles, but only to the mapping of the engines. In other cases, hydrogen was mixed with natural gas even in a percentage of 30% and more substantial interventions were required.

Whereas in case of blend with natural gas both fuels are stored together, in case of blend with liquid fuels two different tanks are required and the mixing occurs inside the cylinder.

In addition to the emissions point of view, the advantages of using blends compared to hydrogen alone can be understood from the point of view of economic acceptability for the user, since these blends offer costs (per litre of gasoline equivalent) reduced compared to hydrogen, from the point of view of acceptability regarding safety issues, and from the expansion of the user base with the same amount of hydrogen distributed.

While on the one hand there would remain the need to create a production infrastructure for hydrogen, on the other hand using blend would allow to avoid a new distribution infrastructure. In fact, the current natural gas distribution network could be used for natural gas/hydrogen mixtures up a defined hydrogen content. However, in the case of gasoline/hydrogen blends, a distribution system based on small cylinders of metal hydrides could be used to carry out distribution to refuelling stations with trucks, similar to what happens today for liquid fuels. Furthermore, thanks to the possibility of using the mixtures in existing internal combustion engines, it would not be necessary to replace the vehicles, but simply to modify them. All this means reducing the bulk of the investment to the production infrastructure alone, postponing the rest of the investment required to switch to an entirely hydrogen system for a second time.

In maritime sector, dual fuel diesel/hydrogen tug will be tested during next years (see Section 6.1.5).

5.1.3 Gas turbines

Hydrogen, being a gaseous fuel, can be used in place of traditional fuels in current turbomachinery-based electricity generation systems. Although the combustion of hydrogen with oxygen only generates water, burning it in the air also produces nitrogen oxides.

At the reference cycle level, the hydrogen/air cycles are similar to the cycles that can be obtained starting from another fuel (e.g. natural gas), but due to secondary effect (irreversibility in balance of plant of gas turbine) feeding a turbine with hydrogen instead of methane generates an increasing in energy efficiency (around 2%) and a reduction in energy consumption [17].

A first experiment on the use of 100% hydrogen in a gas turbine consisted of a 15 MW plant which entered service in 2010 in Fusina (Italy) by ENEL (Figure 5.10). The hydrogen used was a waste product of ethylene cracking operated in the nearby plants of the petrochemical centre of Porto Marghera. The

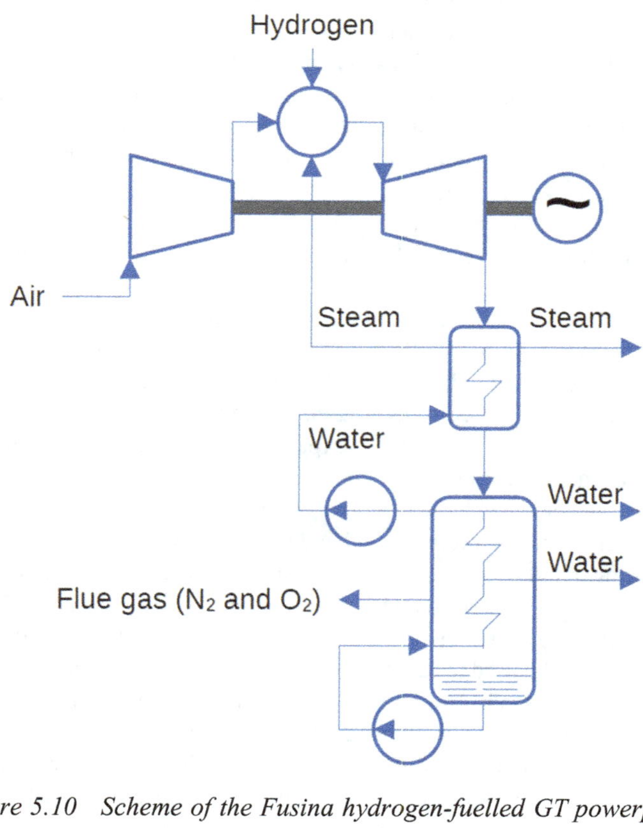

Figure 5.10 Scheme of the Fusina hydrogen-fuelled GT powerplant

combustion of hydrogen in the air took place in a 12 MW turbogas unit and the heat of the high-temperature exhaust fumes was recovered by generating steam for a nearby coal plant, obtaining another 3 MW [18,19].

In Port Island (Kobe) in Japan a 100% hydrogen 1 MW gas turbine was running in a combined heat and power configuration. Thanks to 'dry low NO_x combustion technology' developed by Kawasaki a reduction in environmental impact and an increase in efficiency have been obtained [20,21].

Blending feeding of hydrogen with natural gas is an option for many original equipment manufacturers (GE, Ansaldo, Mitsubishi, Siemens) and many new plants are declared running with different percentage of hydrogen in the blend. For example, tests with a hydrogen volumetric content ranging from 5% up to 100% have been carried out by GE [22].

5.1.4 Direct steam generation

The combustion of hydrogen in air involves only the production of water vapour and nitrogen oxides and, consequently, has considerable environmental advantages compared to the combustion of other fuels in the air. However hydrogen gives certainly its best when it reacts in a stoichiometric mixture with pure oxygen since, in this case, only water vapour is generated (5.9).

This result makes it interesting to use hydrogen when heat is required in the form of steam, especially at high temperatures. In fact, if heat is required for different uses at low temperatures, it is certainly more efficient to obtain it through cogeneration systems, heat pumps or by recovering waste heat where available.

Generating steam by direct combustion of a stoichiometric hydrogen/oxygen mixture is, on the other hand, a system without emissions and with unitary theoretical efficiency and real efficiency very close to unity. The adiabatic combustion temperature is higher than 3000 °C, and therefore a suitable and specially designed combustor is required: the direct steam generator.

Although the prototypes produced are essentially devices of aerospace derivation linked to the development of the Shuttle and Buran engines, the idea has much more distant origins. A first patent of this generator is, in fact, dated 1924 by Ernest E. Tucker [23]. More recently other patents have been presented by Hans J. Sternfeld, Josef Reinkenhof and Heinrich Lambrecht [24,25] and by A.N. Gryaznov and S.P. Malyshenko [26] as well as by other authors.

The new interest for this device arose from the studies carried out at DLR (Deutschen Zentrums für Luft- und Raumfahrt) indicating the possibility of using it in power plants in energy storage systems and peak load coverage, in steam supply networks for industrial use and in steam microgeneration [27–29].

Generally speaking, a direct steam generator (Figure 5.11) is a cylindrical combustion chamber at one end of which enter hydrogen and oxygen which burn generating steam at high temperatures. Along the cylinder, which can be cooled externally with water (or steam at a low temperature), there is a progressive introduction of water (or steam at a low temperature).

The prototypes made gave partially satisfactory results, but the duration of the tests carried out with hydrogen was rather modest. At IVTAN (High Temperature

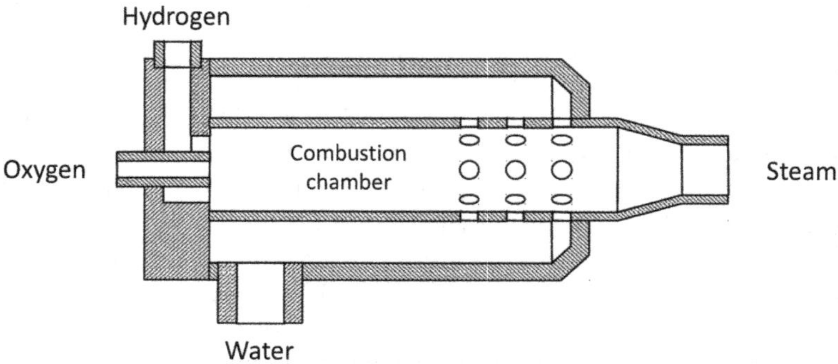

Figure 5.11 Scheme of a direct steam generator

Institute of the Russian Academy of Science) a prototype with a thermal power of 25 MW, a vapour pressure of 10 MPa and a temperature of 1000 °C was tested more than 20 years ago [30]. The only critical area seems to be that where combustion takes place due to the high temperatures developed.

For this purpose, the DLR prototype provided in sequence:

- an ignition chamber with a strong excess of hydrogen as a temperature moderating agent;
- a combustion chamber in which oxygen was added until reaching the stoichiometric ratio and water as a moderating agent of the temperature;
- an evaporation chamber in which the steam was made homogeneous.

In 2007, Carl W. Glasgow, Robert C. Olson and Edward J. Stockton presented a patent with a very different concept to obtain the same result [31]. In this case the flame is directed towards a mass of liquid water below and the reactor is not refrigerated from the outside. An interesting experiment was carried out on the prototype: the level of the liquid water mass was raised until the flame was completely immersed. A solution of this kind obviously entails a complete confinement of the flame and therefore minimises the problems of resistance of the materials to high temperatures. Unfortunately, as was to be expected, in these conditions the flame was not sufficiently stable.

To solve the problem of the too high temperature in the area close to the reactants' injection, another patent was presented in 2009 by G. Spazzafumo [32]. The concept is to provide for the premixing of water to the reactants in such a way that the heat is removed directly from the reaction zone thanks to the evaporation of the water. In practice, it is a question of replacing the nitrogen in the air, as a moderating agent of the temperature, with spray water which, thanks to the phase transition, should have a moderating effect even higher than that of nitrogen.

A similar concept was developed at Clean Energy Systems llc [33,34]. Starting from an oxycombustor fed with natural gas or biogas they can produce a stream of CO_2 and steam at high temperature (up to 1760 °C) without pollutant. In a specific configuration they can also provide a system to separate the stream of CO_2 and steam. They claimed that the oxycombustor is capable to operate also as a direct steam generator.

Recently at IVTAN restarted experimental activity with a smaller (150 kW) unit [35].

Table 5.3 shows a comparison between the basic characteristics of these steam generators and traditional boilers.

Converting hydrogen into electricity makes sense only if there is an actual excess of hydrogen compared to the local demand and a simultaneous electricity deficit in the local market. Although fuel cells are currently the most efficient system for, the economic aspect could overturn the comparison in favour of systems based on turbomachinery as the scale factor becomes important for the latter. It is therefore probable that the market will be divided between applications based on fuel cells for small and medium-sized users, up to a few tens of MW, distributed throughout the territory and large power plants based on turbomachinery [36]. In order for this scenario to occur that the efficiency of turbomachinery powerplants is not too lower than that of systems based on fuel cells.

Having a direct steam generator, one can think of carrying out internal combustion steam cycles, which is impossible using fuels other than hydrogen.

Hydrogen, if burned in air, is, in fact, a 'whatever' fuel in the sense that the result is a gas, devoid of any harmful substance apart from nitrogen oxides, but still an incondensable gas at room temperature. However, when burned with oxygen in a stoichiometric ratio, it is a 'unique' fuel since the reaction product is pure water in a gaseous state. This allows the adoption of steam thermodynamic cycles which can have characteristics that are decidedly different from normal steam thermodynamic cycles.

Table 5.3 Comparison between a traditional steam generator and a direct one [36]

Type	Traditional	Direct
Fuel	Gas, oil, coal	Hydrogen
Oxidant	Air	Oxygen
Equivalence ratio	>1.1	<1.1
Steam maximum temperature [°C]	535–600	1000–1500
Combustor specific thermal power [MW/m^3]	0.1–1	1,000,000
Specific volume of units [m^3/MW]	200–300	0.01
Heat transfer methodology	Allothermic	Mixing
Generator wall temperature [°C]	600–850	200 (water film cooling) 500–1000 (steam cooling)
Temperature difference of exhaust gases and lowest cycle temperature [°C]	100	No exhaust gases at all
Combustion efficiency	0.9–0.94	0.98–0.995
Starting time [s]	$>10^4$	<50
Transition time for 50% load changing [s]	500	<1
Pollutants	CO_2, NO_x, SO_x, Ash	None

The first thermodynamic cycles of this type were developed with a view to use hydrogen as a storage system for the energy produced in excess during empty night hours, weekends and, in general, periods with reduced demand [27]. In this case, in fact, the electrical energy can be easily converted, by electrolysis, into hydrogen and oxygen already compressed to values up to 20 MPa. However, hydrogen could also be produced from different sources and with different processes with or without co-production with oxygen. Gas compression could be provided during periods of lower electricity demand in order to reduce costs.

All cycles of this type have in common the use of at least one direct steam generator. This device is in fact capable of improving the efficiency of a steam plant for the following reasons:

- combustion is internal to the working fluid and therefore there are no stack losses; obviously there are losses connected to the discharge of the combustion products, but the discharge takes place at the condensation temperature;
- in a traditional steam system, the temperature of the steam is necessarily lower than the temperature of the walls of the heat exchangers and therefore the boiler limits the maximum temperature of the cycle; in a direct steam generator used as a mixture superheater, the temperature of the walls is much lower than that of the steam and, therefore, the maximum temperature of the cycle can be very high, with a consequent increasing in efficiency;
- superheating can be carried out near the turbine rather than in the boiler and this allows the adoption of a number of superheating systems higher than the 2–3 normally adopted in a traditional system;
- reaction time is very fast and transient in turbine operation can be reduced at seconds [37].

The most critical aspect is the correct dosage of the reagents: a non-stoichiometric mixture in fact involves the permanence, inside the vapour, of one of the two reagents in the gaseous phase and, therefore, the presence of non-condensable substances. If the excess reagent is hydrogen, the possible leakage of air in the depression section can cause the formation of explosive mixtures, while in the case of oxygen, corrosion problems increase.

Thanks to the direct steam generator, the upper limit of the cycle temperature is moved from the steam generator to the turbine. This means that one can think of steam cycles pushed up to the temperatures typical of turbogas cycles. Of course, it is necessary to adopt an appropriate cooling system for the blades. The gas turbines of H-series use steam in a closed circuit to cool the blade from the inside and therefore should be suitable. This system allows, therefore, to operate at 1500 °C as happens for turbogas systems [38].

In the framework of the Japanese project WE-NET the development of hydrogen-fuelled turbines capable to operate at 1700 °C was planned [39,40]. Actually Mitsubishi Hitachi Power Systems developed a J-series gas turbines that operate at a turbine inlet temperature of 1600 °C. However these turbines are an evolution of the G-series and therefore adopt air-cooling for combustors instead of steam cooling and are not suitable for a pure steam cycle. Furthermore when they will be put on the market (2025), they will be capable to operate with a mix of 30% hydrogen and 70% natural gas.

Although in the case of systems powered with hydrogen and oxygen the fluid is unique, in some cases we can speak of combined systems due to the existence of a topping cycle and a bottoming one. In fact, the steam generated, that is the combustion product of the topping cycle, is discharged to the outside, although this occurs downstream of the condenser rather than the chimney. And the heat still available downstream of the expansion of the topping cycle is transferred to the bottoming cycle. Different types of combined hydrogen/oxygen cycles have been studied: for the sake of brevity, only some of the most significant are considered here.

The Graz cycle [41], also called Jericha cycle, consists of two cycles (Figure 5.12), a topping Brayton (points 7-8-3-4), with discharge of the combustion products (points 10-13-15-16), and a bottoming Hirn (points 15-22-23-3-4-10-13).

The two cycles therefore have a part (points 9-3-4-5) in common with each other and with the combustion products. The fluid operating in the two cycles in fact joins downstream of the compression of the topping cycle (point 8) and of the expansion in the high-pressure turbine of the bottoming cycle (point 23) and separates again during the desuperheating phase going in part (point 6) towards the second desuperheater and partly (point 10) towards the low-pressure steam turbine. The combustion products, or more precisely a corresponding flow rate of steam, are also sent to the low-pressure steam turbine to be then condensed and discharged downstream of the condensate extraction pump.

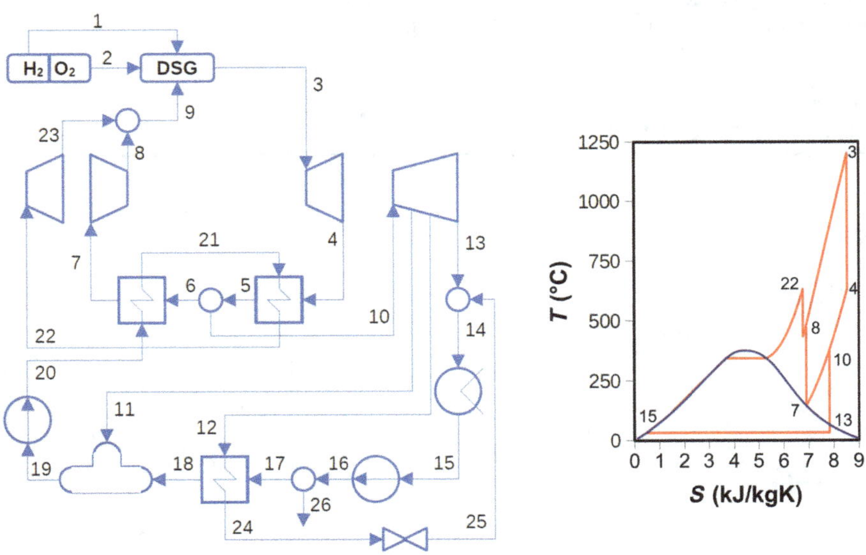

Figure 5.12 Graz cycle and plant scheme

Simulating the Graz cycle with the following data [42]:

- Brayton
  - maximum pressure 5 MPa;
  - compression ratio 12.5;
  - maximum temperature variable from 1200 °C to 1600 °C;

- Hirn
  - condensing pressure 0.005 MPa;
  - evaporation pressure 15 Mpa;
  - minimum temperature difference between hot fluid and cold fluid 20 °C;
  - pressure of the first steam tapping and of the mixture regenerator (degasser) 0.18 MPa;
  - pressure of the second steam tapping 0.03 MPa;

- turbine efficiency 90%;
- compressor efficiency 88%;
- efficiency of the pumps 75%.

the results reported in Table 5.4 have been obtained.

In the framework of the Japanese WE-NET project, the Graz cycle has been modified in various ways in order to achieve even better performance. This has led to the cycle schematically represented in Figure 5.13, called bottoming reheat

Table 5.4 Results of Graz cycle for different maximum temperatures

Maximum temperature	1200 °C	1400 °C	1600 °C
Ratio between flows from direct steam generator and high-pressure turbine	0.47	0.50	0.52
Ratio between flows at compressor and high-pressure turbine	1.94	1.39	1.02
Efficiency (based on HHV)	56.2%	57.8%	59.0%

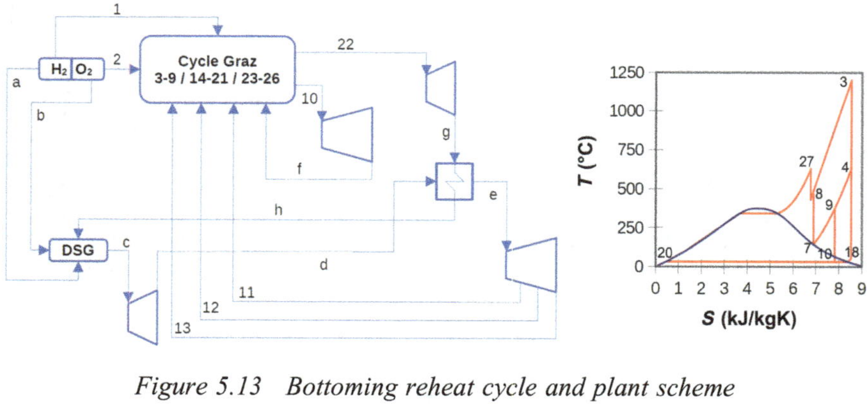

Figure 5.13 Bottoming reheat cycle and plant scheme

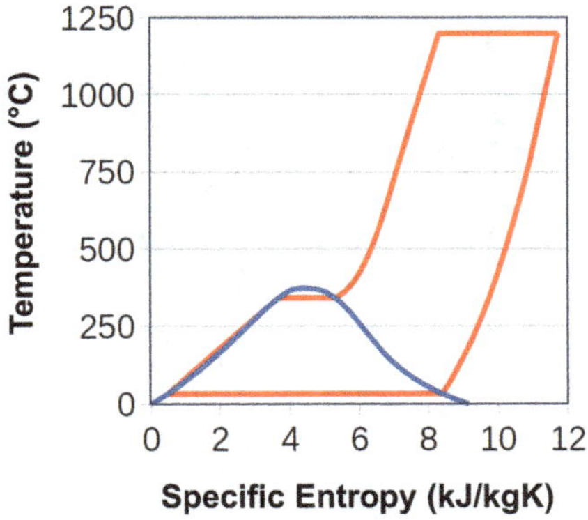

Figure 5.14 Spazzafumo cycle

cycle. Anticipating the low-pressure expansion of the bottoming cycle (point e) with respect to that of the combustion products (point 10), two independent cycles are obtained. This takes place at the expense of a certain system complication since the two cycles are made in two almost entirely separate circuits so that a direct steam generator, two turbines and a surface superheater must be added [43]. With reference to hydrogen HHV, the Graz cycle allows to obtain an efficiency of up to 60% at 1700 °C, while the bottoming reheat cycle allows an increase of about 2% in absolute terms.

A critical element of these systems could be the steam compressor as it operates near the saturation curve, but careful design should avoid any malfunction. In any case, even a compressor efficiency of 75% instead of 88% as in simulation presented would lead to a reduction in efficiency of less than 1%. On the other hand, the situation of desuperheaters is more problematic since their maximum temperature is higher than 1000 °C.

In Figure 5.14, the Spazzafumo cycle is shown [42,44–47]. It is a steam cycle which however has some characteristics in common with the Ericsson cycle. Both cycles have, in fact, two isotherms (E-F and G-A) and two isobars (B-E and F-G). And the area between the abscissa axis and the F-G transformation represents a totally recoverable amount of heat, just like in an Ericsson cycle.

However, there are some substantial differences:

- the presence of C-D vaporisation, which can only be avoided in hypercritical cycles;
- the presence of G-A condensation which is not only isotherm but also isobaric;
- A-B compression, which corresponds to the compression isotherm of the Ericsson cycle, but in reality it is not isotherm, although it is very close to an isotherm since a liquid is pumped.

While the efficiency of the Ericsson cycle is equal to the efficiency of a Carnot cycle operating between the same extreme temperatures, the efficiency of the Spazzafumo cycle cannot reach such a level unless we consider hypercritical cycles.

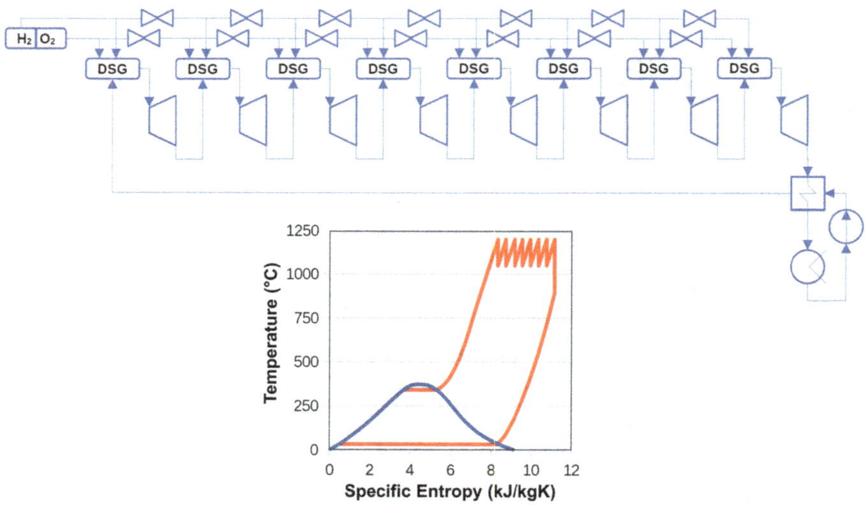

Figure 5.15 Spazzafumo plant scheme and thermodynamic cycle (real)

The limit cycle is followed by a real fluid (water-vapour-gas) which evolves into an ideal machine capable of exchanging heat during an isothermal expansion. Since a real machine is not capable of exchanging significant amounts of heat during an expansion, to obtain an isothermal expansion it would be necessary to continuously inject, during the expansion, high-temperature steam generated directly by the combustion of hydrogen and oxygen. Also this solution is not practicable and therefore the real configuration foresees a para-isotherm expansion realised by alternating mixing re-heaters and turbines (Figure 5.15).

This solution involves an overall non-negligible flow rate variation along the isotherm. It follows that the heat available at the end of the expansion is in excess of the quantity required by the heating and evaporation of the water and superheating of the steam. In order to use all the available heat and reduce the consumption of hydrogen, it is possible to think about preheating the reactants (hydrogen and oxygen) before sending them to the combustion chamber.

Furthermore, since the para-isotherm expansion involves a permanence of the steam leaving the turbines at high-temperature values, a final adiabatic expansion is required to obtain a not excessive temperature at the inlet of the surface regenerator allowing the use of conventional materials.

The variation in flow rate and the introduction of adiabatic expansion have a negative effect on the cycle yield. However, in the same assumptions made for the two previous cycles, the efficiency is around 65%, variable in relation to the number and the characteristics of expansions.

In addition to the cycles related to systems using only hydrogen, there are cycles suitable for systems that use both fossil fuels and hydrogen. Such a combined use can take place either by mixing or separately.

The mixing of hydrogen with other fuels and subsequent combustion in air does not involve significant changes to the Brayton cycle, but only an improvement in combustion with fewer unburned emissions.

The case is different when hydrogen is used 'in series' with fossil fuel to improve the performance of a steam cycle. In this case the combustion of the fossil fuel takes place in air, while that of hydrogen in stoichiometric oxygen. These cycles in which hydrogen and oxygen are used to improve the performance of a traditional steam plant are called 'attached' cycles [36].

In Figure 5.16, a traditional steam plant is shown on the left. Assuming classical values for the design parameters ($p_{\max} = 18$ MPa, $T_{\max} = 550$ °C, $\eta_P = 0.7$, $\eta_T = 0.86$), an efficiency of the thermodynamic cycle of just over 42% is obtained. By replacing the two superheating with just one hydrogen/oxygen stoichiometric mixing superheater and raising the superheating temperature up to 800 °C, an attached system is obtained (shown on the right), which achieves an efficiency close to 46%.

In such a plant, the hydrogen is used only in the superheating phase. Therefore, the heat generated by its combustion is mainly used at a temperature much higher than the average temperature of the cycle. This means that the efficiency of hydrogen use, or marginal efficiency, is higher than the efficiency of the entire cycle. In fact, by comparing the two powerplants keeping constant the thermal power supplied by fossil fuel, the second requires an additional thermal power and offers an additional mechanical power. The flow of hydrogen supplied is responsible for both these additional power and the ratio between these two quantities represents the marginal efficiency:

$$\eta = \frac{W_{mech} + \Delta W_{mech}}{W_t + \Delta W_t} \Rightarrow \eta_{marg} = \frac{\Delta W_{mech}}{\Delta W_t} \tag{5.11}$$

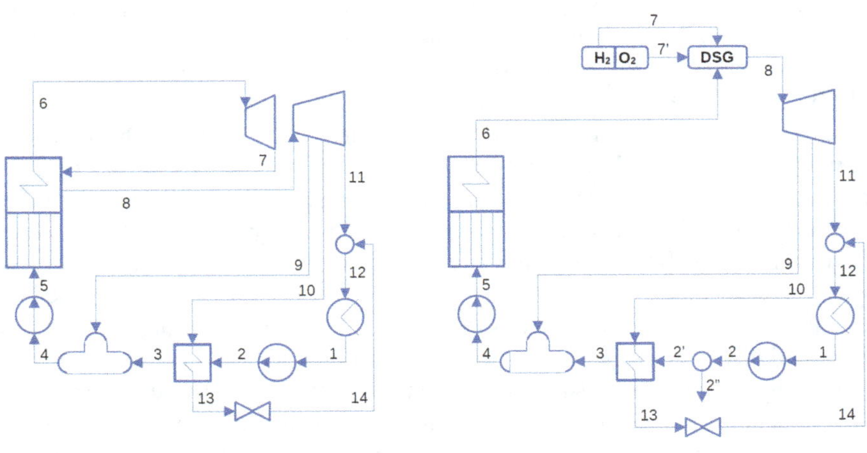

Figure 5.16 Traditional Hirn cycle (left) and 'attached' option (right)

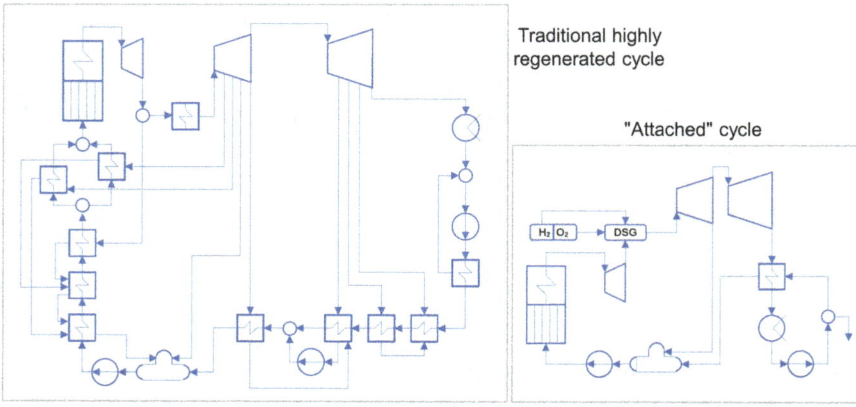

Figure 5.17 Comparison between the complexity of a traditional highly regenerated cycle and the simplicity of an 'attached' cycle

which, in the specific case, is equal to 55%, a value that can be further increased, even above 60%, by adding a second superheater and/or pushing the superheating temperature to higher values.

In addition to the high hydrogen conversion efficiency, a goal of cycles of this type may be the simplification of the system: Figure 5.17 shows on the left the scheme of a traditional powerplant with eight taps and on the right the scheme of the same powerplant modified by eliminating the taps and replacing the surface reheating with a mixing one based on direct steam production. Thanks to the mixing superheating technique, a second reheating could also be introduced upstream of the low-pressure turbine.

At low superheating temperatures, this simplification is not favourable in terms of efficiency: this can be explained by the low temperature at the end of expansion, which involves a low temperature of regeneration. Increasing the superheating temperature also increases the temperature at the end of the regeneration and the efficiency of the second scheme exceeds that of the first, up to approaching 60%.

The mixing re-heating technique could also be used in the repowering of existing powerplants when there is the need to replace the boiler. In this case also, the medium pressure turbine could be replaced. The high pressure turbine operates at the same temperature of the reference case to avoid eventual stress cause by high temperature and high pressure at the same time. The low-pressure turbine operates at low temperature as occurs in the reference case and with the same tappings, so that there is no need for changes in this part of the powerplant. The new medium pressure turbine, however, operates at very high temperature using a mixing re-heater and discharges superheated steam at a still high temperature. Since the inlet temperature in the low-pressure turbine is higher than the temperature of the water at the inlet of the regenerative section consisting of the desuperheaters, it is possible to take advantage of the entire desuperheating at the outlet of the medium pressure turbine leaving unchanged the flow rates of the taps. The increased regeneration

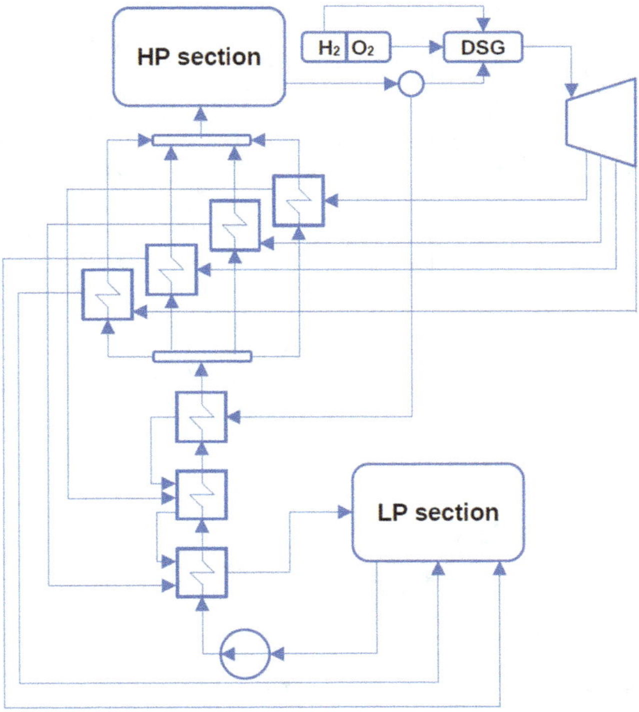

Figure 5.18 Modified medium pressure section

capacity due to desuperheating entails a modification for the medium pressure section of the plant as shown in Figure 5.18. Since the scheme of all the other regenerators remains unchanged, this desuperheating contributes to raising the end of regeneration temperature.

The entropic diagram that follows this definition of the thermodynamic cycle is shown in Figure 5.19 and, for the most part, follows that relating to the reference system. It detaches from it only in the area affected by re-heating, expansion in the medium pressure turbine and subsequent desuperheating, that is, the 'Brayton' part of the cycle. The overall thermodynamic cycle looks a lot like a Graz cycle, but is devoid of the vapour compressor.

In the bar diagrams in Figure 5.20, the theoretical performances related to the repowering of an existing plant are reported in terms of efficiency, temperature at the end of regeneration and power [42]. The bars respectively indicate the reference case and the repowered cases with superheating temperatures of 800 °C (BTr), 1200 °C (MTr) and 1600 °C (LTr).

It is noted that, already at 800 °C of superheating temperature, interesting performances can be obtained. This is due to the good behaviour of the system from the point of view of regeneration. Having kept the regenerative capacity of the taps intact, combined with the availability of high-temperature heat, even allows, in the

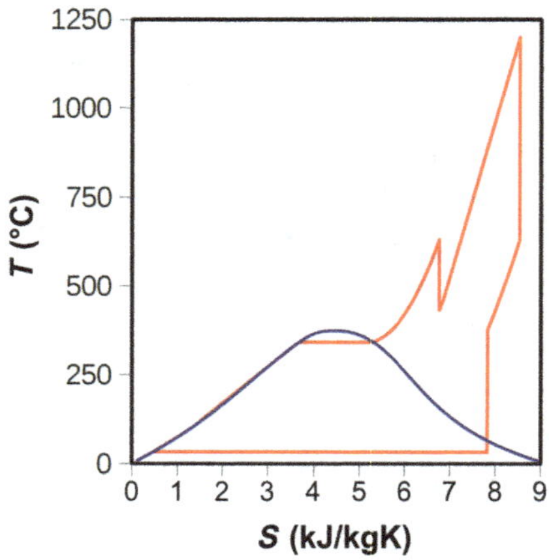

Figure 5.19 Repowered Hirn cycle

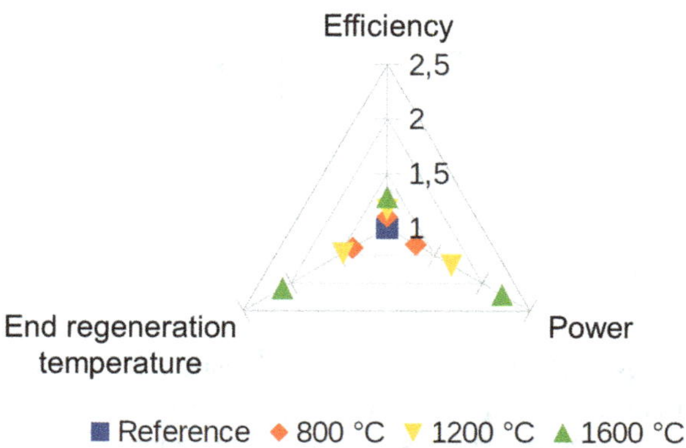

Figure 5.20 Repowered system: efficiency, regeneration end temperature and generated power

case of an superheating temperature of 1600 °C, to push regeneration until the start of the second superheating. This system is therefore powered only by hydrogen and oxygen and the boiler is entirely replaced by the regenerator and the re-heater with mixture. The repowering obtained ranges from 30% to 120%, with an increase in plant efficiency ranging from 18% to almost 50% in relative terms.

5.1.5 Catalytic combustion

For residential applications, the required temperatures are comparatively low and a type of combustion without flame and without generation of nitrogen oxides may be of interest. In fact, hydrogen, in the presence of air, can burn in the absence of flame from temperatures slightly above the ambient temperature up to about 500 °C. To carry out this process it is necessary the presence of a suitable catalyst,

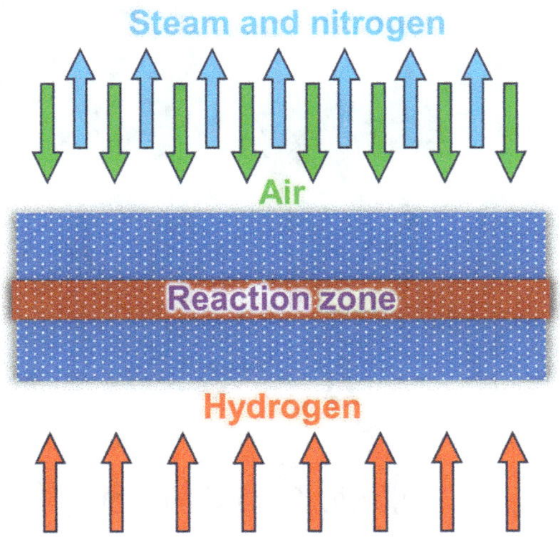

Figure 5.21 Scheme of the catalytic combustion process

usually platinum, dispersed inside a porous matrix in such a way as to have a very extensive reaction surface. Hydrogen and air are supplied separately on the two opposite faces of the porous structure, as shown in Figure 5.21, and they are mixed to form the combustible mixture in a reaction zone inside the porous structure characterised by the presence of the catalyst. The separation between the hydrogen flow and the air is total and, therefore, there is no risk that the reaction goes back to the hydrogen tank since it would not find combustion air.

Temperature and reaction rate can be controlled by regulating the hydrogen flow and, therefore, there are no nitrogen oxides among the reaction products, but only water vapour.

Obviously, at present, the use of hydrogen for purely thermal applications at low temperatures is not reasonable; however, in the long term, it could become so if hydrogen production technologies prove to be more efficient than electricity production technologies with the same primary source available.

5.2 Fuel cells

The fuel cell is a chemical reactor in which, through an oxidation-reduction process with ion transfer, the direct conversion of the chemical energy of the fuel into electrical energy takes place.

The theoretical principles on which this device is based date back to the time of the first studies in the electrochemical field and, in particular, to Sir Humphry Davy who, in 1801, carried out the electrolysis process and to his student, Michael Faraday who, in 1834, published the laws of electrolysis. A few years later, in 1839, Sir William Grove found that the process was reversible and created the first fuel cell, which he called the 'gas voltaic battery' (Figure 5.22).

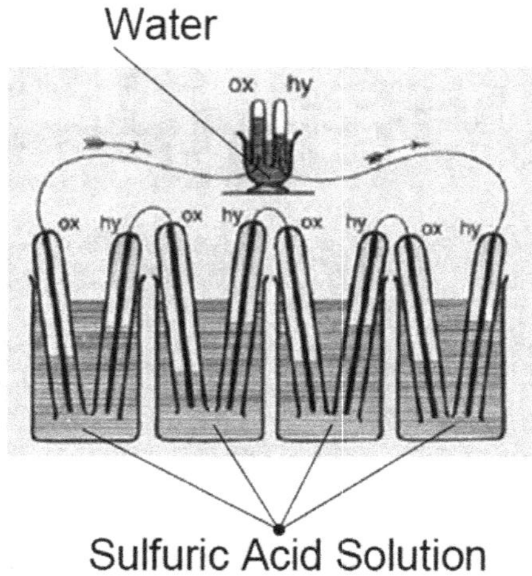

Figure 5.22 Original drawing of gas voltaic battery [48]

At that time the fuel cell had no practical application. Therefore, apart from some sporadic research, it fell into oblivion for about a hundred years. Only in the second half of the 1950s there was a significant revival of interest with the construction of the first membrane fuel cell by W. Thomas Grubb and the construction of an alkaline stack made up of 40 fuel cells for a total power of 5 kW by Francis Thomas Bacon. These cells operated at a temperature of 200 °C–240 °C and at a pressure of 4-5.5 MPa and thanks to these characteristics the voltage was 0.78 V at 800 mA/cm^2).

Although the first application in those years was a 15 kW alkaline stack for an Allis-Chalmers tractor, the sector in which it had the greatest development was that of aerospace missions in the following decade. One of the reasons is undoubtedly identifiable in the availability of hydrogen and oxygen on board. In fact, although the fuel cell is theoretically capable of using any oxidisable substance as a fuel and any reducible substance as an oxidant, a fundamental characteristic is the reactivity of the usable substances and hydrogen is decidedly more reactive than traditional fuels.

Figure 5.23 shows a schematic of an acid electrolyte fuel cell, like that of Grove. The fuel is sent to the anode compartment in which it undergoes the oxidation reaction by releasing electrons that accumulate on the electrode. Through an external user circuit these electrons reach the cathode to participate in the oxidant reduction reaction that feeds the cathode compartment. The two electrodes are separated by an electrolyte through which they diffuse ions which can be negative or positive, depending on the type of electrolyte, basic or acid.

The fundamental components of a fuel cell are therefore the electrodes and the electrolyte.

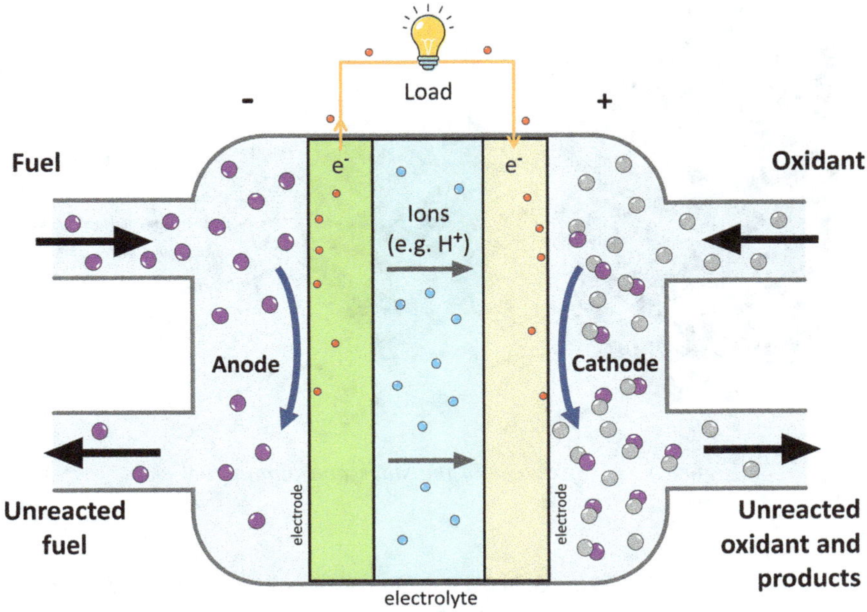

Figure 5.23 Fuel cell working principle

The electrodes must be characterised by good electronic conductivity and mechanical and corrosion resistance. Generally, the fuel is gaseous and poorly miscible with the electrolyte: for this reason the electrode must have a porous structure (Ø=10–100 μm) which increases the contact surface between the different phases. The electrode is therefore divided into three zones: two external ones which are inactive due to the clear predominance, respectively, of electrolyte and gas, and a central one in which the electrode surface is covered with a thin layer of electrolyte, sufficient to favour the transport of both dissolved reagents and ionic species. It follows the usefulness of thin electrodes that allow to reduce the ohmic losses and the dimensions of the inactive zones. Furthermore, the electrodes must have good catalytic properties to facilitate the kinetics of the electrode half reactions: while at high temperatures the electrode material itself has an adequate catalytic activity, at low temperatures the use of specific and very expensive materials could be required and therefore the catalyst is used in minimal quantities limited to the reaction zone. The catalyst should possess qualities of stability and duration and be insoluble in the electrolyte. Finally, in the case of liquid electrolyte, the electrodes must be covered with a coating capable of generating capillary forces of sufficient intensity to prevent flooding.

Conversely, the electrolyte must have bad electronic conductivity and good ionic conductivity. It must also hinder the direct diffusion of the reagents.

The power provided by the single fuel cell is very low and for this reason multiple fuel cells are stacked together. The geometry of the stack is generally of

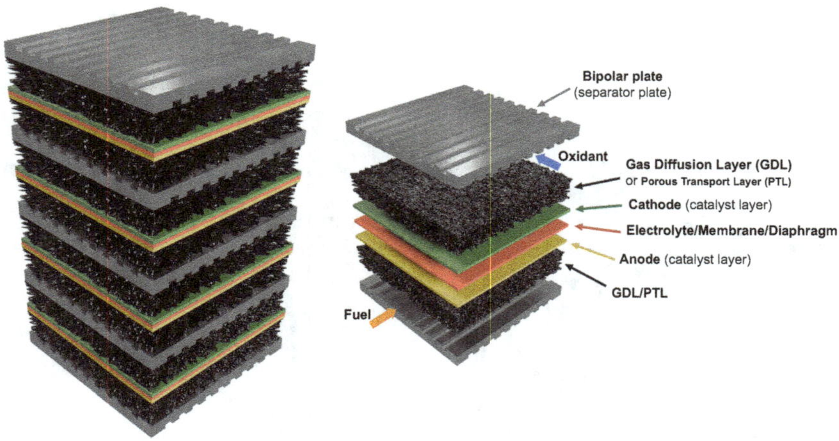

Figure 5.24 Structure of the stack and planar fuel cell

the planar bipolar type (Figure 5.24). The contiguous cells are separated by a plate which acts as an electrical interconnection and on which the fuel and oxidiser distribution pipes are located. In contact with the separator plate there are the anode of one cell and the cathode of the other. Finally, the electrolyte is located between the two electrodes of the same cell.

With reference to the path of the gases along the electrodes, a stack can be co-current, counter-current, cross-flow, serpentine and, in the case of circular cells, also spiral. With respect to the cells, the gas supply and exhaust manifolds can be internal or external and, in the latter case, they can possibly be integrated in the separator plate.

Although the bipolar planar scheme is very simple and offers a reduced electrical resistance, a tubular geometry was initially developed for the high-temperature cells which has the advantage of a better seal against gases, a rather critical feature when the temperature range is wide. The single cell is made up of four coaxial cylindrical layers: from the inside to the outside there is a support tube which also acts as an air collector, the cathode, the electrolyte and the anode. The individual cells are arranged in arrays, electrically interconnected and surrounded by fuel. In order to increase the density, tubular cells with a rectangular section have also been designed.

5.2.1 Energy balance of the fuel cell

Let's initially consider the case of a reversible system. Indicating with U the internal energy, with p the pressure and with v the volume, the enthalpy change H of a system is:

$$\Delta H = \Delta U + p \cdot \Delta v \tag{5.12}$$

and, indicating heat with Q and work with L, replacing the expression of the internal energy given by the first law of thermodynamics, we obtain:

$$\Delta H = Q - L + p \cdot \Delta v \tag{5.13}$$

Considering the fuel cell as an isothermal and isobaric system, indicating with T the temperature, with S the entropy and with L_e the electrical work, we can write:

$$\Delta H = T \cdot \Delta S - L_e - p \cdot \Delta v + p \cdot \Delta v = T \cdot \Delta S - L_e \tag{5.14}$$

Finally, indicating with G the free energy:

$$\Delta G = \Delta H - T \cdot \Delta S = -L_e \tag{5.15}$$

Considering a generic chemical reaction in which A_i are the reactant species and B_j the species produced, respectively with stoichiometric coefficients α_i and β_j:

$$\sum \alpha_i A_i \Rightarrow \sum \beta_j B_j$$

the enthalpy change connected to the reaction is given by the difference between the sum of the enthalpies of the products and the sum of the enthalpies of the reactants. The change in entropy is similarly obtained. Indicating with C_p the thermal capacity at constant pressure and assuming it not dependent on the pressure, we can write:

$$\Delta H(T) = \Delta H^\circ + \int \Delta C_p(T) dT$$

$$\Delta S(T) = \Delta S^\circ + \int \frac{\Delta C_p(T)}{T} dT$$

and therefore also obtain the free energy change.

Ultimately, the fuel cell is fed with a continuous flow of fuel associated with a certain chemical energy ΔH which is converted generating a quantity of electrical energy ΔG and a quantity of thermal energy $T\Delta S$. While the enthalpy and free energy changes connected to the oxidation of a fuel are always negative, the entropy variation can be either negative (e.g. in the case of direct oxidation of hydrogen) or positive (e.g. in the case of direct oxidation of solid carbon) [49]. In the first case, it is necessary to dispose of the heat generated, while in the second case the system absorbs heat from the environment and converts it into electricity [50].

The ratio between the energy available to be converted into electrical work, i.e. the free energy ΔG, and the chemical energy ΔH supplied to the cell is called thermodynamic (maximum) efficiency:

$$\eta_t = \frac{\Delta G}{\Delta H} \tag{5.16}$$

Under standard conditions the change in enthalpy related to the reaction of formation of water in the liquid state is 285.8 kJ/mol and the corresponding change in free energy is 237.1 kJ/mol. Therefore a reversible fuel cell powered by

Table 5.5 Enthalpy and free energy changes, reversible voltages and thermal efficiencies for some reactions (adapted from [51])

Reaction	ΔH° (kJ/mol)	ΔG° (kJ/mol)	E° (V)	η_t (%)
$H_2 + \frac{1}{2}O_2 \Rightarrow H_2O_{(l)}$	−285.234	−237.304	1.23	83
$C + \frac{1}{2}O_2 \Rightarrow CO$	−110.510	−137.343	0.71	124
$C + O_2 \Rightarrow CO_2$	−393.693	−394.572	1.02	100
$CO + \frac{1}{2}O_2 \Rightarrow CO_2$	−283.057	−257.230	1.33	91
$CH_3OH + \frac{3}{2}O_2 \Rightarrow CO_2 + 2H_2O_{(l)}$	−726.941	−702.829	1.21	97
$CH_4 + 2O_2 \Rightarrow CO_2 + 2H_2O_{(l)}$	−891.241	−818.782	1.06	92

hydrogen and pure oxygen has a thermal efficiency of about 0.83 considering HHV. As stated above, the thermal efficiency of a solid carbon and oxygen fuel cell is higher than unity. Table 5.5 shows enthalpy and free energy changes, reversible voltages and thermal efficiencies for some reactions.

The free energy change connected to the reaction can be expressed as:

$$\Delta G = \Delta G^\circ + R \cdot T \cdot \ln \frac{\prod a_{B_j}^{\beta_j}}{\prod a_{A_i}^{\alpha_i}} \tag{5.17}$$

with R gas universal constant, a with activity of the single chemical species and G° standard free energy (at atmospheric pressure) which, indicating with K the equilibrium constant, results:

$$\Delta G^\circ = -R \cdot T \cdot \ln K \tag{5.18}$$

Denoting with n the moles of electrons per mole of the reactant that makes them available, with F the Faraday constant (equal to the charge of one mole of electrons, 96,487 C/mol) and with E the cell potential, the electrical work is also given by:

$$L_e = n \cdot F \cdot E \tag{5.19}$$

Therefore, it results:

$$E = -\frac{\Delta G}{n \cdot F} = E^\circ + \frac{R \cdot T}{n \cdot F} \cdot \ln \frac{\prod a_{A_i}^{\alpha_i}}{\prod a_{B_j}^{\beta_j}} \tag{5.20}$$

with:

$$E^\circ = -\frac{\Delta G^\circ}{n \cdot F} = \frac{R \cdot T}{n \cdot F} \cdot K \tag{5.21}$$

This relationship is known as the Nernst equation and the potential E represents the open-circuit cell voltage. For the pressures at which the fuel cells operate, it is generally permissible to replace the activities with the partial pressures of the gaseous species. It follows that:

$$E = E^{\circ} + \frac{R \cdot T}{n \cdot F} \cdot \ln \frac{p_{H_2} \cdot \sqrt{p_{O_2}}}{p_{H_2O}} \tag{5.22}$$

and finally passing to the molar fractions:

$$E = E^{\circ} + \frac{R \cdot T}{2 \cdot n \cdot F} \cdot \ln p + \frac{R \cdot T}{n \cdot F} \cdot \ln \frac{x_{H_2} \cdot \sqrt{x_{O_2}}}{x_{H_2O}} \tag{5.23}$$

The standard potential at 298 K for a cell fed with hydrogen and oxygen is equal to 1.229 V if the water produced is in the liquid state, while it drops to 1.18 V if the water produced is in the gaseous state since 0.049 V correspond to the free energy change in standard conditions connected to the evaporation of water. As the temperature increases, this value drops by approximately 0.02 V every 100 °C.

The second term of the Nernst equation shows how a pressure increase is favourable to the open-circuit potential. The concentrations of the reagents are also of fundamental importance and the best performances are obtained with pure hydrogen and oxygen. The presence of other substances leads to a reduction in potential, particularly in the case of the products, as the denominator term increases. In the case of diluents, however, the effect is only to reduce the values of the molar fractions and, in this case, the consequence could be either a reduction or an increase in the open-circuit potential. The role of each anodic gas is reported in Table 5.7 (Section 5.2.2). For example, in phosphoric acid fuel cells (PAFC), the presence of nitrogen into the air would lead to a reduction in both the mole fraction of oxygen and that of the water produced, which results in an increase of the argument of the logarithm and, consequently, of the value of the open-circuit potential.

High-temperature cells are also capable of converting carbon monoxide and hydrocarbons and therefore the Nernst equation should also contain the terms relating to the reactants and products of the related conversion reactions. However, the progressive consumption of hydrogen and the kinetics favour, respectively, the water gas shift:

$$CO + H_2O \Rightarrow CO_2 + H_2 \quad \Delta H^{\circ} = -40.8 \ \text{kJ/mol} \tag{5.24}$$

and reforming reactions:

$$CH_4 + H_2O \Rightarrow CO + 3H_2 \quad \Delta H^{\circ} = 206.3 \ \text{kJ/mol} \tag{5.25}$$

and therefore the direct conversion reactions are of little significance.

Moving on to a real system, it is necessary to take into account the irreversibility which obviously have a negative effect and which result in a progressive voltage drop, compared to that given by the Nernst equation, as the current density increases. This voltage reduction is called polarisation and is attributable to three different components:

- ohmic polarisation, due to the internal resistance of the fuel cell;
- activated polarisation, due to the energy required to break the molecular bonds of the reactants and those between the reactant and catalyst atoms;

- concentrated polarisation, due to the limitation of the mass transport speed which manifests itself in the slow diffusion of the gases into the electrodes and of the ions into the electrolyte: the consumption of the reactants and ions in the vicinity of the electrode generates, inside the cell itself, a concentration cell which generates a counter-electromotive force.

The internal resistance R consists of a component relating to the resistance to the passage of ions (essentially inside the electrolyte) and a component relating to the resistance to the passage of electrons (essentially within the electrodes and interconnections).

Since the electrolyte also behaves according to Ohm's law, the ohmic polarisation varies linearly with the current. To make this polarisation uniform to the other two which are defined as a function of the current density i, a quantity defined ASR (area specific resistance) and corresponding to the product of the resistance times the cell surface is used:

$$\Delta V_o = I \cdot R = \mathrm{i} \cdot ASR \tag{5.26}$$

The dominant effect at low temperature is that of the ionic resistance of the electrolyte which however decreases as the temperature increases, inversely to what happens for the resistance to the passage of electrons in the electrodes and in the interconnections. Resistance is a function of numerous parameters related to geometry, materials, manufacturing techniques and operating conditions and is a fundamental parameter because in the normal operating range the ohmic polarisation is the dominant one.

Similarly to what happens in chemical reactions, also in electrochemical ones there is an energy barrier that must be overcome by the reacting substances. Despite the complexity of the reaction mechanism, for electrochemical reactions that have an activated bias higher than 50–100 mV, the dependence of the activated bias on the current density is expressed with good approximation by means of an empirical relationship called Tafel's equation:

$$\Delta V_a = \frac{R \cdot T}{\alpha \cdot n \cdot F} \cdot \ln \frac{i}{i_0} \tag{5.27}$$

in which α is the electron transfer coefficient, i the current density and i_0 the exchange current density.

The Tafel equation is therefore linear in a logarithmic diagram (Figure 5.25) and the slope of the line is called the Tafel slope.

The concentrated polarisation is due to the limitation of the mass transport speed which manifests itself in the slowness of the diffusion of the gases in the electrodes and of the ions in the electrolyte. The proceeding of the reaction in fact creates an inhomogeneity of concentration caused by the consumption of the reactants and ions and by the simultaneous generation of products in the vicinity of the electrode: this fact is comparable to the creation of a concentration cell which originates a counter electromotive force. In general, due to the different diffusivity between hydrogen and vapour, the limiting factor is the removal of the latter.

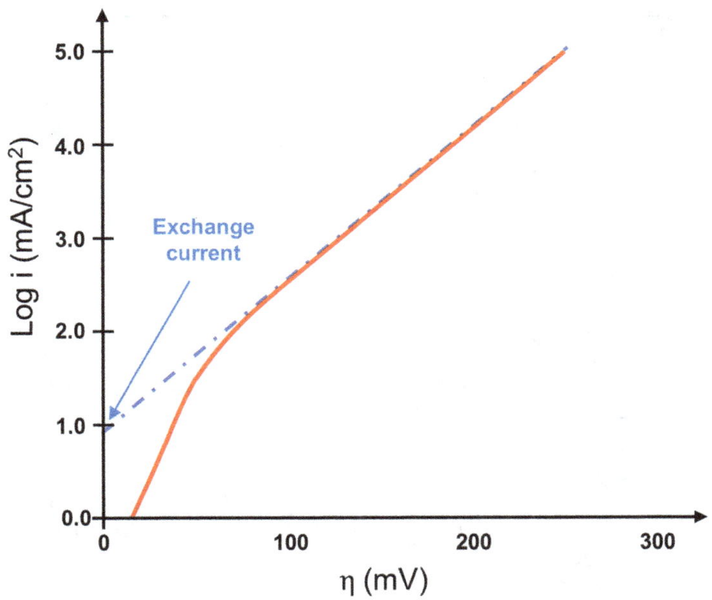

Figure 5.25 Activated polarisation (adapted from [50])

The effect of the concentrated bias is however negligible as long as the cell does not work near the limiting current density i_L, a zone in which the voltage drop rises very rapidly, effectively blocking the direct conversion process.

The trend of the ion concentration inside the electrolyte can be approximated with an 'equivalent profile' as indicated in Figure 5.26. This corresponds to define a diffusion layer inside which there is a linear variation of the concentration and outside which there is no variation of the concentration. The relationship that links the thickness of the diffusion layer to the current density is given by Fick's first law of diffusion:

$$i = \frac{n \cdot F \cdot D \cdot (c_i - c_b)}{d} \tag{5.28}$$

where:

- D is the diffusion coefficient of the reactant ions;
- ci is the concentration of the reactant ions at the electrolyte/electrode interface;
- c_b is the concentration of the reactant ions in the central area of the electrolyte;
- d is the thickness of the diffusion layer.

The limit value of the current density can be obtained by reducing to zero the concentration on the electrode surface:

$$i_L = \frac{n \cdot F \cdot D \cdot c_b}{d} \tag{5.29}$$

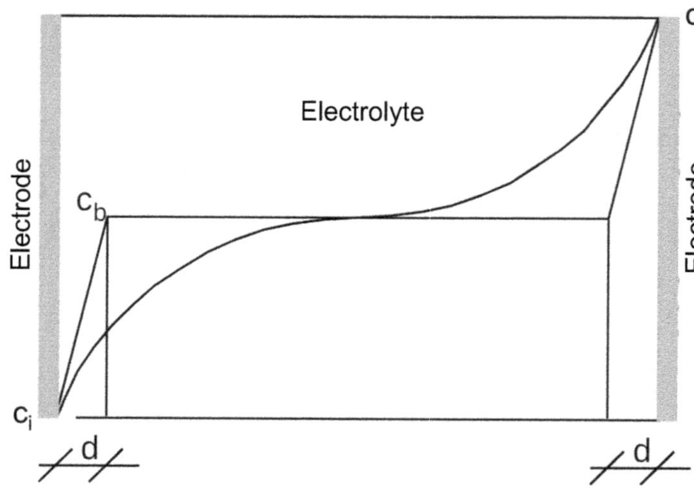

Figure 5.26 Trend of the concentration of ions inside the electrolyte (equivalent profile)

Combining the last two relations, we obtain:

$$\frac{c_i}{c_b} = 1 - \frac{i}{i_L} \tag{5.30}$$

The Nernst equation applied to the electrolyte in open-circuit conditions is:

$$E = E^\circ(T) + \frac{R \cdot T}{n \cdot F} \cdot \ln c_b \tag{5.31}$$

while in closed circuit conditions, we have:

$$E = E^\circ(T) + \frac{R \cdot T}{n \cdot F} \cdot \ln c_i \tag{5.32}$$

The back electromagnetic force (emf) generated by a concentration cell is:

$$\Delta V_c = \frac{R \cdot T}{n \cdot F} \cdot \ln \frac{c_i}{c_b} \tag{5.33}$$

from which:

$$\Delta V_c = \frac{R \cdot T}{n \cdot F} \cdot \ln\left(1 - \frac{i}{i_L}\right) \tag{5.34}$$

And therefore the cell voltage is:

$$V = E - \Delta V_o - \Delta V_a - \Delta V_c$$

$$= E - ASR \cdot i - (a + b \cdot \ln i) - \frac{R \cdot T}{n \cdot F} \cdot \ln\left(1 - \frac{i}{i_L}\right) \tag{5.35}$$

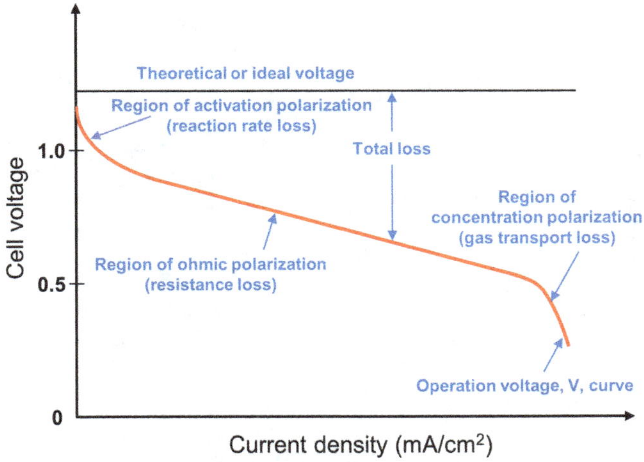

Figure 5.27 Overall effect of polarisations (adapted from [50])

and Figure 5.27 shows the overall effect of the polarisations on the voltage value as the current density varies. The activated polarisation, being a logarithmic function, rises rapidly at low current densities and then stabilises, while the concentrated polarisation has an inverse trend. The result is that the dependence of the overall polarisation on the current density is, by usual values of the latter, linear.

An approximation that is often made starts from the consideration that the concentrated polarisation is negligible far from the limiting current density. For this reason, since the operating conditions suggested by the manufacturers are always far from this limit, the term relating to concentrated polarisation is often omitted and the previous expression is simplified to:

$$V = E - ASR \cdot i - (a + b \cdot \ln i) \tag{5.36}$$

Furthermore, the activated polarisation is strongly linked to the temperature and becomes almost negligible at high temperatures. Therefore, in the case of high-temperature fuel cells, the following is also used:

$$V = E - ASR \cdot i \tag{5.37}$$

Ultimately, the electrical energy (in direct current) produced by the cell is:

$$L_e = L_{rev} - I \cdot \Delta V = n \cdot F \cdot V \tag{5.38}$$

The temperature negatively affects the value of the open-circuit voltage, but has an overall beneficial effect on the polarisations. In fact, in addition to the already highlighted scarce influence of the activated polarisation, at high temperature the ohmic polarisation is lower due to the reduction of the ionic resistance of the electrolyte which, generally, predominates on the increase of the resistance of the electrodes and of the interconnections. The effect of temperature is qualitatively represented in Figure 5.28.

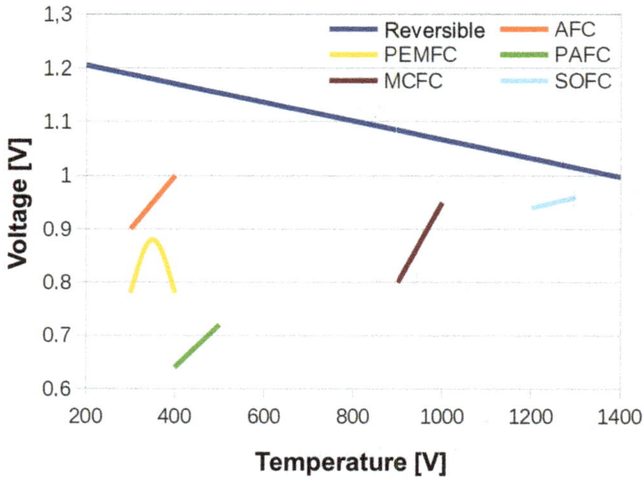

Figure 5.28 Temperature influence

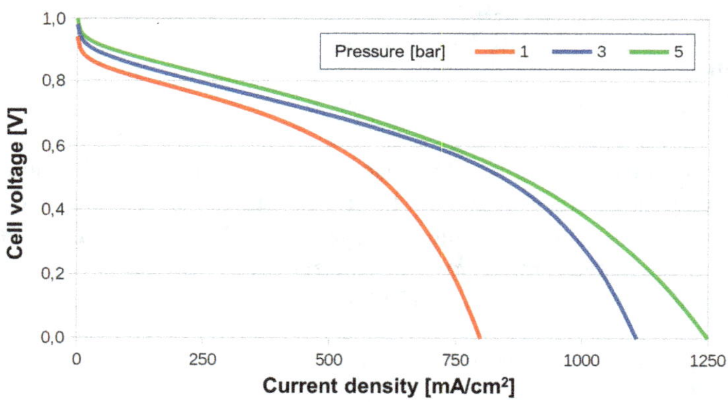

Figure 5.29 Polarisation curves at different pressures

Figure 5.29 shows the polarisation curves at different operating pressures for a proton exchange membrane fuel cell (PEMFC) fed with hydrogen and air and operating at 70 °C with fuel utilisation equal to 0.83. As the operating pressure increases, the cell potential increases and the operating range extends towards higher values of current density. However, this effect quickly diminishes as the pressure increases and, in the case taken as an example, the differences in the passage from 0.3 to 0.5 MPa are almost negligible.

The global efficiency of a fuel cell is given by the product of three different efficiencies:

$$\eta_{FC} = \eta_t \cdot \eta_V \cdot \eta_I \tag{5.39}$$

In addition to the thermal efficiency already encountered in the discussion of the reversible system, there are:

- η_V voltage efficiency, given by the ratio between the effective cell voltage and the Nernst cell voltage:

$$\eta_V = \frac{V}{E} \tag{5.40}$$

- η_I current efficiency, given by the ratio between the effective current and the theoretical current corresponding to the fuel flow rate:

$$\eta_I = \frac{I}{n \cdot F \cdot \dot{m}_f} \tag{5.41}$$

The latter term can also be expressed as the product between the Faraday efficiency η_F and the fuel utilisation φ_f. Faraday efficiency takes into account the different reactions that can lead to the oxidation of a fuel and is, therefore, unitary if only one type of reaction is possible, while it is less than unity if different reactions can occur with different value of the number of electrons involved. However, it does not differ much from unity and, therefore, we can set $\eta_I \approx \varphi_f$. The fuel utilisation is given by the ratio in moles between the flow rate of fuel consumed and that of fuel fed into the cell:

$$\varphi_f = \frac{\Delta \dot{m}_f}{\dot{m}_f} \tag{5.42}$$

The fuel utilisation must take into account not only hydrogen, but also other substances that can originate hydrogen, through water gas shift (5.24) or reforming (5.25) reactions, like carbon monoxide and methane. Therefore, it is expressed through different formulations depending on the type and operating temperature of the fuel cell.

At the exit of the different anode and cathode compartments, there are gases of different composition compared to the inlet due to the electrochemical reactions occurring along the electrodes: that is, there will be a reduction in the moles of reactants and an increase in the moles of products. As utilisation increases, these variations increase in magnitude and, therefore, an increase in utilisation involves a reduction in cell voltage and efficiency. Due to the concentration-dependent multiplication factor of the logarithmic term, the negative effect of high utilisation on the reversible potential is felt especially in the case of high-temperature cells. Furthermore, utilisation negatively affects the value of the voltage drops in closed circuit due to the aggravation of the problems of mass transport which influence the concentrated polarisation.

On the one hand, a poor utilisation of the fuel has a negative effect on the value of the current efficiency and, therefore, it is necessary to find a good compromise. On the other hand, a poor utilisation on the oxidant front does not lead to a deterioration in cell performance, but only higher energy supply costs. Therefore, the compromise solution usually tends to high values of fuel utilisation (75%–85%) and lower values for the utilisation of the oxidiser (not more than 50%).

An interesting option for PEMFCs is the so-called dead-end anode. When the fuel fed to the fuel cell is pure hydrogen, thanks to the fact that water is generated at the cathode, there is no contamination of the hydrogen which can be therefore used completely, without discharging at the anode exit. Actually, it is not possible to have absolutely pure hydrogen and to completely avoid cross-flow of water: therefore, the anodic exit is closed, but with the possibility to purge when necessary.

Substituting the expressions of the various efficiencies and simplifying the corresponding terms, the actual efficiency of the fuel cell results:

$$\eta_{FC} = \left(\frac{V \cdot I}{\dot{m}_f} \cdot \Delta H\right) \tag{5.43}$$

Assuming that $\eta_I \approx \varphi_f \approx$cost, the polarisation curves of Figures 5.28 and 5.30 are also representative of efficiency since:

$$\frac{\eta_{FC}}{V} \approx \left(\frac{n \cdot F}{\Delta H} \cdot \varphi_f\right) \approx \text{cost} \tag{5.44}$$

With the same composition of the reagents, the operating parameters on which one can act to improve the efficiency of the fuel cell are those that can reduce the value of the polarisations:

- temperature:
  - favours the mass transport and the reaction speed, decreases the internal resistance since the ionic resistance in the electrolyte decreases more than the electronic resistance in the electrodes increases;
  - also increases corrosion phenomena, electrolyte losses due to evaporation and sintering of the electrodes;
- pressure:
  - increases the reversible voltage and the solubility of gases in the electrolyte.

Multiplying the cell voltage by the current density gives the power density. Figure 5.30 shows the trend of the latter as a function of the current density and the

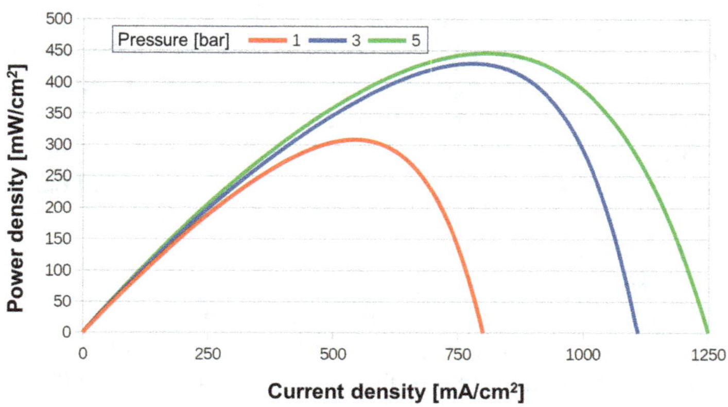

Figure 5.30 Power density vs. current density changing operating pressure

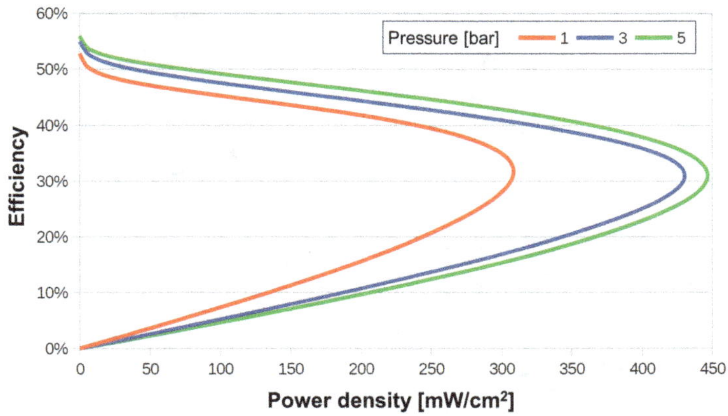

Figure 5.31 Efficiency vs. power density changing operating pressure

variation of the operating pressure. The favourable influence of a pressure increase up to about 0.3 MPa is also confirmed with regard to power.

Since the cell potential decreases almost linearly over much of the cell's operating range, the power density also varies almost linearly until the concentrated polarisation begins to appear. At this point the trend begins to decline until it reaches a maximum beyond which the power drops quite suddenly. As the pressure increases, the maximum shifts towards increasing values of both power density and current density.

Operating at the peak of power density would allow to minimise capital costs. At the same time, this means operating at high current densities and, therefore, at low voltages and low efficiencies. Placing the efficiency as a function of the power density (Figure 5.31), it is also noted that there are two efficiency values for the same power density value: they correspond to different current densities, and, therefore, different fuel consumptions. Furthermore, by operating near the peak of power density, there is a risk of having an unstable behaviour with current oscillations around the peak. Therefore, we tend to operate at current density values lower than those corresponding to the peak of power density, looking for a good compromise between capital costs and operating costs.

With respect to other energy conversion systems, fuel cells give several advantages in terms of efficiency, environmental impact and operations. Thanks also to the fact of carrying out a direct conversion from chemical energy to electrical energy, without passing through the intermediate phases of thermal and mechanical energy, the fuel cell has a higher efficiency than that of conventional systems for energy conversion (Figure 5.32). In fact, the limitations imposed by the efficiency of the Carnot cycle are avoided and mechanical losses are avoided as well.

Furthermore, the fuel cell, unlike other conversion systems, has an efficiency that increases as the load decreases. Taking into account the behaviour of the auxiliaries, it results that in operation at partial loads the system maintains its

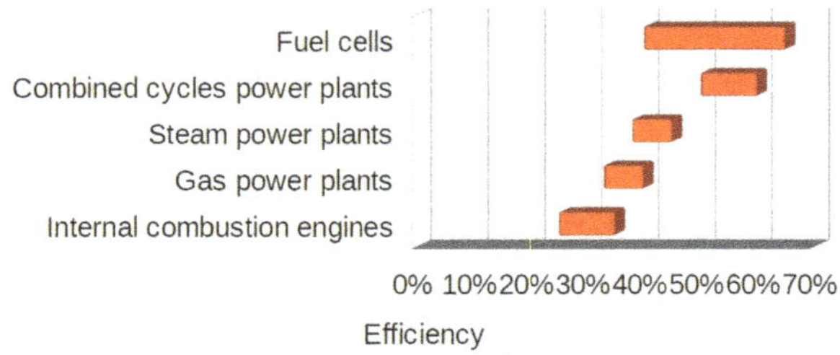

Figure 5.32 Efficiency comparison

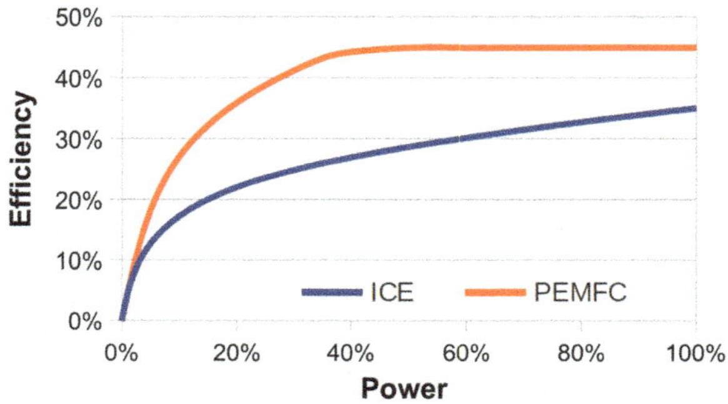

Figure 5.33 Efficiency vs. load

efficiency almost unchanged down to loads varying from 30% to 50% of the nominal one (Figure 5.33).

This is a feature of fundamental importance for vehicular applications, especially in an urban environment when vehicles are subject to frequent transients. The coupling with batteries also makes it possible to reduce the installed power in terms of fuel cells and to recover energy during braking, further increasing the efficiency of the system.

Another peculiar feature is the insensitivity to the scale factor: in fact the cells are assembled in stacks and the stacks can be placed in series and/or in parallel to each other. Therefore, standardised modules of limited size are created with a consequent reduction in costs and delivery, in assembly and start-up time, and the efficiency of the system is equal to that of the single module, except for variations due to the auxiliaries. The efficiency is therefore very high even for minimal powers and this means that small fuel cells can be used for very small-size mobile applications.

As regards the environmental impact, three aspects can be distinguished:

- chemical pollution:
    - the sulphur content of the fuel must be very low to ensure an adequate life span of the cell components and therefore the emissions of sulphur oxides are very low;

  - electrochemical combustion also limits emissions of nitrogen oxides and unburned products;
  - if the fuel is pure hydrogen, the emissions are zero;

- thermal pollution: the heat is discharged into the atmosphere and the impact is therefore more limited than if it were discharged mainly into water as occurs for most of the thermoelectric plants;
- noise pollution: the reduction of moving parts to the auxiliaries only makes systems with fuel cells much quieter than systems with internal combustion engines and/or turbomachinery.

These characteristics allow a great flexibility in the location of a plant, both for the high environmental acceptability and for the non-dependence on the availability of large masses of cooling water.

As already highlighted, there is always a certain production of thermal energy during fuel cell operation that does not participate in the direct conversion process. This thermal energy can sometimes still have an interesting thermal level and can therefore be used for co-generation or for the generation of other electrical energy by means of turbomachinery or, possibly, by other conversion systems, for example by exploiting the thermoelectric effect.

As the operating temperature increases, the flexibility towards the fuel increases and the high-temperature cells are able to use natural gas, synthesis gas produced by gasification of coal and/or biomass, landfill gas and biogas without the need for a very thorough purification. The case of low-temperature cells is different, which requires a drastic reduction of carbonaceous compounds, in particular of carbon monoxide.

Furthermore, fuel cells, thanks also to the absence of mechanical moving parts, require reduced maintenance and can be managed remotely in an automated way.

Modularity also offers a further advantage. While ancillary equipment has to be dimensioned for the final powerplant size, fuel cell stacks can be added gradually, avoiding the advance commitment of large capitals and reducing the economic risk associated with forecasts of energy needs.

The set of listed characteristics means that the construction of a power plant based on this technology may encounter fewer obstacles in the choice of the site and may involve construction and management economies such as to make fuel cells competitive at a specific cost higher than that of turbomachinery. This is especially true thanks to the high yields, even at partial loads and even in small plants.

When it comes to building basic power plants, characterised by large power and full load operation, the focus shifts to nominal yield and plant cost. For fuel cells, the latter is almost linear with the installed power, while for turbomachines the cost of the plant increases less markedly with increasing size. In relation to these facts, it is conceivable that fuel cell systems retain an economic advantage up to sizes of the order of 10–20 MW, while for larger sizes the plants based on specific thermodynamic cycles for hydrogen should guarantee greater profitability.

5.2.2 Types of fuel cells

The classification of fuel cells can be carried out according to various criteria: the one mainly used is based on the type of electrolyte (Table 5.6): alkaline fuel cell (AFC), proton exchange membrane fuel cell (PEMFC), phosphoric acid fuel cell (PAFC), molten carbonate fuel cell (MCFC) and solid oxide fuel cell (SOFC). Recently in the group of AFC was developed the anionic exchange membrane fuel cells (AEMFC). PEMFC are also called polymer electrolyte fuel cell (PEFC) or solid polymer fuel cell (SPFC): acronym PEMFC is preferable to differentiate from the AEMFC which are also based on a polymeric membrane.

Furthermore, a distinction can be made based on the temperature. Low-temperature cells are those operating at a temperature not exceeding 200 °C–250 °C, whereas high-temperature cells are those operating at temperature above 600 °C: the former can possibly be used in cogeneration systems for electricity and heat, while the latter can also provide heat for a combined cycle plant.

A further distinction is that between direct, indirect and regenerative cells. The former are able to use the fuel as it is or to process it internally, while the second require an external treatment of the fuel and the third are able to operate also as electrolytic cells thus regenerating the reagents. The latter present a certain manufacturing difficulty due to the fact that it is difficult to identify catalysts capable of operating satisfactorily in the two opposite directions of reaction.

In some cases, the direct cells, although belonging to the types indicated in Table 5.6, take on specific names. In particular there are DMFCs (direct methanol fuel cells) and DEFCs (direct ethanol fuel cells), which are polymeric membrane

Table 5.6 Classification and main characteristics of fuel cells

Type	AFC	PEMFC	PAFC	MCFC	SOFC
Electrolyte	Potassium hydroxide	Proton exchange membrane	Phosphoric acid	Molten carbonate	Yttria-stabilised zirconia or perovskite
Electrode	Carbon- or nickel-based	Carbon-based	Carbon-based	Nickel-based	Nickel-based or metal cermet
Catalyst	Nickel or silver	Platinum	Platinum	Nickel and nickel oxide	Nickel-based
Temperature [°C]	65–220	40–160	205	650	600–1000
Transferred ion	OH^-	H^+	H^+	CO_3^{--}	O^{--}
Reforming	External	External	External	External or internal	Internal
Refrigerant	Electrolyte	Water or air	Water	Gas	Gas

Table 5.7 Role of any anode gases

Substance	AFC	PEMFC	PAFC	MCFC	SOFC
H_2	Fuel	Fuel	Fuel	Fuel	Fuel
CO	Poison (if Pt used)	Poison	Poison > 0.5%	Fuel (thanks to water gas shift)	Fuel (thanks to water gas shift)
CH_4	Poison	Diluent	Diluent	Diluent or fuel (thanks to reforming)	Fuel (thanks to reforming)
CO_2	Diluent	Diluent	Diluent	Product	Diluent or product
H_2O	Product	Diluent	Diluent	Product	Product
H_2S and COS	Poison	Poison	Poison > 50 ppm	Poison > 0.5 ppm	Poison > 1 ppm

and directly use methanol and ethanol without any external treatment, and IR-MCFCs (internal reforming MCFCs (sometimes reported as DFCs, i.e. direct fuel cells), which directly use natural gas. In addition, DCFCs (direct carbon fuel cells) are being studied, high-temperature cells, but also alkaline, capable of directly using carbon in the solid state introduced into the anodic compartment. Finally fuel cells directly fed with dimethyl ether [52], ammonia and urea [53] and formic acid [54] are under development.

The term perovskite was coined in honour of Lev Perovski, a great mineral collector and minister of the Russian imperial court, and initially attributed to opaque cubic-shaped crystals he found in the Ural Mountains. More generally it indicates an oxide with an ABO_3 structure, that is, similar to that of natural oxide ($CaTiO_3$). Generally element A is an alkaline earth metal (calcium, magnesium, barium, strontium) while element B is a transition metal (cobalt, titanium, manganese, zirconium).

As already mentioned, the various gases that can be found in the anode fuel can play a different role depending on the type of fuel cell. Table 5.7 reports the behaviour of the main substances.

5.2.3 Alkaline electrolyte fuel cells

These cells use an aqueous solution of up to 50% by weight of potassium hydroxide as electrolyte and operate at temperatures between 60 °C and 120 °C. This temperature can be raised up to 250 °C by increasing the concentration by weight of the hydroxide up to 85%. Potassium hydroxide is preferred to other hydroxides due to its higher conductivity.

The hydroxide provides the hydroxyl ion as a charge carrier inside the electrolyte (Figure 5.34) and the electrode half reactions are:

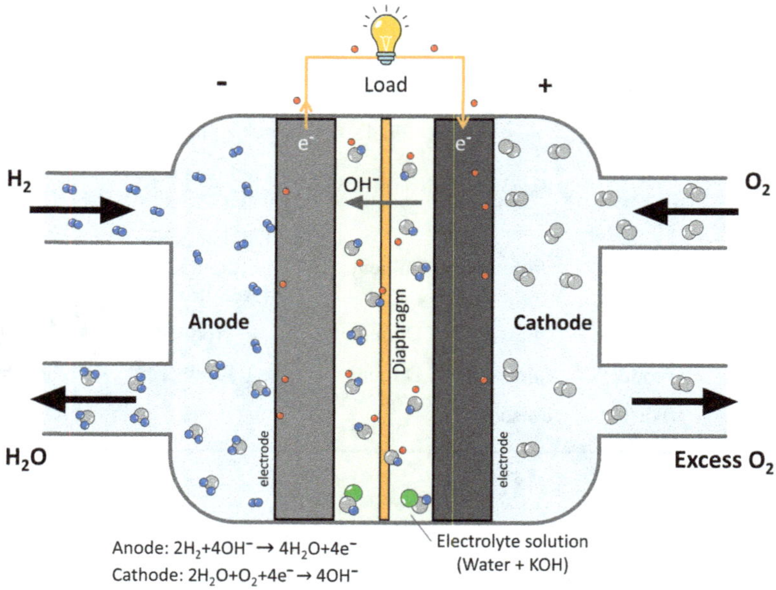

Figure 5.34 Working scheme of an alkaline fuel cell

$$\text{anode: } H_2 + 2OH^- \rightarrow 2H_2O + 2e^-$$
$$\text{cathode: } H_2O + \frac{1}{2}O_2 + 2e^- \rightarrow 2OH^-$$

The water is then produced at the anode and is removed together with the unused hydrogen which is recirculated.

Of course, the alkaline fuel cells can also use platinum; however, a wide range of non-noble catalysts, such as nickel and silver, can be used [55,56]. This is a significant advantage over other types of low-temperature cells which must use noble metals.

However, it is very critical towards the composition of the gases. Even in case that platinum is not used, these cells are generally fed with pure hydrogen and oxygen since the electrolyte reacts in the presence of even modest percentages of carbon dioxide to form potassium carbonate:

$$2KOH + CO_2 \rightarrow K_2CO_3 + H_2O \tag{5.45}$$

Using pure reactants allows very high yields (up to 65%), but also considerable operating costs. Their use is therefore limited to sectors in which efficiency problems are pre-eminent over economic ones (e.g. aerospace industry). In fact, in the 1960s of the last century, the alkaline cells were successfully used in the missions of the Apollo series and following, for example for the Space Shuttle Orbiter.

Another important aspect in this regard is the duration of these cells, about 5000 hours [57]. This is of course more than adequate for aerospace missions, but insufficient for stationary land applications. Since use in mobile land applications is less demanding, 5000 hours may be an adequate lifespan and some prototypes were, in fact, built. Among the first, in addition to the more famous Electrovan of General Motors and Austin A-40 of prof. Karl V. Kordesch, a motorcycle made by Union Carbide Corporation and powered by hydrazine (N_2H_4) deserves a special mention. However, there are currently no mobile application.

In traditional alkaline cells, the liquid electrolyte was retained in an asbestos matrix, while the more recent cells have adopted an electrolyte circulation system which, in addition to cooling the cell itself, maintains all the quantities involved (e.g. the concentration of hydroxide and that of hydroxyl ions) to a level of greater uniformity and, above all, should allow the removal of accumulated impurities and, consequently, allow the use of air as an oxidant. Since the electrolyte, as it circulates, puts all the cells into contact, a design is required that takes into account the possible occurrence of eddy currents capable of reducing performance and damaging the electrodes.

It seems clear that alkaline technology will evolve into anion-exchange membrane (AEM) technology, which will build on the achievements of alkaline fuel cells to improve their performance and in particular lower their costs [57,58]. In AEMFCs, there is no circulation of alkaline solution, which instead soaks the polymer membrane interposed between anode and cathode. Moreover, the concentration of the electrolyte is an order of magnitude lower.

Many challenges have to be addressed before AEMFCs become commercial, including stability, durability (over 1000 h) and low-cost production process. Generally, AEMFCs offer several advantages over liquid-based alkaline systems, such as better CO_2 tolerance, decreased gas crossover, which for FCs results in a higher open-circuit voltage and increased efficiency. Companies like AFC Energy is developing commercial AEMFC but currently it is not known the performance and price.

5.2.4 Proton exchange membrane fuel cells

These cells (Figure 5.35) use a polymeric membrane capable of transferring protons as electrolyte and operate at temperatures usually between 60 °C and 80 °C. This limitation is related to the fact that the membrane must remain humid and, if the temperature were excessive, the water could evaporate faster than it is produced at the cathode, thus dehydrating the membrane. At the same time, the cell must be designed in such a way as to avoid flooding the electrodes.

The protons that are formed at the anode are the charge carriers inside the electrolyte and the electrode half reactions are:

$$\text{anode: } H_2 \Rightarrow 2H^+ + 2e^-$$
$$\text{cathode: } 2H^+ + \frac{1}{2}O_2 + 2e^- \Rightarrow H_2O$$

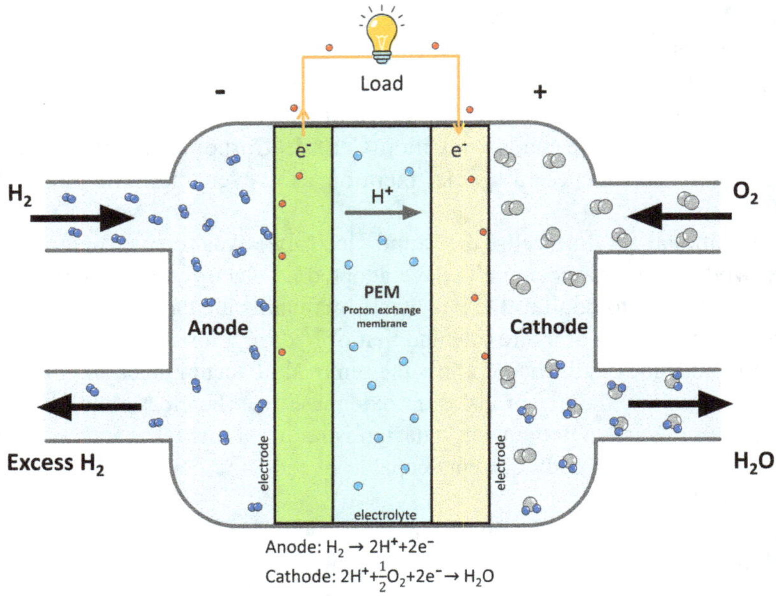

Figure 5.35 Working schematic of an acid fuel cell

The water is produced at the cathode and exits together with the unused oxidiser. The latter can be pure oxygen, in which case it is recirculated at the cathode inlet, or air that is discharged into the atmosphere.

The cell structure is planar with the electrocatalysts in direct contact with the membrane on one side and with a porous and conductive layer that allows the diffusion of the gases which are distributed through conduits obtained in the interconnection plates.

Water is the only liquid present in the cell and this means that corrosion problems are very low.

For a long time the membranes used were produced by extrusion of the molten polymer, such as the Nafion 117 from DuPont, characterised by a thickness of 178 μm. Thanks to a new membrane production method, dispersion cast, Nafion NR-212 and NR-211 allow a lower ohmic resistance due to reduced thickness respectively of 50 μm and 25 μm [59].

The material of these membranes consisted of a fluoropolymer similar to Teflon to which sulphonic acid groups were chemically bonded and was extremely stable from both a chemical and thermal point of view, so much so that it passed tests lasting over 50,000 hours. Up to temperatures of 125 °C, it resisted attacks from even strong acids and bases [50]. The main flaw was the excessive cost. In order to reduce the cost and simplify the production process, the most recent technology foresees that the polymer is converted from non-ionic to ionic and then dissolved in an alcoholic solution to be finally cast in sheets of the desired thickness. A further improvement was achieved by including a support layer to increase

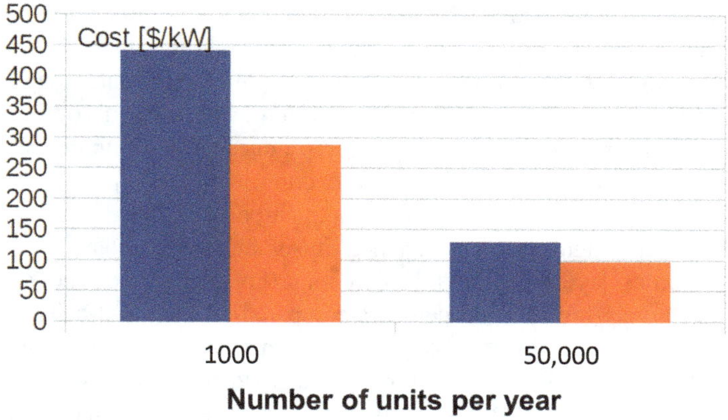

Figure 5.36 Cost comparison of 5 kW and 10 kW systems

the mechanical strength of the membrane. An example of this is the Primea membrane series made by Gore Fuel Cell Technologies. In any case, even the most recent membranes have high costs and it is estimated that only a significant increase in demand can lead to a significant reduction in cost (we are talking about a reduction factor of five for a market of millions of square meters per year) [59].

Figure 5.36 shows the unitary cost for 5 kW and 10 kW systems in case of production of 1000 and 50,000 units per year [59].

The two porous conductive layers (gas diffusion layers) that enclose the membrane are usually made of carbon fibre and include a hydrophobic material (e.g. polytetrafluoroethylene) to avoid flooding of the cathode and to facilitate the removal of the water produced and not evaporated in the cathode gas. In addition to this, these layers have other functions: distributing the gas through the porous structure, conducting electrons and acting as a mechanical support.

Electrocatalysts are directly applied to the membrane or conductive porous layer. The basic material of the electrocatalysts is generally carbon, while the cathode catalyst is made up of platinum, as well as the anodic one if the cell is fed with pure hydrogen.

The presence of platinum entails a disadvantage in terms of capital cost, but also problems of poisoning the catalyst by carbon monoxide. It follows that, in the case of using fuels other than hydrogen, a pre-treatment of the fuel is required which achieves a high degree of purification from carbon monoxide, ammonia and substances containing sulphur and halogens.

In addition to looking for possible substitutes for platinum, to reduce the cost of the electrodes, the use of platinum particles with a size of around 3.5 nm dispersed on a carbon or graphite support has been considered. It seems that this dimension is the optimal one to obtain the maximum specific activity of platinum. In any case, the total platinum load currently used is about 0.1–0.7 mg/cm^2.

As regards the binder used in the electrodes, polytetrafluoroethylene (hydrophobic) has been replaced by perfluorosulphonic acid (hydrophilic) in order to improve the membrane/catalyst contact and reduce the platinum load: with less than mg_{Pt}/cm^2 between the anode and cathode, a current density of 600 mA/cm^2 at 0.7 V was obtained with reforming gas. An example of this technology is the 3M seven-layer Membrane Electrode Assembly which has exceeded 10,000 operating hours with reforming gas and which can be produced in an automated and continuous way with further benefit for cost reduction.

The interconnections can be made of both carbon and metal.

Thanks to the high electrical conductivity of both the bipolar plates and the conductive porous layers and to the high ionic conductivity of the membranes, PEMFCs allow very high current densities compared to other fuel cells, even up to 1000 mA/cm^2 [60]. The power density under normal operating conditions, i.e. with utilisation of reforming gas equal to 85% and of oxidant equal to 40%, is approximately 50 mW/cm^2 at 0.7 V. With hydrogen, it can increase by a factor of ten.

The polymer cells used in the Gemini space program operated at 50 °C and 0.2 MPa and reached a current density of 37 mA/cm^2 at 0.78 V.

Furthermore, PEMFCs have the ability to react quickly to load variations.

The low temperature allows a very rapid start-up, a very interesting feature for use in the transport sector and in the uninterruptible power supply sector. However, this feature also involves problems. First of all, at temperatures below 120 °C, carbon monoxide more easily attacks the platinum of the catalysts and is admissible in concentrations of a few ppm while the carbon monoxide content of a syngas after water gas shift (5.24) is of the order of 1%. Furthermore, low-temperature cells require wider exchange surfaces for refrigeration compared to higher temperature cells. And their use in co-generation is more difficult. Finally, the thermal management of the cell is more complex because of the narrow range of variation.

The water balance is influenced by several parameters (temperature, pressure, current, reagents flow rate) and therefore water management is a critical aspect. Furthermore, two opposite effects occur: the dragging of water from the anode towards the cathode by the protons and the reverse flow caused by the moisture gradient that is formed inside the membrane. This second phenomenon is also influenced by the thickness of the membrane as the return of water towards the anode is facilitated in thin membranes.

To obtain a correct water management, an accurate planning of the distribution of the channels on the bipolar plates is necessary. Furthermore, in case of excessive evaporation, the reagents can be humidified, being careful not to flood the electrodes.

The humidity of the membrane is also important because a dry membrane does not allow perfect adhesion with the electrode and the phenomenon of membrane detachment is practically irreversible. It follows that, even if temporary, an incorrect water management can reduce the life of the cell.

The temperature increase between inlet and outlet is contained within 10 °C through the use of a cooling fluid, usually water, which circulates inside the heat exchangers integrated in the stack.

As already stated in Section 5.2.1, Nernst equation establishes a favourable effect of an increase of the operating pressure for all kind of fuel cells. Moreover, for a PEMFC, an increase in the oxygen pressure also causes a significant reduction in polarisation at the cathode. However, it is necessary to take into account the increased power consumption for compression.

PEMFCs have been tested for over 20,000 hours in the laboratory under conditions similar to those of real applications and have shown a reduction in performance equal to 4–6 mV/h, corresponding to a reduction of 0.67%–1% every 1000 hours. Moving on to complete systems, both vehicular and stationary applications are numerous, but they have rarely exceeded 8000 hours of operation and, in general, with much higher levels of degradation.

While carbon dioxide and hydrocarbons are diluents, carbon monoxide and sulphur are contaminants. Carbon monoxide negatively affects performance already at 10 ppm. Addition of small amounts of oxygen could solve the problem (Figure 5.37); however, in the case of platinum/ruthenium catalysts, the tolerance rises to 200 ppm [50].

The presence of ammonia can generate a deterioration of the membrane. Furthermore, it is good to filter the electrode flows to avoid the presence of particulates, alkali metals and heavy hydrocarbons, substances that can significantly reduce the life of the cell.

To solve some problems of traditional PEMFC, several research works have been conducted in order to increase the operating temperature (120 °C–200 °C). Compared to traditional PEMFCs, they offer faster reaction kinetics, higher tolerance to impurities, and better water management. Furthermore, since the heat is available at higher temperatures, they can be used for co-generation of electricity and heat, similar to what happens with PAFCs.

Higher temperatures also help to improve the performance since an increase in temperature causes the decrease of the ohmic resistance of the electrolyte and accelerates the kinetics of the electrode reactions. Experimental data suggest a

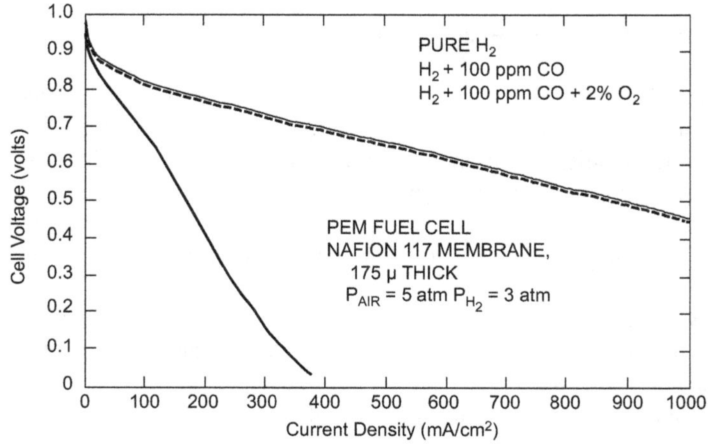

Figure 5.37 Effect of CO and O_2 at the anode of a PEMFC [50]

voltage gain in the range of 1.1–2.5 mV for each degree (°C) of temperature increase.

Obviously, the high temperature also brings problems yet to be resolved. In particular, because of limitation due to the vapour pressure of water into the membrane and the risk of dehydration of the latter, to allow the operating increase of the cell a different kind of membrane is necessary. This goal involves the creation of new membranes because Nafion would require pressures of over 1 MPa to avoid dehydration. Polybenzimidazole requires much less water than Nafion and, therefore, withstands these temperatures better and allows for much easier water management. The disadvantage is that the membrane must be impregnated with phosphoric acid and, therefore, problems similar to those of PAFCs arise. Furthermore, there are issues connected to durability, performance, and also electrode fabrication and cost since current high-temperature technology requires a higher load of platinum [61].

5.2.5 Phosphoric acid fuel cells

The first fuel cell had a sulphuric acid solution as an electrolyte. Nowadays the category of liquid acid electrolyte cells is represented only by phosphoric acid cells since the stability of this acid is high compared to that of other acids and this means that it can be operated at higher temperatures (150 °C–220 °C). This allow to reduce the platinum and, at the same time, to make less critical the phenomenon of the poisoning of electrocatalysts by carbon monoxide, which is tolerated in concentrations up to 1% about. Phosphoric acid is used in concentrations up to 100% thus minimising the vapour pressure of the water and making easier its management. The acid is usually held in a silicon carbide matrix.

As in the case of fuel cells with proton exchange membrane, protons are formed at the anode as charge carriers inside the electrolyte and also the electrode half reactions are identical:

$$\text{anode: } H_2 \Rightarrow 2H^+ + 2e^-$$
$$\text{cathode: } 2H^+ + \frac{1}{2}O_2 + 2e^- \Rightarrow H_2O$$

Using reformed natural gas, the efficiency varies between 33% and 38%, but by recovering the heat produced, a co-generation system can be created or, by using suitable working fluids, additional electricity can be generated. When using a gas derived from fossil fuels, biomass or waste, it is very likely that the concentration of contaminants (CO, COS and H_2S) exceeds that tolerated by the anode: operating at 200 °C the limit for carbon monoxide is 1%, while the total sulphur volume must not exceed 20 ppm.

The electrodes, which use platinum as a catalyst, are made of carbon particles bonded by polytetrafluoroethylene (PTFE). The mass of catalyst is 0.1–0.25 mg/cm^2 for the anode and 0.5 mg/cm^2 for the cathode. This difference is due to the slowness of the oxygen reduction reaction. The mass of the binder ranges from 30% to 50% of the total and allows to obtain a porous electrode which is supported on a porous graphite substrate. Graphite is a fairly inexpensive material with good electronic conductivity and good corrosion resistance. However, above a cell voltage value of about 0.8 V, carbon corrosion and platinum dissolution phenomena can occur.

To improve cathode performance by reducing the amount of platinum, electrodes with platinum alloys and transition metals (Ti, Cr, V, Zr, Ta) have been experimented with some success. In addition, organic macrocycles were also considered which showed an activity comparable to that of platinum, but present stability problems.

The bipolar plates can also contain the channels that carry the reagents to the electrodes and the channels that remove the products (and any diluents) from the electrodes.

The cooling of the cells can take place either by means of a liquid refrigerant or by means of a gaseous refrigerant. In both cases, the refrigeration channels are located in special plates that are inserted every five cells. The use of air as a coolant allows for a much simpler and cheaper system, even if the liquid coolant requires much smaller channels.

The power density is 0.31 W/cm^2.

These cells have been developed especially for stationary applications and hundreds of batteries of this type have been produced and tested mainly in the USA and Japan for a total power approaching 100 MW and with systems that are mainly of size between 50 and 200 kW, but they also reached 5 MW in a couple of plants and even 11 MW at the Goj plant, in Japan. The Goj plant operated at around 0.8 MPa, but most of the acid cells built operated at atmospheric pressure. In the face of an improvement in performance, there is, in fact, a greater system complexity which is justified only for high powers.

Tens of units have exceeded 30,000 hours of operation and some even 40,000 hours without even needing to top up the electrolyte despite the losses due to evaporation and with an average degradation of performance equal to 2 mV every 1000 hours.

While still the most affordable type of fuel cell, development has almost stalled and there have been no major improvements in at least 25 years. This is due to the widespread opinion that PEMFCs have better prospects in terms of both efficiency and cost, as well as being more compact and therefore suitable for mobile applications.

5.2.6 Molten carbonate fuel cells

These cells use a mixture of molten alkali metal carbonates as electrolyte and operate at temperatures of 600 °C–700 °C. This temperature range is limited downwards by the melting temperature of the carbonates, which is around 550 °C, and upwards by problems of corrosion and mechanical stability of the materials that make up the different components of the cell.

Given the particular type of electrolyte, the charge-carrying ion inside is the carbonate ion and the electrode half-reactions are (Figure 5.38):

$$\text{anode: } H_2 + CO_3^{-\,-} \Rightarrow H_2O + CO_2 + 2e^-$$

$$\text{cathode: } CO_2 + \frac{1}{2}O_2 + 2e^- \Rightarrow CO_3^{-\,-}$$

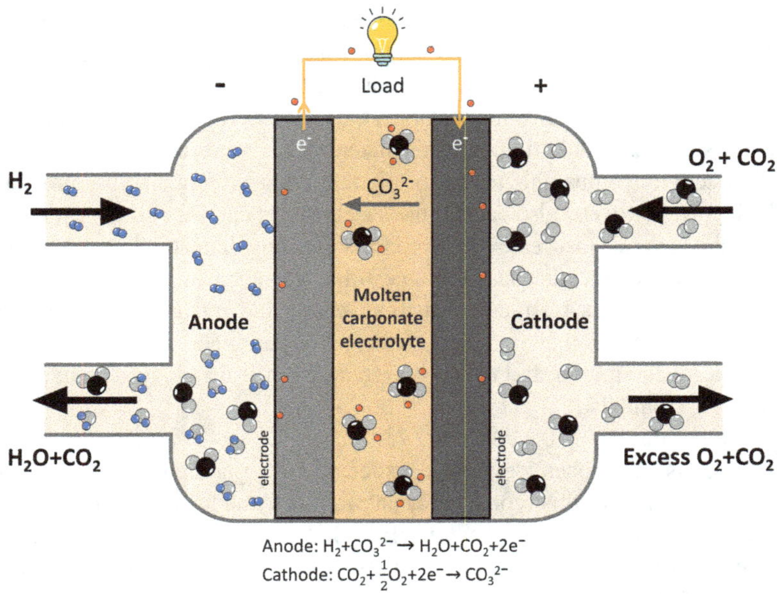

Figure 5.38 Working schematic of a molten carbonate fuel cell

The water is then produced at the anode together with the carbon dioxide. At the cathode, however, carbon dioxide is consumed in the same quantity. From these reactions, it is evident that the ideal composition of the reactants is 100% H_2 for the fuel and 33% O_2 plus 67% CO_2 for the oxidant. For economic reasons, however, processed natural fuels are used which therefore have a very variable composition linked to both the primary fuel and the process undergone, while a mixture of air or oxygen and combustion products is used as an oxidant. The concentration of carbon dioxide in the air is not sufficient and, usually, recirculation of the exhausted anodic is used, which is very rich in carbon dioxide.

Since the molar fractions of carbon dioxide at the anode and at the cathode are different, the Nernst equation must take this substance into account as a cathodic reagent and as an anodic product and therefore takes the form:

$$E = E^\circ(T) + \frac{R \cdot T}{2 \cdot n \cdot F} \cdot \ln p + \frac{R \cdot T}{n \cdot F} \cdot \ln \frac{x_{H_2} \cdot x_{CO_{2(C)}} \cdot \sqrt{x_{O_2}}}{x_{H_2O} \cdot x_{CO_{2(A)}}} \tag{5.46}$$

At the operating temperatures of these cells, the water gas shift reaction (5.24) is not particularly favoured. However, hydrogen is consumed continuously, while water and carbon dioxide are produced at the anode with equal continuity and in equal measure. Therefore, the equilibrium is constantly shifted towards the products and carbon monoxide can be considered as a fuel since continuously reacts with steam to form hydrogen. Methane, on the other hand, is a diluent that passes through the anodic compartment without reacting.

The electrolyte is a mixture of alkali metal carbonates ($LiCO_3$, KCO_3) in the molten state, retained in a lithium and aluminium oxide matrix ($LiAlO_2$) which is inert and constitutes 40% by weight of the electrolyte. The thickness of the electrolyte is kept in the range of 1–2.5 mm to reduce internal voltage drops.

The cathode is made of 70%–80% porous nickel oxide with an average diameter of the pores of 10 μm; the anode is made of 60%–70% porous sintered nickel with an average pore diameter of 5 μm. No particular catalysts are required and this allows to produce relatively inexpensive electrodes and increases the versatility towards the fuel.

They reach efficiencies ranging from 45% to 55%, but it is possible to recover the waste heat to power a bottoming cycle and reach overall efficiencies in the 55%–65% range. The high resistance of the interconnections and the cathode limit the power density to 100–200 mW/cm^2.

Molten carbonate cells are suitable for applications that do not require a quick start or particular compactness and lightness (stationary and marine applications).

The role played by carbon dioxide is fundamental for this type of cell. Its presence in the electrode compartments is in fact of primary importance: always at the cathode, as it is essential for the semi-reaction that produces the carbonate ion, and at the anode in the case of the presence of carbon monoxide in the fuel gas, as it is essential for hinder the Boudouard reaction which leads to the formation of solid carbon:

$$2CO \Rightarrow CO_2 + C_{(s)} \quad \Delta H^\circ = -159.7 \ \text{kJ/mol} \tag{5.47}$$

In fact, the carbon formed is deposited on the anode, obstructing the pores and thus preventing contact between the reagents and the electrolyte. In this reaction, the carbon dioxide tends to shift the equilibrium towards the formation of monoxide and therefore preserves the electrode and allows a regular functioning of the cell. Since the set of reactions that occur along the anode involves a progressive consumption of carbon monoxide and a continuous production of carbon dioxide, it is sufficient for the following condition to be verified at the entrance to the anode compartment:

$$x_{CO_2} \geq \frac{x_{CO}^2}{K_{eq}} \tag{5.48}$$

where K_{eq} is the equilibrium constant of the reaction and x indicates the mole fraction of compound.

If the combustible gas does not comply with this condition at the entrance to the anodic compartment, it is possible to change its composition using one of the different methods proposed for this purpose. The main ones are:

- addition of water vapour to the combustible gas;
- partial recycle of anodic products.

The recycle of the anodic products, poor in hydrogen and carbon monoxide and rich in water and carbon dioxide, acts directly on the condition to be verified, whereas the addition of water vapour obtains indirectly the same effect by pushing the water gas shift reaction (5.24) and therefore entailing a consumption of monoxide and a production of dioxide.

Both of these methods involve an impoverishment of the anodic gas since they increase the overall flow rate without contributing proportionally to increasing the combustible fraction: therefore, the hydrogen mole fraction is reduced and this implies a lowering of the cell voltage.

At the cathode, the carbon dioxide is only consumed and it is therefore essential to recover the anodic products, rich in this substance, and send them to the cathode. Before they reach it, however, combustion must be completed to avoid the presence in the cell of hydrogen/oxygen mixtures which are dangerous at operating temperatures. Furthermore, unless the fuel sent to the anode is rich enough in carbon, the amount of carbon dioxide that is recirculated is insufficient since the cathodic utilisation of the latter cannot be too high. If necessary, partial recirculation of the exhausted cathode is used in such a way as to bring the carbon dioxide content at the cathode inlet to an adequate value.

An interesting alternative is the use of MCFC for carbon dioxide concentration. Rather than using air as an oxidant, it is possible to use chimney exhaust fumes which contains a significant amount of carbon dioxide in addition to the oxygen corresponding to the air excess required by combustion. This concept has two advantages: recycle from anode to cathode is not necessary and the carbon dioxide at the anodic exit is much more concentrated than the original carbon dioxide at the cathodic inlet. Assuming, for example, the following composition of the chimney exhaust fumes used as an oxidant: 73.75% N_2, 15% H_2O, 7.5% CO_2 and 3.75% O_2, and CH_4 as a fuel, with a fuel utilisation of 0.85, the anodic exhaust would have approximately the following composition: 69.65% CO_2, 27.85% H_2O and 2.5% CH_4. Burning the residual methane the final composition would be approximately 58% CO_2, 26% H_2O and 16% N_2. Comparing the dry fumes, carbon dioxide results concentrated by a factor close to nine, from 8.8% to 78.4%.

A significant problem in the design of systems with molten carbonate cells is the thermal regulation of the electrode flows and of the cell itself since it is necessary to remain sufficiently above the solidification temperature of the electrolyte.

In the case of anode gas, the problem generally arises when the supply is carried out with hydrogen and it is necessary to equip the system with a heat exchanger that provides for its preheating. If, on the other hand, synthesis gas is the fuel fed, it is usually supplied by a gasification or reforming process and, therefore, the temperature could be too high.

The cathode gas is usually also entrusted with the function of refrigerant fluid and, therefore, the cathode balance is more complex as it could be the result of the confluence of three flows: the oxidant, the total recirculation of the anodic exhaust and the partial recirculation of the exhausted cathode. Since the predominant function of cathodic recirculation is the thermal one, it follows that the composition of the cathodic supply cannot be adjusted to the optimal one. For this reason, an alternative system was proposed [62,63], however never applied. The fundamental concept is that refrigeration must be entrusted to a flow that does not generate any composition variation in the flows optimised from a chemical point of view. The ideal solution consists in adopting a recirculation of a gas operating in a section of the plant that is not the site of chemical reactions. The flow rate that flows in this recirculation is thus found to operate in an open circuit, but which can be considered closed and which allows the transfer of thermal energy to the outside.

A particular application of MCFCs is CO_2 concentration. An exhaust gas stream from a power plant can be dried and mixed to the cathodic air. Therefore, the fuel cell benefits from the higher concentration of carbon dioxide, which crosses the electrolyte and join the anodic flow. The latter, when dried, contains only carbon dioxide and minimum amounts of residual unreacted gases [64,65].

5.2.7 Solid oxide fuel cells

Solid oxide fuel cells operate at very high temperatures because only in this way the solid electrolyte is conductive. However, although the first solid oxide fuel cells needed 1000 °C, the current temperature required are in the range 650 °C–800 °C and a possible target for new high-conductivity electrolytes and nanostructured electrodes is as low as 350 °C [66–68].

The electrode half reactions are therefore (Figure 5.39):

$$\text{anode: } H_2 + O^{--} \Rightarrow H_2O + 2e^-$$
$$\text{cathode: } \frac{1}{2}O_2 + 2e^- \Rightarrow O^{--}$$

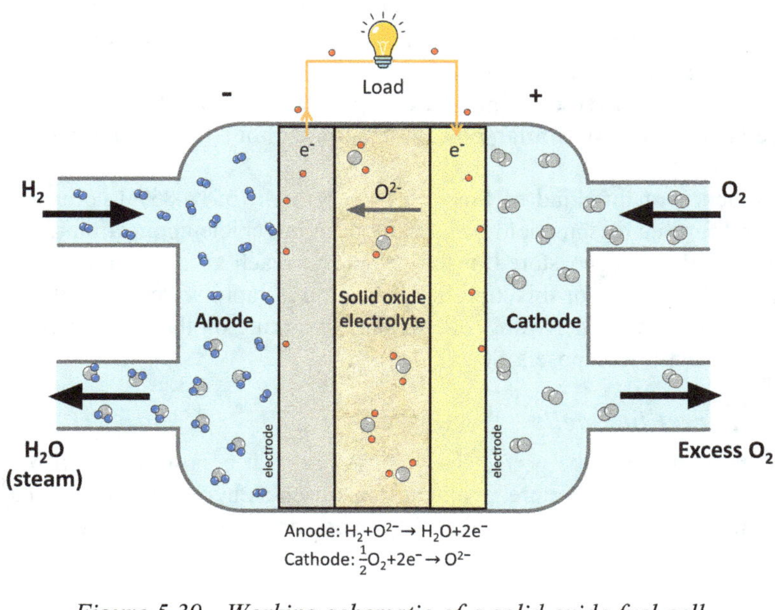

Figure 5.39 Working schematic of a solid oxide fuel cell

In some cases, low-temperature solid oxides, for example Gd-doped barium cerate, exchange protons rather than oxygen ions. This mechanism was already studied more than 30 years ago [69] and also in recent years [70] because this kind of fuel cells is considered very promising.

The first solid oxide cells had a tubular structure because the wide thermal excursion posed serious sealing problems and the tubular structure allows to keep the anodic and cathodic flows structurally separate. With the lowering of the operating temperature, it was possible to move to a planar structure as for all other types of fuel cells.

The electrolyte most commonly used is zirconium dioxide stabilised with diyttrium trioxide and at the operating temperatures of the cell it is a non-porous solid. The anode is a cermet of zirconium dioxide with cobalt or nickel, while the cathode is made of lanthanum trioxide and manganese doped with strontium. The ceramic materials of which the cell is made make it immune from corrosion problems and the absence of liquids prevents electrolyte losses and problems of flooding the electrodes. The high temperature can, however, create corrosion problems in other components of the cell (e.g. in the metal interconnects).

The high temperature also allows the direct conversion of carbon monoxide:

$$\text{anode}: CO + O^{--} \Rightarrow CO_2 + 2e^-$$

however, given the continuous consumption of hydrogen and the continuous production of water, the water gas shift reaction (5.24) remains favoured despite the high temperatures and carbon monoxide is mainly converted indirectly.

As with molten carbonate cells, it is possible to use a broad spectrum of fuels directly and it is possible to recover heat at high temperatures both for cogeneration and to power a subjected cycle, but there is also the considerable advantage of not having to manage the carbon dioxide.

Furthermore, these cells have a compactness comparable to that of polymer cells thanks to a similar power density without the need for expensive catalysts.

Efficiency of this kind of fuel cell is very high (65%–85%) compared with others and for this reason could be usefully used in vehicle application to decrease the amount of hydrogen stored in the vehicle to reach an acceptable mileage. A problem to be solved for this kind of application is rapid warm-op of the system. Kraftwerk Group together with Volkswagen is working on this, developing a mini tubular element to compose a SOFC [71]).

5.2.8 Direct fuel cells

Although hydrogen is the perfect fuel for fuel cells, it is nowadays still produced from fossil fuels. Furthermore, synthetic fuels used as hydrogen carriers may also be available in the future. This makes very interesting research works in the field of direct fuel cells, i.e. capable of using a fuel directly without converting it previously into hydrogen outside the stack.

Direct methanol fuel cells (DMFC)

This type of fuel cell [72] is derived from PEMFCs and was developed with the aim of overcoming the difficulties imposed by the storage of hydrogen in mobile systems, both vehicular and non-vehicular. In fact, methanol has the same H/C ratio as methane, but is a liquid fuel in ambient conditions. As the fuel treatment is not necessary, the system is very simplified and compact.

The performances achieved are 180–250 mA/cm^2 with cell voltages of 0.25–0.4 V, i.e. with power density equal to 40–100 mW/cm^2 which is still far from that, equal to 200 mW/cm^2, assumed as a target for marketing. The low cell voltage is attributable to the migration of methanol or its decomposition products from the anode towards the cathode, a phenomenon which involves, in addition to the reduction of fuel use, the contamination of the cathode electrocatalyst. The methanol that crosses the membrane without participating in the electrochemical reaction can reach values equal to 30%–50% of what is, instead, electrochemically oxidised. To reduce this phenomenon, it is possible to resort to diluted methanol, but this choice involves a reduction of the current density.

Water also exhibits considerable passage across the membrane.

The required platinum load is very high, from 3 to 5 mg/cm^2, although experimental cells with a significantly lower load have been made.

The operating temperature was increased from 60 °C to 100 °C.

These problems mean that, at the moment, DMFCs are not considered suitable for stationary and vehicular applications. However, there is some interest in portable applications with powers below 1 kW.

Internal reforming MCFC

Thanks to the high working temperature, MCFC can be designed with catalysts capable to perform internal reforming of methane (5.25). Therefore methane reacts with steam producing carbon monoxide and hydrogen. And the continuous consumption of hydrogen causes the reaction to progress continuously and the methane is progressively and almost completely consumed notwithstanding the temperature is lower than that typical of external reforming. It follows that internal reforming cells are more efficient since they also use methane as a fuel.

Moreover, reaction (5.48) is strongly endothermic and can use part of the heat generated by the fuel cells, reducing the need for cooling.

Other direct fuel cells

Various fuels can be used to directly power a fuel cell: they are being studied with the aim of having alternatives to the use of hydrogen, which can avoid the complications associated in particular with its storage and transport. This is why direct conversion systems based on ethanol [73], dimethyl ether [74–77], ammonia [78–80], urea [81–83] and formic acid [84,85] are being studied.

There is also some interest in fuel cells that directly consume coal as fuel (Direct Carbon Fuel Cells) [49,51], which are interesting as a more efficient alternative to electricity production through steam cycles.

Finally, a very recent research proved that methylcyclohexane can be used as a fuel for SOFC [86].

5.2.9 Reversible fuel cells

A reversible fuel cell is a device capable to operate either as an electrolyser or a fuel cell, i.e. capable to convert chemical energy into electricity and also electrical energy into chemical energy. It is the equivalent of a hydro-pumping system which accumulates energy when it is in excess and then returns it when needed. The difference is that a reversible fuel cell store energy by means of hydrogen, and chemical storage is much more compact than potential energy storage. For such a reason eversible fuel cells have attracted increasing attention in recent years, although their efficiency is lower than that of electrolysers and fuel cells [87].

In 2023, a specific EU funding project, REFLEX (Reversible solid oxide Electrolyser and Fuel cell for optimised Local Energy miX), was finished to test a first application of this system in a field application [88].

5.2.10 Biofuel cells

Most organic substances can be subjected to combustion generating heat that can be used to produce electricity. Using biofuel cells, it is possible to produce electricity directly.

Biofuel cells can be indirect or direct. In the first case (Figure 5.40), a biofuel feeds a traditional fuel cell [89]. Therefore, it consists of a bioreactor coupled to a fuel cell based on inorganic catalysts. In the second case, the biological process is integrated into the anode compartment of the fuel cell where an organic fuel (e.g. glucose) is oxidised by enzymes and generates positive ions and electrons. Such enzymes can be isolated (enzymatic fuel cells) or belonging to whole microorganisms (microbial fuel cells).

The use of whole microorganisms allows to avoid the processes of isolation and purification of the enzymes, while the use of isolated enzymes allows to improve the efficiency since it reduces the number of reactions to those useful for the process.

In general, direct biofuel cells are suitable for very small applications. A particular case is that of the so-called abiotic biofuel cells, which means that a biofuel is converted by using non-biological catalysts. This kind of fuel cells could allow very particular uses also in the medical sector for activating microelectronic implantable devices [90].

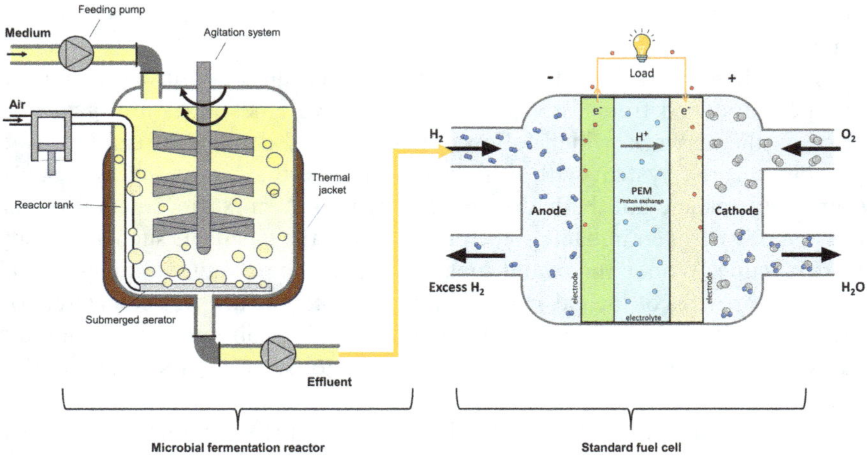

Figure 5.40 Indirect biofuel cell

5.3 Power to fuel

The current energy system is mainly based on fossil fuels and only a minority of them are used to produce electricity. Conversely, the transition to non-fossil energy sources involves, with current technologies, an unbalanced production towards electricity hourly request, hence the need to convert part of this electricity into fuels.

The simplest and sufficiently mature technology to produce a fuel using electricity is the electrolysis of water which allows to obtain hydrogen. Although TRL of hydrogen technology is very high in many cases, there are at the date not so many companies able to realise single core electrolyser with a size of 10 MW or more, with long life guarantee: this is a limit for large-scale deployment of power-to-X approach coupled with a lack of sufficient renewable energy.

Hydrogen allows to:

- decarbonise end uses, being a zero-emission energy carrier;
- compensate for the random production from renewable energy sources;
- improve flexibility and stability of the electric grid;
- interconnect the electric grid with the gas network.

The hydrogen produced from electricity surplus can be used to store energy and regenerate electricity or as a fuel or as a feedstock for industry: we speak, respectively, of power-to-power, power-to-fuel and power-to-X, where X can be substituted by many chemicals like methanol, ammonia and so on, not necessarily used as fuel.

Unless you can use hydrogen directly on site, there is the necessity to distribute it. As already described in Section 4.7, large-scale hydrogen distribution is still an issue to be solved. The simplest way is hydrogen injection into the natural gas pipelines which is possible within certain percentages depending on national regulations [91,92].

A problem arises from the difference between the intermittent amount of hydrogen produced from RES and the irregular volumes of natural gas transported by the network, which follows the end users' demand. It can be solved by equipping the system with a hydrogen storage that functions as a bag to ensure the constancy of the hydrogen content into the mixture. In addition, the injection of hydrogen impacts some properties of the gas itself such as density, calorific value, Wobbe index and speed of sound. The change in the chemical-physical properties of the gas in turn directly affects the utilities connected to it in terms of efficiency and safety. If the end user is unable to use this mixture and requires pure hydrogen, it is necessary to separate the gases before distributing.

Hydrogen can also be used to produce synthetic fuels more or less similar to the already used fuels derived from fossil sources. In this case, they can be distributed using the already existing infrastructures and can be used into the existing engines, furnaces, boilers, and so on, without or with minor changes.

5.3.1 Power to methane

The reverse reaction with respect to the decomposition of methane is called hydrogasification and is, of course, a process that aims to directly produce methane:

$$C + 2H_2 \Rightarrow CH_4 \quad \Delta H^\circ = -74.95 \text{ kJ/mol} \tag{5.49}$$

Obviously, the carbon source for hydrogasification can be made up of coal or biomass, which also contains a significant amount of oxygen, in addition to other secondary elements. In this case, hydrogen has the dual function of combining with carbon to form methane and extracting oxygen forming water.

The same result can be obtained starting from carbon dioxide by means of the Sabatier reaction:

$$CO_2 + 4H_2 \Rightarrow CH_4 + 2H_2O \quad \Delta H^\circ = -165 \text{ kJ/mol} \tag{5.50}$$

Actually, such a reaction is not direct [93,94]. The reverse water gas shift reaction reduces carbon dioxide to carbon monoxide:

$$CO_2 + H_2 \Rightarrow CO + H_2O \quad \Delta H^\circ = 41 \text{ kJ/mol} \tag{5.51}$$

which is finally converted to methane:

$$CO + 3H_2 \Rightarrow CH_4 + H_2O \quad \Delta H^\circ = -206 \text{ kJ/mol} \tag{5.52}$$

Figure 5.41 shows the conceptual scheme of the system.

Carbon dioxide can be obtained by capture based on absorption from the exhaust gases of thermoelectric plants, refineries and industries such as steel and cement factories. A liquid sorbent, usually mono- or di-ethanolamine, is used to separate the CO_2 from the gas stream and then it is regenerated by heating or depressurisation. This is the most mature carbon capture technology and can

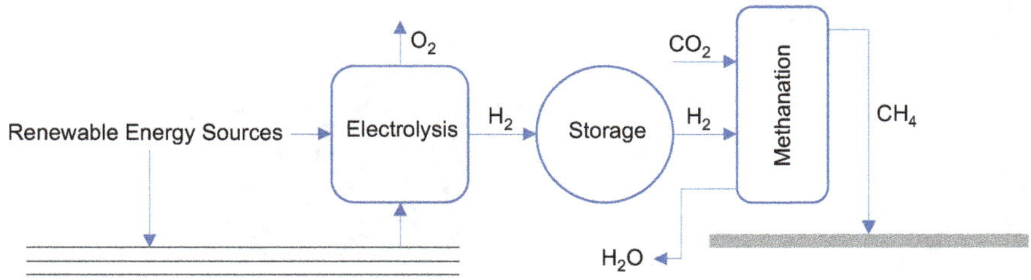

Figure 5.41 Conceptual scheme of the system power to methane

achieve a high level of carbon dioxide removal. Other available technologies are physisorption, which uses activated carbon or metal oxides, and membrane-based separation methods.

Carbon capture from air with solid sorbents or with aqueous basic solutions is available, but the air CO_2 content is very small and these techniques are not yet competitive [95] and high energy consuming.

Biogas also has a high content of carbon dioxide, from 30% to 50%, while the rest is already converted into methane possibly with small amounts of sulphur and silicon compounds [96,97]. Such a biogas can be used, possibly after purification, for methanation, without traditional upgrading consisting in CO_2 separation.

The gas produced through the methanation process has, after condensation of water vapour, chemical-physical properties comparable to those of natural gas and is defined as a substitute natural gas (SNG) or as the simplest e-fuel after pure hydrogen from electrolysis. It is also sometimes called synthetic natural gas, but it is an inappropriate definition as the terms 'natural' and 'synthetic' are in general antithetical. Thanks to its properties, the SNG can be introduced into natural gas transportation and distribution networks without limitations [98,99], provided that the quality requirements imposed by the Transmission System Operator are met. Furthermore, having a higher energy density than hydrogen, it allows you to store larger quantities of energy more easily [100].

After direct injection of hydrogen, production of a SNG and its injection into the existing natural gas pipelines is maybe the best current option to store large amounts of renewable energy on a seasonal basis and therefore to obtain a vast decarbonisation of the energy system. Figure 5.42 shows a comparison with flywheels, batteries, compressed air energy storage, pumped hydro storage and hydrogen.

Due to the high investment and management costs, the current cost of the SNG is higher than that of natural gas [91,102,103]. However, while the cost of the SNG is gradually decreasing, the cost of natural gas is characterised by strong variability on average upwards.

The electrolyser has a fairly fast dynamics that allows rapid adaptation to the variation of available electricity. The dynamics of the methanation reactor, however, is slower and therefore does not allow direct coupling. Moreover, the optimisation of the plant could imply a different speed between production by the electrolyser and

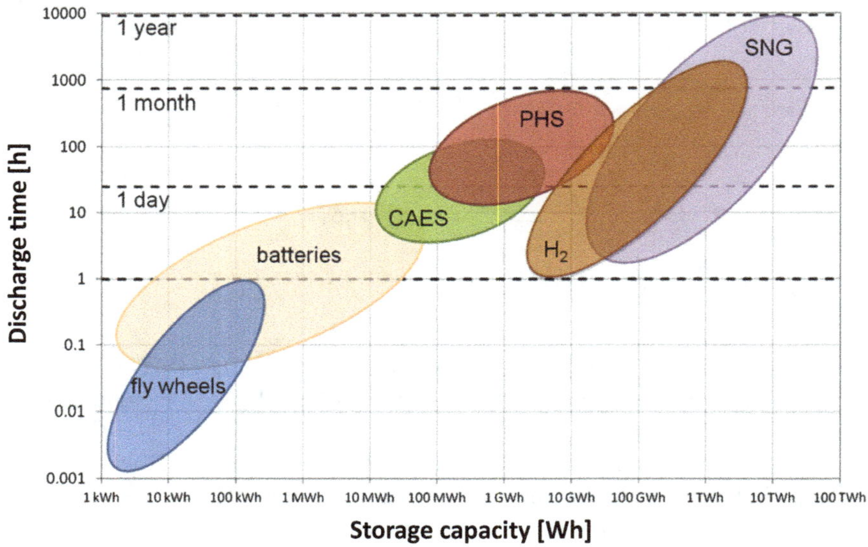

Figure 5.42 Comparison among different energy storage systems [101]

consumption by the methanator [104]. It follows the need for temporary storage of hydrogen which can also contribute to increasing the hours of operation at full load of the plant by reducing the production costs of the SNG [105].

The methanation process can be carried out both chemically and biologically.

In the first case, it is necessary to resort to the use of catalysts because the process is favoured by low temperatures which slow down the reaction kinetics. Usually, solid catalysts are used deposited on a porous support which has both the purpose of providing adequate mechanical resistance and the purpose of creating an extensive active surface. Among the selective catalysts for the methanation reaction, the one with the greatest activity is ruthenium, followed by nickel; other elements with a good activity are rhodium, palladium, platinum and molybdenum [106,107]. Nickel is the most used being the least expensive. In particular, the typical unsupported catalyst is Raney nickel, whereas in case of supported catalyst, nanoparticles of nickel are deposited on porous substrates of alumina (Al_2O_3), silica (SiO_2) or zeolites. Since nickel is poisoned by sulphur compounds, in case of methanation of a biogas or a syngas it is necessary to reduce the sulphur content to less than 1 ppm [108,109]. The control of the temperature inside the methanation reactor is also important in order to prevent the sintering of the catalyst [91].

Since methanation reactions are characterised by a reduction in the number of moles, the formation of methane is favoured at high pressures. For a stoichiometric mixture of hydrogen and carbon dioxide, the influence of temperature and pressure on the equilibrium conversion is shown in Figure 5.43.

Typical temperature and pressure values are in the range 250 °C–400 °C and up to 30 bar, respectively [110,111].

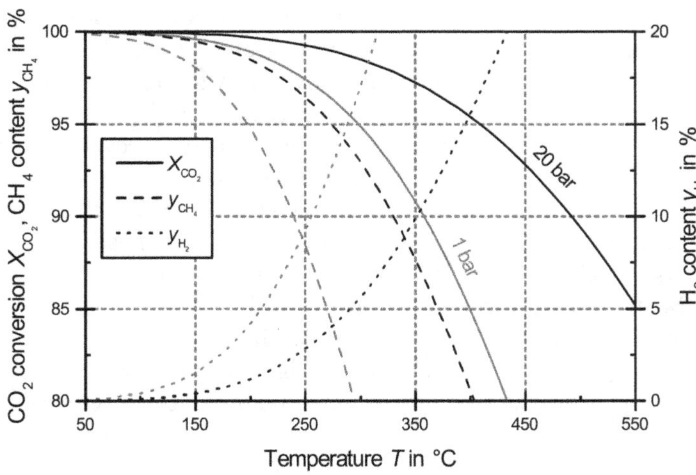

Figure 5.43 CO_2 conversion and H_2 and CH_4 content, at equilibrium with stoichiometric feeding

The progress of the conversion also depends on the composition of the reagent gas which is sometimes represented by the parameter:

$$F = \frac{m_{H_2} - m_{CO_2}}{m_{CO} + m_{CO_2}}$$

When we talk about methanation of carbon dioxide, however, we refer directly to the ratio between carbon dioxide and hydrogen. The stoichiometric ratio is 1:4. Under-stoichiometric hydrogen involves a not complete conversion of carbon dioxide, whereas over-stoichiometric hydrogen is favourable but with a very limited effect when the temperature is contained within 300 °C [112].

Reactors are classified as follows:

- two-phase
  - in a fixed bed
    - adiabatic
    - isotherms
  - fluidised bed
- three-phase
- structured

Those currently used are fixed bed reactors in which the catalyst is kept in a well-defined space inside the reactor [113]. Among these, the most used are the adiabatic ones that operate at pressures of 20–100 bar [114] and temperatures of 240 °C–700 °C [91]. The wide range of temperatures is due to the fact that adiabaticity raises the temperature favouring the kinetics and disfavouring the chemical equilibrium: hence the need to use more reactors in series intercooled in order to obtain a progressive advancement of the reaction at decreasing temperature levels.

Feeding an adiabatic reactor with a methane-free mixture results in a very high reaction temperature which is dangerous for the catalyst [115]. This is the case of a mixture of pure hydrogen and carbon dioxide fed to the first reactor. The most important technology for adiabatic fixed bed reactors is TREMP (Topsøe Recycle Energy-efficient Methanation Process) which uses a partial recycle of the reactor products at the inlet in such a way that an adequate amount of methane is contained into the gas fed to the first reactor [116,117] to decrease the temperature reaction.

The process is schematised in Figure 5.44. Downstream of each reactor, the gas temperature is brought back within the range of 250 °C–350 °C. The catalysts used (MCR-2X, MCR4) are stable at high temperatures allowing the reactors to operate up to 700 °C with a maximum pressure of 30 bar. The heat recovered at this temperature allows the production of superheated steam at high pressure (100 bar, 540 °C).

Table 5.8 reports an example for the composition and characteristics of the gas in the various passages starting from a synthesis gas obtained from biomass gasification.

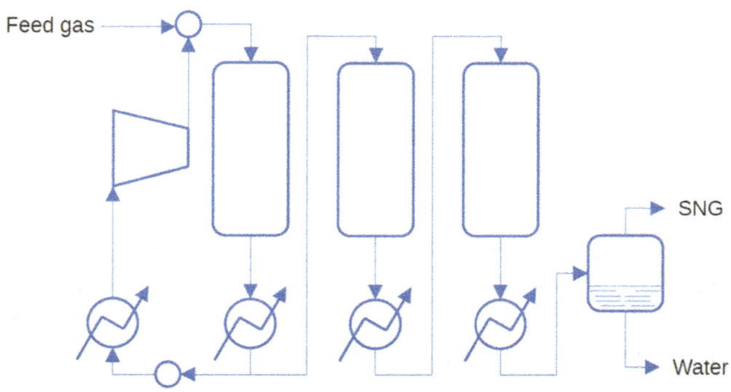

Figure 5.44 TREMP process scheme

Table 5.8 Characteristics of flows in the TREMP process [118]

	Feed	**R1$_{in}$**	**R1$_{out}$**	**R2$_{out}$**	**R3$_{out}$**	**SNG**
T (°C)		300	604	451	303	23
p (MPa)	2.730	2.720	2.710	2.705	2.700	2.700
Flow (m^3/h)	535	1416	1255	348	334	119
			Gas composition (vol%)			
H_2	65.45	36.87	20.96	8.10	1.77	3.11
CO	9.84	4.28	1.17	0.00	0.00	0.00
CO_2	8.96	6.13	4.46	2.07	0.95	1.67
CH_4	11.30	28.12	37.44	44.35	47.28	82.95
H_2O	0.00	19.19	29.82	38.84	43.06	0.10
N_2	4.4	5.41	6.15	6.64	6.93	12.16

Other processes do not use any recirculation. For example, the RMP process which presents a multi-stage quenching configuration with six intercooled fixed bed adiabatic reactors. Figure 5.45 shows an outline of the process.

The syngas enters the methanation unit distributed between the first three reactors. The temperature control in the first reactor is carried out by mixing a steam flow with the incoming syngas. The pressure can vary between 4.5 and 77 bar, while the maximum temperature is close to 800 °C. The addition of steam penalises the production of methane in favour of that of carbon dioxide.

The Imperial Chemical Industries (ICI) process, shown in Figure 5.46, is similar, but much simpler as it is based on only three methanation reactors. Also in this case, the steam is added to the incoming syngas to keep the temperature of the first reactor under control [118].

The gas produced by these processes without recirculation often has a high carbon dioxide content. For this reason, Foster Wheeler developed the VESTA process depicted in Figure 5.47 and based on the ICI process [119]. The removal of acid gases upstream of the water gas shift (5.24) and the removal of carbon dioxide downstream of the last reactor are characteristics of this process.

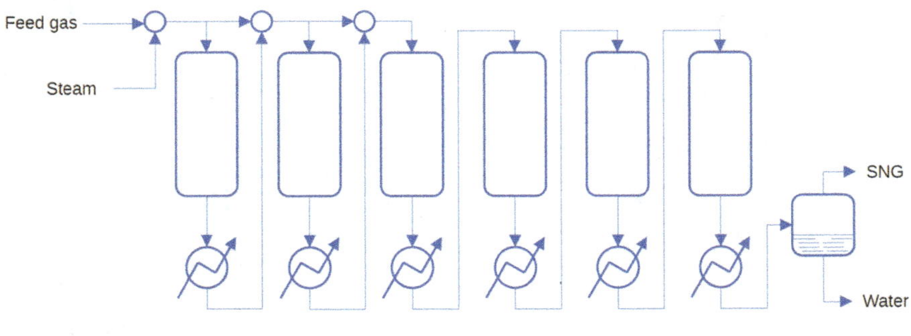

Figure 5.45 Scheme of the RMP process

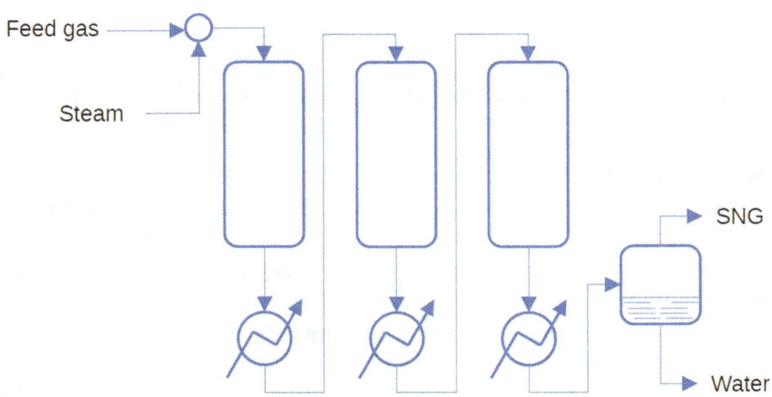

Figure 5.46 Scheme of the ICI process

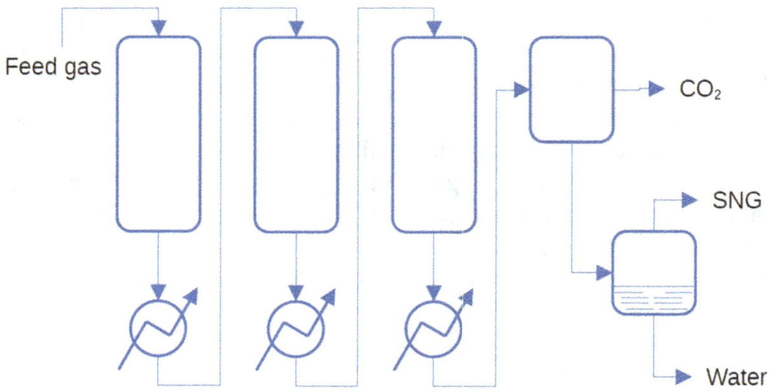

Figure 5.47 Scheme of the VESTA process

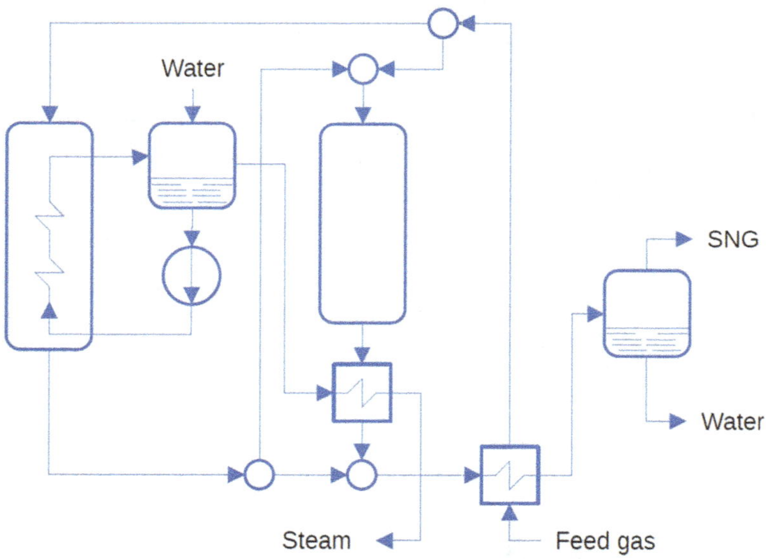

Figure 5.48 Scheme of the Linde process with isothermal reactor

The configuration of the isothermal process is much simpler and involves the use of a tubular fixed bed reactor which, however, is constructively more complex and therefore more expensive [91,120]. It also has greater maintenance needs [121]. The operating temperature is obviously lower than in the case of adiabatic reactors, usually in the range 250 °C–300 °C: this reduces the heat recovery efficiency, unless there is the need for low-temperature heat [118,122,123]. The operating pressure is also lower, usually between 1 and 20 bar.

Linde developed a process with two reactors in series (Figure 5.48): the first isothermal and the second adiabatic.

A longer life span of the catalyst and lower reaction costs are further advantages of isothermal reactors [91]. Figure 5.49 shows a reactor with a circular distribution of the tube bundles for steam generation, incorporated in the catalyst bed.

Fluidised bed reactors are characterised by approximately isothermal temperature profiles and the ease of removing or adding catalyst without the need to stop the process. However, the fluidisation of the catalyst particles presents some problems. The gas velocity, on the one hand, must be high enough to maintain fluidisation, while, on the other hand, it must not be too high to avoid elutriation of the catalyst [118]. Furthermore, the friction generated by fluidisation can lead to a reduction in the life of the reactor itself due to the high mechanical stress on the walls and the risk of catalyst deactivation [109].

Although this concept has been studied since the 1950s, only one commercial plant based on fluidised bed reactors has been built, named COMFLUX process, constructed in 1977 and commercially operated at the Ruhrchemie plant site in Oberhausen (Germany) [118,119,124].

A flexible solution of methanation based on a fluidised bed one step reactor was also proposed to better follow methanation process driven by non-programmable renewables [125].

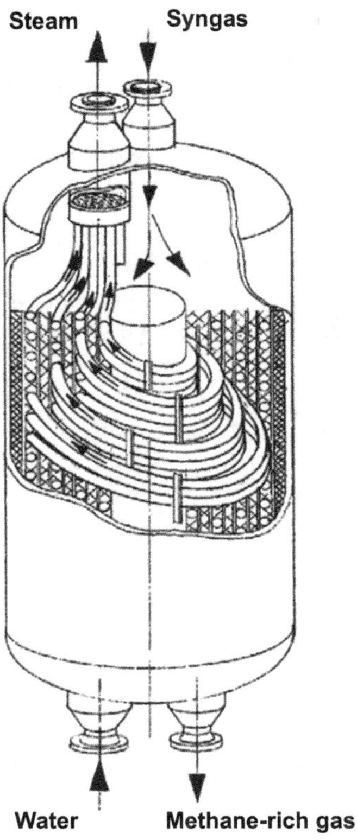

Figure 5.49 Example of an isothermal reactor

The concept of three-phase methanation, however, dates back to the 1970s and 1980s. The introduction of a heat transfer liquid facilitates the dispersion of the heat released during the Sabatier reaction, allowing to obtain an almost isothermal temperature profile in the reactor [126]. However, the technological feasibility of three-phase reactors is still in the experimental phase.

Finally, structured reactors consist of a metal body crossed by channels whose internal surface is coated by the catalyst [122]. Heat is then transferred very quickly and evenly [91,126]. The main obstacles to the diffusion of this technology lie in the difficulty of depositing the catalyst and its replacement in case of deactivation [127].

CO_2 conversion is favoured at low temperatures and high pressures. The operating window of nickel and ruthenium-based catalysts is between 250 °C and 450 °C. Figure 5.50 shows that, for temperatures lower than 350 °C and pressures higher than 5 bar, the CO_2 conversion is always higher than 95%. Increasing the temperature, a conversion of more than 95% can be obtained by operating at increasing pressures.

Figure 5.51 shows that the operating pressure is less important from the point of view of carbon dioxide conversion since the marginal conversion increase is decreasing as the pressure increases. However, pressure is important to increase the methane purity as the impact on hydrogen and carbon oxides is significant.

A 6 MW power to gas plant was built for Audi by Etogas in 2016 in Werlte (Germany) and is currently the largest in the world [128,129]. It includes three alkaline electrolysers of 2 MW_{el} each powered by a wind farm of four turbines of 3.6 MW_{el} located in the North Sea. The carbon dioxide comes from a nearby bio-methane production plant. Methanation is carried out by means of a 325 Nm^3/h SNG isothermal fixed bed reactor. The plant does not work in stationary conditions, but dynamically follows the availability of electricity supplied by the wind farm, i.e. for about 4000 h/year, so the annual production of SNG is around 1000 t/year.

As already mentioned, the variability of the electricity available for the production of hydrogen makes it necessary its storage so that the methanation reactors can

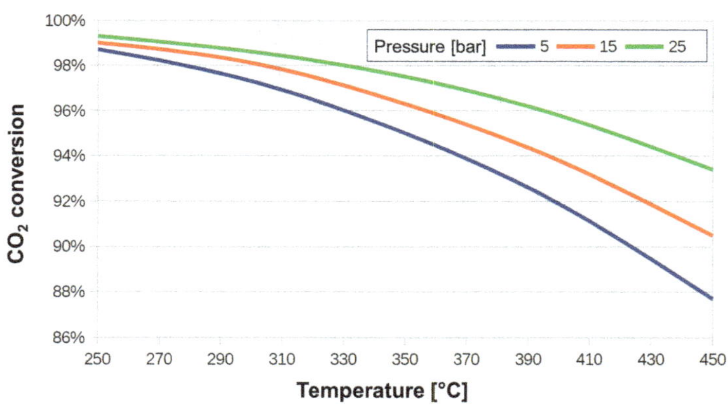

Figure 5.50 Carbon dioxide conversion vs temperature and pressure with stoichiometric feeding

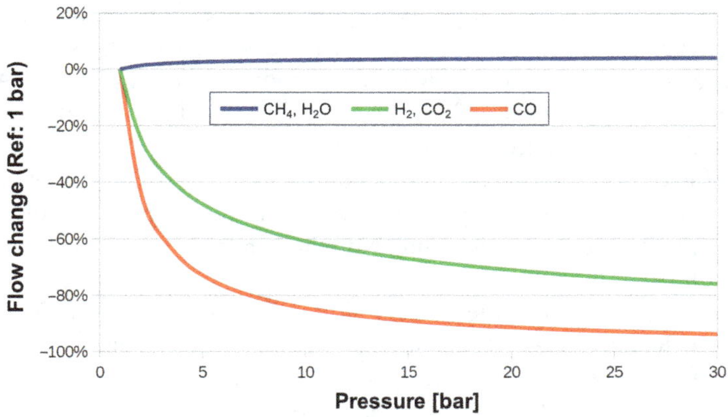

Figure 5.51 Percentage change in the concentration of hydrogen, methane, carbon oxides, and water in the SNG at equilibrium (feed $CO_2/H_2=4$) for a one-step process (T=300 °C)

operate in stationary conditions. To eliminate, or at least reduce, the complexity and associated costs, the reactors need to have the ability to operate dynamically. Whereas some authors argue that load changes do not have a major effect on the catalyst [91,130,131], there are others reporting the risk of cracking or sintering [109].

Fixed bed reactors seem unsuitable for such behaviour and should not be operated below 40% of rated load, while three-phase reactors can drop to 25% of rated load [131]. When the availability of hydrogen falls below these thresholds, it is necessary to switch to stand-by mode. The permanence of pure carbon dioxide inside the reactor is critical for the catalyst: therefore, after a shutdown, it is necessary to remove the residual carbon oxides by flushing the reactor either with hydrogen or with an inert gas [132]. Another important issue related to shutdowns is the duration: for a short shutdown is more convenient to keep the reactor in temperature, whereas for a long shutdown the energy required for a cold start-up could be smaller.

The alternative to the catalytic process is the biological one. The last phase of the anaerobic digestion process is methanogenesis in which the methanogenic microorganisms decompose the acetic acid produced in the previous phases and generate methane both directly and by combining carbon dioxide and hydrogen [133,134].

$$\begin{gathered} C_2H_4O_2 + H_2O \Rightarrow CH_4 + H_2CO_3 \text{ (acetoclastic methanogenesis)} \\ H_2CO_3 \Rightarrow H_2O + CO_2 \\ CO_2 + 4\ H_2 \Rightarrow CH_4 + 2\ H_2O \text{ (hydrogenotrophic methanogenesis)} \end{gathered} \tag{5.53}$$

As can be seen, this last reaction of the process is analogous to (5.50). The difference is that the former is favoured by the presence of catalysts, while this one is made possible by the presence of enzymes that are present in a group of micro-organisms (*Archaea*) able to produce methane when in mesophilic or thermophilic conditions (20 °C–70 °C). Hydrogen is the energy source for these microorganisms, whereas oxygen is toxic to the them. For such a reason, some oxidised compounds are needed, like the carbon oxides which are among the most diffused. Although the process is anaerobic, traces of oxygen are tolerated and some impurities such as nitrogen and sulphur are also tolerated [91,135].

The efficiency of the process can be improved by increasing the mass transfer coefficient and the gas solubility. This can be achieved by agitation of the water solution and by increasing the pressure, with the drawback of increasing the power consumption [91,136,137].

The most diffused kind of reactor for biological conversion is the Continuously Stirred Tank Reactor (CSTR), but in case of conversion of carbon dioxide and hydrogen also different types of reactors (fixed-bed and trickled-bed) are suitable and, although still at an experimental level, they promise to be more efficient [138].

An important feature is the ability to operate at low loads: while adiabatic fixed bed reactors should not drop below 40%, biological reactors can thrust down to 10% [91].

Biological methanation is generally characterised by temperature and pressure lower than those of catalytic methanation: 20 °C–70 °C and 1–3 bar for biological and 250 °C–700 °C and 1–100 bar for catalytic. Also GHSV ranges are much lower for biological methanation (up to 300 h^{-1}) with respect to catalytic methanation (thousands h^{-1}). Tolerance to impurities is much higher for *Archaea* than for catalysts. Another advantage of the biological process is the need of only one reactor, whereas in the catalytic process this is possible only in case of isothermal reactor which is not the most common. However, the use of adiabatic reactors allows a better heat recovery [123].

Fixed-bed catalytic methanation is a mature technology, whereas almost all the alternatives are still at a lab or pilot-scale level [91,139].

The main cost in a power to methane plant is the electrolyser, whereas the contribution of methanation reactors is rather low [91]. An important option to reduce the global cost is dynamic operation of the methanation reactor since its higher cost is compensated by the lower cost of the hydrogen storage.

The cost of SNG depends on electricity and carbon costs and it is extremely variable inside the range 0.04–0.60 €/kWh [99,140–144].

Some options for system integration of SNG production with or without power generation have been discussed by Spazzafumo and his colleagues [145–148]. Since the methanation is a strong exothermic process, the main problem of system integration is the heat recovery. The mentioned papers report efficiencies in the range 52%–74%, based on HHV. Bargiacchi also carried out a Life Cycle Assessment of these layouts, considering global warming potential, acidification potential and non-renewable cumulative energy demand [148]. Comparing natural gas and SNG production from biomass and renewable electricity, the latter showed a significantly lower impact under global warming potential and non-renewable cumulative energy demand, whereas a higher impact resulted under the acidification footprint, mainly because of the mining of catalyst metals used for electrodes.

5.3.2 Power to methanol

Methanol is an alcohol industrially produced by hydrogenation of carbon oxides:

$$CO + 2H_2 \Rightarrow CH_3OH \quad \Delta H^\circ = -90.7 \text{ kJ/mol} \tag{5.54}$$

$$CO_2 + 3H_2 \Rightarrow CH_3OH + H_2O \quad \Delta H^\circ = -49.5 \text{ kJ/mol} \tag{5.55}$$

It is a very important raw material in the chemical industry. It can be used as a fuel either in internal combustion engines, pure or blended with gasoline, or in DMFC [72]. Moreover, it can be used as a hydrogen carrier since hydrogen can be obtained from methanol steam reforming (reverse of (5.55)).

Its storage and transportation are easy because it is liquid at ambient conditions. It is toxic and hazardous like other common petroleum-based fuel [149].

Although traditional methanol production is based on fossil fuels, green methanol can be produced from green hydrogen and biomass or captured CO_2.

Just as for methane, methanol synthesis is favoured by low temperature and high pressure. Typical values are 200 °C–300 °C and 50–100 bar with a copper-based catalyst [150,151]. However, higher temperatures can be adopted to favour the reaction kinetics with the result that a lower final conversion is obtained. Actually this is not a real issue since it is possible to recycle the unconverted carbon oxides and hydrogen as shown in the schematic layout of a methanol production plant of Figure 5.52. In this way, very high conversion levels can be obtained with small losses due to periodical purging to eliminate the unwanted substances accumulated.

Catalysts can be poisoned by various substances. For example, a Cu/ZnO-based catalyst can be deactivated by a gas containing 0.05 ppm of sulphuric acid, 1 ppb of hydrochloric acid, few ppb of metal carbonyl, 0.25 mg/Nm^3 of alkalis, 1.0 mg/Nm^3 of tar and 0.1 mg/Nm^3 of particles [152]. However, since poisoning substance are usually removed upstream of the catalysis bed, the catalyst can be deactivated only by sintering.

Catalysts like MK-151 Fence by Haldor Topsøe are able to maintain higher values of activity and selectivity even close to the end of their life [153,154].

Almost all hydrogen and carbon oxides can be converted thanks to the liquid state of methanol: unconverted reactants can be easily separated and recycled

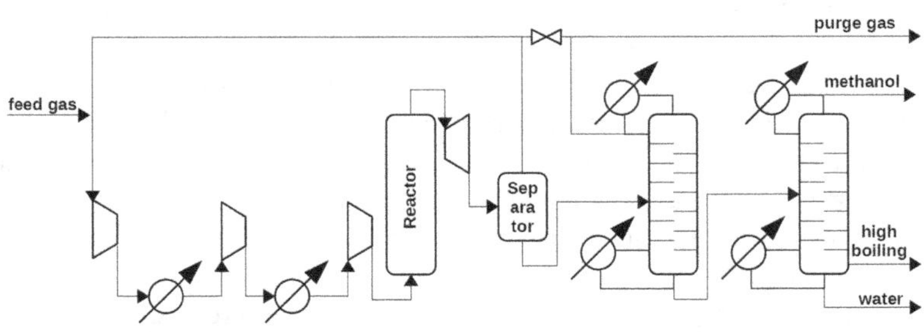

Figure 5.52 Flowsheet for a methanol production plant [149]

limiting the losses to periodical purge to eliminate impurities. The American Society for Testing and Materials classifies commercial methanol in A and AA purity grades [155]. To obtain the required purity, further stages of gas removal by flashing and distillation could be needed [150,152].

Scale economy is very important: a 10,000 t/d plant costs only about the triple of a 2000 t/d plant.

Reactors can be classified in the following way [152]:

- fixed bed reactors (FBR):
  - adiabatic, used in series with inter-cooling, has the advantages of easy scale-up, no mechanical stress of catalyst, and defined residence time, but the disadvantages of a lower heat transfer, the danger of hotspots and the several pressure vessels and piping;
  - polytropic:
    - quench, with two separate beds and injection of cold feed reactant gas in between, has the advantages of no mechanical stress of catalyst and improved temperature control, but the disadvantage of a large catalyst volume;
    - gas-cooled converter (GCC), with catalyst filled into shell or tube side and refrigerated by cold gas;
- steam raising converter (SRC), quasi-isothermal with catalyst refrigerated by water evaporation, has a very efficient heat recovery, good heat transfer and temperature control, but is expensive requiring a large number of tubes;
- fluidised bed reactors, isothermal, with heat exchange carried out by a submerged coil, which guarantees a high heat transfer coefficients, and a uniform temperature distribution; the disadvantages are a non-uniform residence time, the attrition of catalyst, the erosion of internals and the problematic scale-up;
- liquid-phase reactors, isothermal with catalyst suspended in liquid hydrocarbons, used as heat transfer medium, which results in excellent heat transfer performance, but also in high mechanical stress of catalyst.

Commercial reactors are commonly classified on the basis of flow direction: axial or radial. Lurgi SRC is the most common axial reactor, whereas Haldor Topsøe is the most common radial reactor. Several data about axial or radial flow commercial reactors are shown in Tables 5.9 and a comparison of some characteristics is shown in Figures 5.53–5.57.

Radial flow reactors are characterised by a lower pressure drop, which means lower power needed for recycled gas recompression. Isothermal Methanol Casale (IMC) process is one of the most advanced technologies for methanol production: it is based on an axial-radial flow reactor.

Other types of reactors have been proposed [156–160]; however, they have not been demonstrated on an industrial scale.

The range of temperature for methanol production from carbon dioxide is 230 °C–270 °C, similar to that suitable for production from syngas. Although the

Table 5.9 Characteristics and operating parameters of main commercial reactors

	Licensor	Flow	Type	Catalyst location	Heat exchanger	Stages	Max capacity (t/d)
Lurgi	Lurgi	Axial	SRC	Tube-side	Tubular	1	1500–2200
Super-converter	MGC and MHI	Axial	GCC/SRC	Double-pipes	Tubular	1	Not available
Mega Methanol	Lurgi	Axial	GCC and SRC	Shell-side	Tubular	2	5000–10,000
LPMeOH	Air products	Axial	Slurry	Shell-side	Tubular	1	Low
IMC	Methanol Casale	Mixed	SRC	Shell-side	Plate	1	7000–10,000
Variobar	Linde	Radial	SRC	Shell-side	Tubular	1	4000
MRF-Z	Toyo (TEC)	Radial	SRC	Shell-side	Bayonet	1	5000
Haldor Topsøe	Haldor Topsøe	Radial	Adiabatic	Fixed bed	Intercooler	2-4	10,000

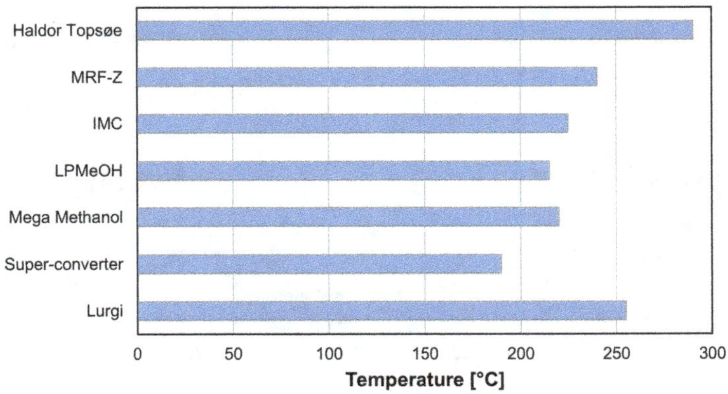

Figure 5.53 Outlet temperature

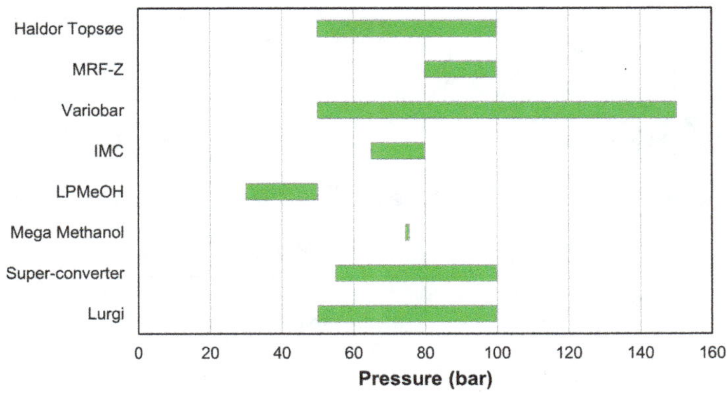

Figure 5.54 Pressure range

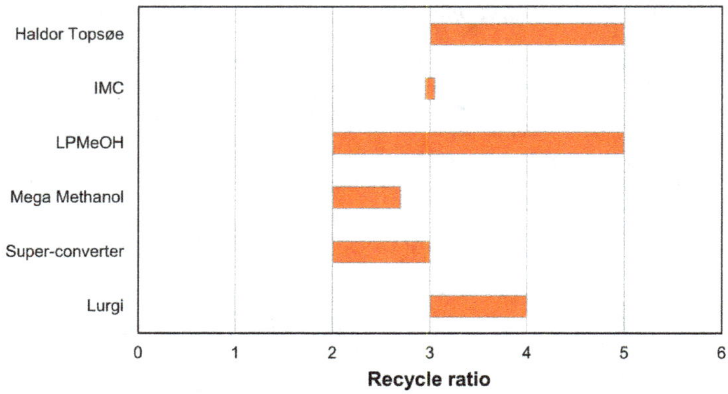

Figure 5.55 Recycle ratio range

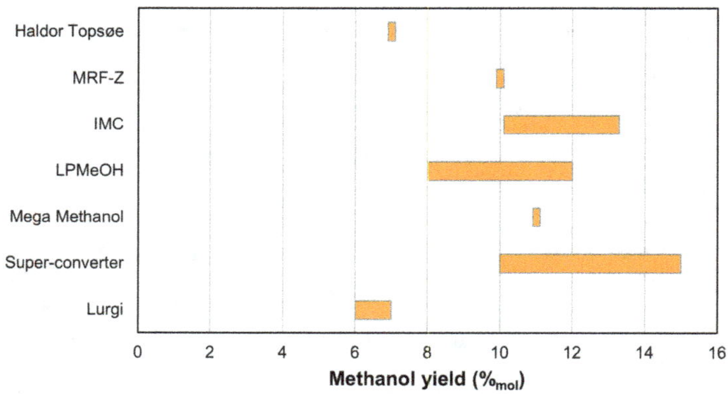

Figure 5.56 Methanol yield range

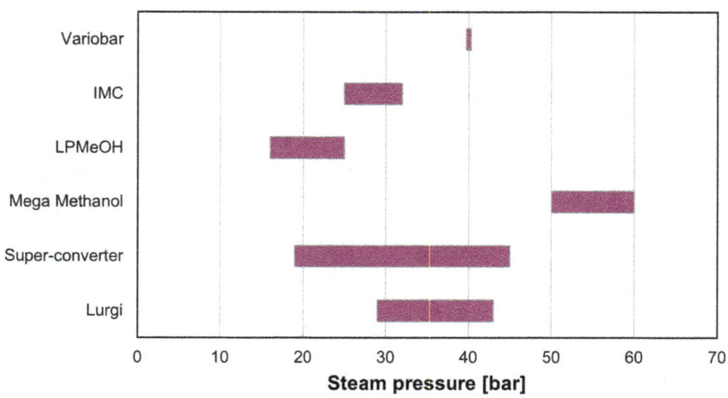

Figure 5.57 Steam pressure range

use of syngas results in a higher methanol production (84.5% vs. 63.0%–63.7%), the use of carbon dioxide has the advantage of a lower presence of impurities which results in cost saving for methanol distillation. In particular, there are no paraffins and ketones, the amounts of higher alcohols, dimethyl ether, and esters are significantly reduced [152]. Moreover, methanol production from carbon dioxide allows to capture it from sources with a great concentration like chimneys of power plants and industries, thus reducing climate-altering gases emissions into the atmosphere. A disadvantage is the need to oxidise one more atom of oxygen and this requires an extra mole of hydrogen (5.55). This involves also a much greater presence of water inside the reactor, and it is necessary to develop specific catalysts [161–164].

Biological conversion from carbon dioxide to methanol does not exist or, at least, has not yet been discovered. However, some bacteria, called methanotrophic bacteria, consume methane for their metabolism and produce methanol, although not a final product. Therefore, this process could be useful to obtain a liquid fuel after having produced methane from carbon dioxide.

Methanotrophic bacteria live in the same environments of methanogenic bacteria being that they are mesophilic too and take advantage from the methane production by methanogenic. The environment is anaerobic; however, methanotrophic bacteria can find oxygen from some compounds or can live at the aerobic-anaerobic interface.

Methanotrophic bacteria metabolism is very complex:

- methane monooxygenase enzymes catalyse the conversion of methane into methanol;
- methanol dehydrogenase enzymes catalyse the conversion of methanol into formaldehyde;
- formaldehyde dehydrogenase enzymes catalyse the conversion of formaldehyde into formate;
- and finally formate dehydrogenase enzymes catalyse the conversion of formate into carbon dioxide.

Therefore, to obtain methanol it is necessary to inhibit the second process of the chain.

Contrary to the traditional energy-intensive process, which needs two steps (methane reforming, 5.25, and methanol synthesis, 5.54), biological methanol production occurs in a single step [165]:

$$CH_4 + \frac{1}{2}O_2 \Rightarrow CH_3OH \quad \Delta H^\circ = -126.2 \text{ kJ/mol} \tag{5.56}$$

at atmospheric pressure, low temperature and almost neutral acidity [166]. Unfortunately, this technology is still at laboratory scale.

5.3.3 Power to other synthetic fuels

Dimethyl ether (DME) is the simplest ether (CH_3OCH_3). It is a colourless, not toxic and highly flammable gas which can replace LPG (liquefied petroleum gas) or

Table 5.10 Physical properties of conventional diesel and DME [52]

	Unit	Diesel	DME	Reference
Molecular formula	[–]	$C_{14}H_{30}$	CH_3OCH_3	[167]
Oxygen content	[wt%]	<1	34.8	[168]
Density	[kg/l]	0.856	0.661*	[167]
Lower heating value	[MJ/kg]	41.66	28.62	[167]
Cetane number	[–]	51–54	55–60	[168]
Kinematic viscosity (40 °C)	[mm^2/s]	3	0.184	[168]
Normal boiling point	[°C]	125–400	−24.9	[167]

*Density in liquid state at p = 1.013 bar and T = −25 °C.

diesel fuel with minor modifications and some important advantages [52]. Table 5.10 shows a comparison between characteristics of diesel fuel and DME. The disadvantage of DME is the lower density and energy content due to the presence of oxygen.

DME can also be used as a hydrogen carrier [169] or directly in fuel cells [74–77]. It is liquid at atmospheric pressure and −24.9 °C or at 25 °C and 0.6 MPa [167,170]. Compared to other green fuels, DME seems to offer some interesting advantages: the highest conversion efficiency, and the lowest cost and well-to-wheel greenhouse gas emissions [167,171]. Furthermore, DME can be distributed using the existing LPG infrastructure [167,172].

Current world production is based on dehydration of methanol produced from synthesis gas mainly obtained from fossil fuels. After distillation of methanol, DME is obtained through the exothermic reaction below [173]:

$$2CH_3OH \Leftrightarrow CH_3OCH_3 + H_2O \quad \Delta H^\circ = -23.5\ \text{kJ/mol} \tag{5.57}$$

Low temperatures are favourable and typical values are in the range 220 °C–360 °C with a methanol conversion up to 70%–85% [162,174]. The low temperature implies slow reaction kinetics and requires the use of catalysts, typically solid weakly acidic catalysts such as γ-Al_2O_3 [170,174]. Pressure has no influence on the advancement of the process; however, it is useful to reduce the volume of the reactor and can reach 2 MPa [174–177].

Although the main products in the outlet gas are DME and water, considerable amount of methanol and minor amounts of by-products (carbon monoxide, hydrogen and hydrocarbons) has been reported by some researchers [174,178–180].

Among different production pathways, Figure 5.58 also shows a direct route consisting of a direct methanol synthesis and subsequent dehydration of DME in a single reactor [52,173]. This process involves different reactions: methanol synthesis (5.54), methanol dehydration (5.57) and water gas shift reaction (5.24). The global reaction is:

$$3CO + 3H_2 \Leftrightarrow CH_3OCH_3 + CO_2 \quad \Delta H^\circ = -90.4\ \text{kJ/mol} \tag{5.58}$$

The advantages of the direct synthesis consist in a favourable advancement of the reactions, due to the methanol and water direct consumption, and in the reduction of the number of reactors [181–183]. However, the product at the exit of the reactor includes also carbon dioxide which results in a more complex purification process.

Finally, there is a third route to produce DME, that is producing methanol starting from hydrogen and sequestered carbon dioxide used as a 'green' carbon source (Figure 5.58). The CO_2-based methanol production has the advantage of a strongly reduced amount of impurities with respect to the traditional process starting from syngas [184,185]. The main issue is that the reverse water gas shift reaction (5.51) produces additional water which inhibits reactions in which water is produced. This could be solved carrying out that reaction in a separate reaction converting CO_2 to CO and extracting water. This allows a high yield, but also complicates the plant layout [186].

Since removal of water is extremely beneficial both for direct and indirect processes, some alternative concepts are under development: reactive distillation [187,188], sorption-based and membrane-based water removal [189]. Reactive distillation consists of a combined reaction and continuous products removing allowing a full conversion of reactants. Sorption-based water removal aims to the in-situ adsorption of water for example from zeolite particles. Zeolite is continuously fed and withdrawn to be regenerated in a separate reactor or alternatively a batch process adsorption/regeneration can be adopted. Membrane-based water removal is based on a selective membrane which allows water to exit from the reactor directly during the process.

Ammonia is a colourless gas with a boiling point of −33.3 °C, but it is also highly miscible with water. In its concentrated form, it is caustic and hazardous. However, it is a widely used product, for example as a fertiliser as well as for the synthesis of pharmaceutical and cleaning products.

The direct reaction of nitrogen and hydrogen:

$$N_2 + 3H_2 \Rightarrow 2NH_3 \quad \Delta H^\circ = -92 \ \text{kJ/mol} \tag{5.59}$$

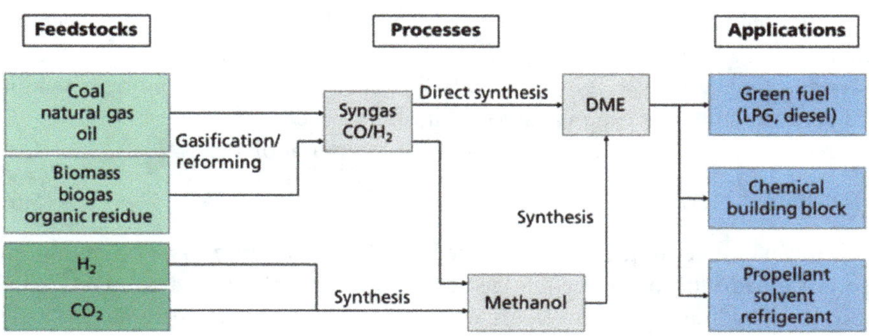

Figure 5.58 Overview over DME production pathways for different feedstocks [52]

is exothermic; however, it is not spontaneous due to the strong triple bonds between the nitrogen atoms and requires temperatures higher than 800 °C, which is unfavourable to chemical equilibrium. Only some kinds of bacteria, thanks to an enzyme called nitrogenase, can break such bonds at room temperature [190,191].

The conventional process to obtain ammonia is the Haber–Bosch process in which the use of a solid catalyst based on iron, potassium, calcium and aluminium oxides allows an heterogeneous reaction at a temperature in the range 350 °C–550 °C and at a pressure in the range 15–30 MPa. Even at this temperature, equilibrium is unfavoured and the advancement of reaction is limited to 25%–35%, so that it is necessary to recover and recycle unreacted reagents [192]. Purity of reactants is very important since traces of sulphur and chlorine compounds affect irreversibly the catalyst reducing its lifetime.

Alternative electrochemical processes could have advantages in term of conversion level and energy consumption, but they are still in a research phase [193–195]. An innovative electrochemical approach uses catalytic nitrogen activation which allows to produce ammonia at rates close to equilibrium at a temperature of 250 °C. Previous experiments were carried out using phosphate electrolytes [196,197]. To avoid anode degradation at current densities exceeding 50 mA cm^{-2}, Sagara and Kubota recently used molten alkaline electrolytes. At the time of writing this book, the article has not yet been published except in the form of a short abstract [198].

The current ammonia industry uses hydrogen produced from fossil fuels. Due to its very large market, it has significant impact on the environment. To produce 'green' ammonia, it is thus necessary to produce 'green' hydrogen, but also 'green' nitrogen. Contrary to hydrogen, nitrogen is abundant in the atmosphere and needs only to be separated from other air components. Its separation form air is mainly based on a cryogenic process, whereas for small production also membranes or pressure swing adsorption can be used [199]. However, all these processes use electric power so that it is necessary to avoid the use of fossil fuels to generate such electricity.

At high temperature (850 °C–950 °C), the ammonia synthesis reaction (5.59) inverts and ammonia decomposes into hydrogen and nitrogen in the presence of nickel as a catalyst. Therefore, considering its very high hydrogen mass concentration, about 17.6%, ammonia is considered as a promising hydrogen carrier. Decomposition could occur also at lower temperatures, but this requires expensive catalysts.

Ammonia could be used in direct ammonia fuel cells. Thanks to high operating temperature, a SOFC can decompose ammonia in nitrogen and hydrogen, using the latter as a fuel. However, it is not easy to find catalysts offering long-term stability and external ammonia decomposition still offers better performance [78].

Ammonia could also burn in air at a concentration of 15%–27%:

$$4NH_3 + 3O_2 \Rightarrow 2N_2 + 6H_2O \tag{5.60}$$

However, due to its low flame speed and high resistance to auto-ignition, the use combined with other fuels is necessary, for example with a relatively small amount of hydrogen. Moreover, its corrosive nature requires to remove copper-based components. Another issue arises from nitrogen oxides production.

Although power to ammonia could be a good means to store surplus electrical power, ammonia plants are designed for continuous operation, which is a great challenge considering fluctuating behaviour of renewable energy source. An intermediate hydrogen storage is fundamental to buffer the energy fluctuations avoiding to stop the plant. The hydrogen storage represents an important cost of a power to ammonia plant; however, the cost of a power to ammonia plant is mainly affected by the cost of the electrolyser [200]. Therefore, the electrolyser should operate at the possible highest average operating capacity. When the main task of the plant is ammonia production, the size of the electrolyser should be small and the surplus electricity should be used for other purposes. An option to minimise the hydrogen production and storage costs is to combine different renewable energy sources, like for example solar and wind [201]. Another option to reduce costs is electric power import from the grid when no renewable energy sources are available [202], but it is important that such an imported energy be 'green' through Power Purchase Agreement of green electricity.

At high pressures, 13–25 MPa, urea can be obtained from liquid ammonia and carbon dioxide (Bazarov reaction). The process is carried out in two steps. A first strong exothermic reaction allows to obtain ammonium carbamate:

$$2NH_3 + CO_2 \Leftrightarrow [NH_4][H_2NCO_2] \tag{5.61}$$

which can be decomposed into urea and water:

$$[NH_4][H_2NCO_2] \Leftrightarrow CO(NH_2)_2 + H_2O \tag{5.62}$$

Urea is a solid highly soluble in water widely used as a fertiliser thanks to the higher nitrogen content.

A very interesting feature of urea is that electrolysis of urea aqueous solutions requires a much lower voltage (0.37 V) with respect to water electrolysis [203]. Therefore, also urea can be considered as a hydrogen carrier.

Moreover, urea can be directly used in alkaline fuel cells [81].

Finally, another interesting synthetic product is formic acid (HCOOH), the simplest carboxylic acid. It is a colourless liquid with flammability limits of 12%–38% and is commercially available at a concentration of 85% [54]. It is corrosive, but not toxic. Its density at 20 °C is 1.22 g/cm^3 and melting and boiling points are, respectively, 8.4 °C and 100.8 °C.

The direct process from carbon dioxide and water to formic acid should be preferable, but is disfavoured. A large review of research works carried out in such a direction is reported in [54], including the electrochemical direct process:

$$\text{Cathode: } 2CO_2 + 4H^+ + 4e^- \Rightarrow 2HCOOH$$

$$\text{Anode: } 2H_2O \Rightarrow O_2 + 4H^+ + 4e^-$$

Considering indirect processes to obtain formic acid, the most widely used is methyl formate hydrolysis which consumes carbon monoxide and water and uses methanol as a recyclable reactant. A first reaction generates methyl formate from methanol and carbon monoxide:

$$CH_3OH + CO \Rightarrow HCO_2CH_3 \tag{5.63}$$

then methyl formate is hydrolysed producing formic acid and regenerating methanol:

$$HCO_2CH_3 + H_2O \Rightarrow HCOOH + CH_3OH \tag{5.64}$$

Therefore, the global reaction consumes only carbon monoxide and water:

$$CO + H_2O \Rightarrow HCOOH \tag{5.65}$$

which could be obtained from the reverse water gas shift reaction (5.51).

This reaction can be reverted supplying heat making formic acid a hydrogen carrier, since hydrogen can be finally obtained thanks to the water gas shift reaction (5.24):

$$HCOOH \Rightarrow CO + H_2O \Leftrightarrow CO_2 + H_2 \tag{5.66}$$

Furthermore, using catalysts like platinum and ruthenium, formic acid can be directly decomposed into hydrogen and carbon dioxide. Therefore, formic acid can be used as a liquid hydrogen carrier adapting existing infrastructures used for fossil fuels distribution [204]. Its hydrogen content is 4.34% by weight and due to gravimetric density, its energetic density is comparable with 700 bar commercial pressure cylinder one [54]. After distribution, hydrogen can be obtained and used, for example, inside a PEMFC to generate electric power [205].

Besides its use as a hydrogen carrier, formic acid could be used in direct formic acid fuel cells [84,85]. However, there are still some issues to be solved before obtaining commercial direct fuel cells.

A process gaining increasing interest is electrolytic CO_2 reduction which can be achieved both with alkaline or acid electrolytic cells and allows to obtain different liquid or gaseous products (Figure 5.59). Here are some examples of reduction half-reactions [206]:

$$CO_2 + 8H^+ + 8e^- \Rightarrow CH_4 + 2H_2O \tag{5.67}$$

$$CO_2 + 6H_2O + 8e^- \Rightarrow CH_4 + 8OH^- \tag{5.68}$$

$$CO_2 + 6H^+ + 6e^- \Rightarrow CH_3OH + H_2O \tag{5.69}$$

$$CO_2 + 5H_2O + 6e^- \Rightarrow CH_3OH + 6OH^- \tag{5.70}$$

$$2CO_2 + 12H^+ + 12e^- \Rightarrow C_2H_4 + 4H_2O \tag{5.71}$$

$$2CO_2 + 8H_2O + 12e^- \Rightarrow C_2H_4 + 12OH^- \tag{5.72}$$

$$CO_2 + 2H^+ + 2e^- \Rightarrow HCOOH \tag{5.73}$$

However, although along the last decades some researchers tested different catalysts for example for direct electrochemical reduction of carbon dioxide to

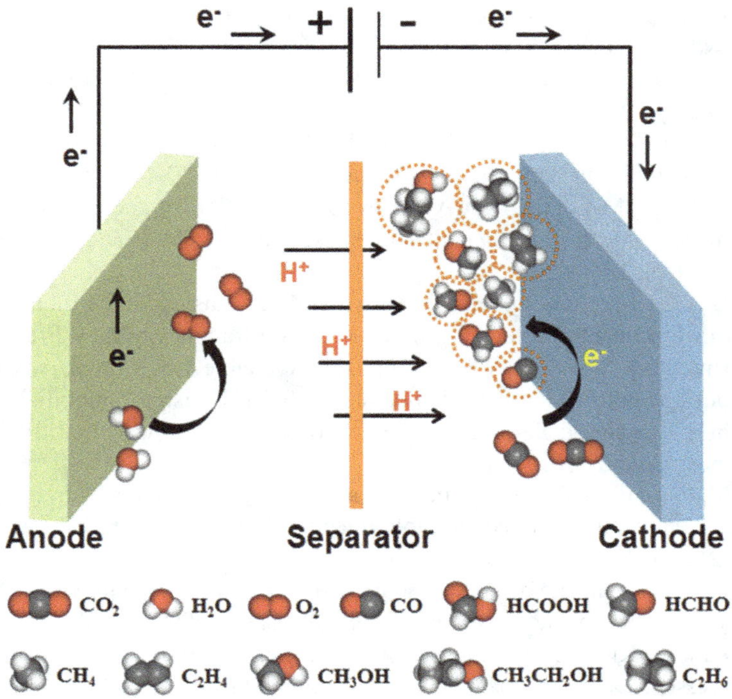

Figure 5.59 Co-electrolysis [206]

methanol was tried [207–211], results are not yet cost competitive. In any case, electrolytic CO_2 reduction also allows to obtain hydrogen and carbon monoxide which can be used for a separate synthesis process [212].

References

[1] Eichlseder H, Grabner P, and Heindl R. Hydrogen internal combustion engines. In: Stolten D (ed.). *Hydrogen and Fuel Cells: Fundamentals, Technologies and Applications*. 1st edn. Weinheim: Wiley-VCH; 2010. pp. 811–30.

[2] Wikimedia Commons, François Isaac de Rivaz (1752–1828). File: Rivaz Engine.jpg – Wikimedia Commons [Internet]. 1804 patent in the Republic of Valais. 2011 [cited 2023 Nov 1]. Available from: https://commons.wikimedia.org/wiki/File:Rivaz_Engine.jpg.

[3] Fondazione Barsanti e Matteucci. The history of the invention – The internal combustion engine [Internet]. [cited 2023 Nov 1]. Available from: https://www.barsantiematteucci.it/en/history-of-the-invention/.

[4] CNBC. Toyota commissions Yamaha Motor to develop hydrogen-fueled V8 engine [Internet]. 2022 [cited 2023 Nov 1]. Available from: https://

www.cnbc.com/2022/02/22/toyota-commissions-yamaha-motor-to-develop-hydrogen-fueled-engine.html.

[5] Furuhama S. Hydrogen engine systems for land vehicles. *Int J Hydrogen Energy*. 1989;14(12):907–13.

[6] Petkov T, Veziroğlu TN, and Sheffield JW. An outlook of hydrogen as an automotive fuel. *Int J Hydrogen Energy*. 1989;14(7):449–74.

[7] DeLuchi MA. Hydrogen vehicles: an evaluation of fuel storage, performance, safety, environmental impacts, and cost. *Int J Hydrogen Energy*. 1989;14(2):81–130.

[8] BMW Group. Design e mobilità sostenibile: I disegni delle future stazioni di rifornimento di veicoli a idrogeno in mostra alla Bovisa – Press release [Internet]. 2002 [cited 2024 Aug 31]. Available from: https://www.press.bmwgroup.com/italy/article/detail/T0020471IT/design-e-mobilità-sostenibile:-i-disegni-delle-future-stazioni-di-rifornimento-di-veicoli-a-idrogeno-in-mostra-alla-bovisa?language=it.

[9] Mathur HB, and Das LM. Performance characteristics of a hydrogen fuelled S.I. engine using timed manifold injection. *Int J Hydrogen Energy*. 1991;16(2):115–27.

[10] Pede G. L'utilizzo dell'idrogeno nei motori a combustione interna – Dispensa a cura dell'Ing. Giovanni Pede, ENEA, 8 Maggio 2008. ENEA; 2008.

[11] Park C, Kim Y, Oh S, *et al.* Effect of fuel injection timing and injection pressure on performance in a hydrogen direct injection engine. *Int J Hydrogen Energy*. 2022;47(50):21552–64.

[12] Norbeck JM, Heffel JW, Durbin TD, Tabbara B, Bowden J, and Montano MC. *Hydrogen Fuel for Surface Transportation*. Warrendale, PA: SAE International; 1996.

[13] Verhelst S, and Wallner T. Hydrogen-fueled internal combustion engines. *Prog Energy Combust Sci*. 2009;35(6):490–527.

[14] Rotary Performance Inc. RX7.com. Rotary displacement – Rotary performance technical articles [Internet]. 2004 [cited 2023 Nov 1]. Available from: http://www.rx7.com/techarticles_displacement.html.

[15] Jaura AK, Ortmann W, Stuntz R, Natkin B, and Grabowski T. Ford's H2RV: an industry first HEV propelled with a H_2 fueled engine – a fuel efficient and clean solution for sustainable mobility. *SAE Tech Pap*. 2004;2004(724).

[16] Bengt R. Malmö hydrogen and CNG/hydrogen filling station and hythane bus project. *In: 16th World Hydrogen Energy Conference 2006, WHEC 2006*. 2006. pp. 1566–75.

[17] Lopez-Ruiz G, Castresana-Larrauri J, and Blanco-Ilzarbe JM. Thermodynamic analysis of a regenerative Brayton cycle using H_2, CH_4 and H_2/CH_4 blends as fuel. *Energies*. 2022;15(4):1508.

[18] Balestri M, Benelli G, Donatini F, Arlati F, and Conti G. Enel's Fusina hydrogen-fed power generation plant. In: 2007 *Int Conf Clean Electr Power, ICCEP '07*. 2007. pp. 456–63.

[19] ENEL. ENEL: first hydrogen-fuelled power now on-line in Venice [Internet]. 2009 [cited 2024 Aug 30]. Available from: https://corporate.

enel.it/en/media/explore-press-releases/press/2009/08/enel-first-hydrogen-fuelled-power-now-on-line-in-venice-.

[20] NEDO. World's first heat and electricity supplied in an urban area using 100% hydrogen – towards establishing optimal energy control technology in local communities [Internet]. 2018 [cited 2023 Nov 2]. Available from: https://www.nedo.go.jp/english/news/AA5en_100382.html.

[21] Kawasaki. World's first successful technology verification of 100% hydrogen-fueled gas turbine operation with dry low NO_x combustion technology improving power generation performances to realize a hydrogen society. *New Energy Industrial Technology Development Organizaion.* 2020.

[22] GE Gas Power. Hydrogen-fueled gas turbines [Internet]. 2023 [cited 2023 Nov 2]. Available from: https://www.ge.com/gas-power/future-of-energy/hydrogen-fueled-gas-turbines.

[23] Tucker EE. *Oxyhydrogen Steam Generator* – United States Patent No. US1483917A of Feb, 19, 1924 [Internet]. United States; US1483917A, 1924. Available from: https://patents.google.com/patent/US1483917A/en. Accessed 17 December 2024

[24] Sternfeld HJ, Reinkenhof J, and Lambrecht H. *Steam Generator – United States Patent No. US4377067A [Internet]*. United States; US4377067A, 1983. Available from: https://patents.google.com/patent/US4377067A/en. Accessed 17 December 2024

[25] Sternfeld HJ, Reinkenhof J, and Lambrecht H. *Steam Generator – United States Patent No. US4474140A [Internet]*. United States; US4474140A, 1984. Available from: https://patents.google.com/patent/US4474140A/en. Accessed 17 December 2024

[26] Gryaznov AN, and Malyshenko SP. *Steam Generator* – Russian Patent No. RU2079684C1 [Internet]. Russia; RU2079684C1, 1997. Available from: https://patents.google.com/patent/RU2079684C1/en?oq=+RU2079684C1 +. Accessed 17 December 2024

[27] Winter C, and Nitsch J. *Hydrogen as an Energy Carrier: Technologies, Systems, Economy.* 1st edn. In: Winter C, and Nitsch J (eds). Berlin: Springer; 1988. 380 p.

[28] Sternfeld HJ, and Heinrich P. A demonstration plant for the hydrogen/oxygen spinning reserve. *Int J Hydrogen Energy*. 1989;14(10):703–16.

[29] Haidn OJ, Fröhlke K, Carl J, and Weingartner S. Improved combustion efficiency of a H_2/O_2 steam generator for spinning reserve application. *Int J Hydrogen Energy*. 1998;23(6):491–7.

[30] Volkov AG, Bebelin IN, Gryaznov AN, and Malyshenko SP. Development and testing of experimental H_2/O_2 steam generators. In: *XI World Hydrogen Energy Conference, June 1996, Stuttgart (Germany)*. Stuttgart, Germany; 1996. pp. 1913.

[31] Glasgow CW, Olson RC, and Stockton EJ. *Dynamic Combustion Chamber* – International Patent No. WO2007109017A3 [Internet]. World Intellectual Property Organization International Bureau; WO2007109017A3, 2007.

Available from: https://patents.google.com/patent/WO2007109017A3/en. Accessed 17 December 2024

[32] Spazzafumo G. *Generatore diretto di vapore con premiscelazionedi acqua ai reagenti (Direct Steam Generator with Premixing of Water and Reactants)* – Italian Patent No. ITMS20090004A1 (UIBM0001395389) [Internet]. Italy: Ufficio Italiano Brevetti e Marchi; ITMS20090004A1, 2009. Available from: https://patents.google.com/patent/ITMS20090004A1/it. Accessed 17 December 2024

[33] Clean Energy Systems. Technology – oxy-fuel combustion [Internet]. 2023 [cited 2023 Nov 2]. Available from: https://www.cleanenergysystems.com/technology.

[34] Clean Energy Systems. Direct steam gas generators [Internet]. 2023 [cited 2023 Nov 2]. Available from: https://www.cleanenergysystems.com/direct-steam-gas-generators.

[35] Dunikov DO, Borzenko VI, and Schastlivtsev AI. Compact and fast-response 150 kW hydrogen-oxygen steam generator. *Int J Hydrogen Energy*. 2022;47(84):35897–902.

[36] Malyshenko SP, Gryaznov AN, and Filatov NI. High-pressure H_2/O_2-steam generators and their possible applications. *Int J Hydrogen Energy*. 2004;29 (6):589–96.

[37] Schastlivtsev AI, and Borzenko VI. Hydrogen-oxygen steam generator applications for increasing the efficiency, maneuverability and reliability of power production. In: *Journal of Physics: Conference Series, The International Conference "Problems of Thermal Physics and Power Engineering" (PTPPE-2017) 9–11 October 2017, Moscow, Russian Federation*. IOP Publishing; 2017. pp. 012213.

[38] POWER Magazine, and Palatel S. MHPS secures first order for hydrogen-capable J-Series gas turbines [Internet]. 2020 [cited 2023 Nov 2]. Available from: https://www.powermag.com/mhps-secures-first-order-for-hydrogen-capable-j-series-gas-turbines/.

[39] Chiba M, Arai H, and Fukuda K. International and national program and project Hydrogen Energy Technology Development in Japan: new Sunshine Program. In: *XI World Hydrogen Energy Conference, June 1996, Stuttgart (Germany)*. Stuttgart, Germany; 1996. pp. 13.

[40] Yamawaki S, Maya T, and Yasu S. Study of film cooling with shaped holes for a hydrogen combustion turbine blade. In: *XI World Hydrogen Energy Conference, June 1996, Stuttgart (Germany)*. Stuttgart, Germany; 1996.

[41] Jericha H, Wilplinger H, and Perz E. A novel high efficiency steam cycle. *In: ASME Cogen Turbo I, Montreaux*. 1987.

[42] Spazzafumo G. An overview of thermodynamic cycles based on direct stoichiometric combustion of hydrogen and oxygen. *J Sci Ind Res (India)*. 2003;62(1):71–80.

[43] Sugisita H, Mori H, and Uematsu K. A study of thermodynamic cycle and system configurations of hydrogen combustion turbines. *Int J Hydrogen Energy*. 1998;23(8):705–12.

[44] Cicconardi SP, Jannelli E, Perna A, and Spazzafumo G. A steam cycle with direct combustion of hydrogen and oxygen and an isothermal expansion. In: Saetre TO (ed). *Hydrogen Power: Theoretical and Engineering Solutions – HYPOTHESIS II, August 1997, Grimstad (Norway)*. Dordrecht: Springer; 1998. pp.505–10.

[45] Cicconardi SP, Jannelli E, Perna A, and Spazzafumo G. A steam cycle with an isothermal expansion: the effect of flow variation. *Int J Hydrogen Energy*. 1999;24(1):53–7.

[46] Cicconardi SP, Jannelli E, Perna A, and Spazzafumo G. A steam cycle with a mixed quasi-isothermal/adiabatic expansion. In: *Proceedings of HYPOTHESIS III, Hydrogen Power Theoretical and Engineering Solutions International Symposium, July 1999, Sankt Petersburg (Russia)*. Sankt Petersburg, Russia; 1999.

[47] Cicconardi SP, Jannelli E, Perna A, and Spazzafumo G. Parametric analysis of a steam cycle with a quasi-isothermal expansion. *Int J Hydrogen Energy*. 2001;26(3):275–9.

[48] Sir William Groove, Wikimedia Commons (Public domain image). 1839 – William Grove Fuel Cell [Internet]. 1839 [cited 2024 Aug 30]. Available from: https://upload.wikimedia.org/wikipedia/commons/c/ce/1839_William_Grove_Fuel_Cell.jpg.

[49] Cao D, Sun Y, and Wang G. Direct carbon fuel cell: fundamentals and recent developments. *J Power Sources*. 2007;167(2):250–7.

[50] EG&G Technical Services Inc., NETL. *Fuel Cell Handbook* (7th edn) [Internet]. Morgantown, West Virginia; 2004. Available from: https://www.osti.gov/servlets/purl/834188. Accessed 17 December 2024

[51] Giddey S, Badwal SPS, Kulkarni A, and Munnings C. A comprehensive review of direct carbon fuel cell technology. *Prog Energy Combust Sci*. 2012;38(3):360–99.

[52] Semmel M, Ali RE, Ouda M, Schaadt A, Sauer J, and Hebling C. Power-to-DME: a cornerstone towards a sustainable energy system. In: Spazzafumo G (ed). *Power to Fuel: How to Speed Up a Hydrogen Economy. Academic Press*; 2021. pp. 123–51. Accessed 17 December 2024

[53] Perl A. Power to ammonia and urea. In: Spazzafumo G (ed). *Power to Fuel – How to Speed Up a Hydrogen Economy*. 1st edn. New York: Academic Press; 2021. pp. 153–67.

[54] Chatterjee S, Dutta I, and Huang K-WW. Power to formic acid. In: Spazzafumo G (ed). *Power to Fuel – How to Speed Up a Hydrogen Economy*. 1st edn. New York: Academic Press; 2021. pp. 169–210.

[55] Bidault F, and Kucernak A. A novel cathode for alkaline fuel cells based on a porous silver membrane. *J Power Sources*. 2010;195(9):2549–56.

[56] Fazil A, and Chetty R. Synthesis and evaluation of carbon nanotubes supported silver catalyst for alkaline fuel cell. *Electroanalysis*. 2014;26(11): 2380–7.

[57] Ferriday TB, and Middleton PH. Alkaline fuel cell technology – a review. *Int J Hydrogen Energy*. 2021;46(35):18489–510.

[58] Dekel DR. Review of cell performance in anion exchange membrane fuel cells. *J Power Sources*. 2018;375:158–69.

[59] Olabi AG, Wilberforce T, Alanazi A, *et al.* Novel trends in proton exchange membrane fuel cells. *Energies*. 2022;15(14):4949.

[60] Suárez C, Iranzo A, Toharias B, and Rosa F. Experimental and numerical investigation on the design of a bioinspired PEM fuel cell. *Energy*. 2022; 257:124799.

[61] Rosli RE, Sulong AB, Daud WRW, *et al.* A review of high-temperature proton exchange membrane fuel cell (HT-PEMFC) system. *Int J Hydrogen Energy*. 2017;42(14):9293–314.

[62] Cicconardi SP, Lensi R, and Spazzafumo G. Il bilanciamento termico di un impianto combinato a carbone con celle a combustibile. In: *Atti del 2° Convegno Nazionale Gruppi combinati, prospettive tecniche ed economiche, 1988, Firenze, Maggio*. Florence, Italy; 1988. pp. 361–74.

[63] Cicconardi SP, Lensi R, and Spazzafumo G. Coal fed combined power plant with fuel cells. *In: Proceedings of 23rd Intersociety Energy Conversion Engineering Conference, July-August 1988, Denver, vol 2*. Denver; 1988. pp. 239–44.

[64] Campanari S, Chiesa P, and Manzolini G. CO_2 capture from combined cycles integrated with molten carbonate fuel cells. *Int J Greenh Gas Control*. 2010;4(3):441–51.

[65] Wee JH. Carbon dioxide emission reduction using molten carbonate fuel cell systems. *Renew Sustain Energy Rev*. 2014;32:178–91.

[66] Shim JH, Chao CC, Huango H, and Prinz FB. Atomic layer deposition of yttria-stabilized zirconia for solid oxide fuel cells. *Chem Mater*. 2007;19 (15): 3850–4.

[67] Wachsman ED, and Lee KT. Lowering the temperature of solid oxide fuel cells. *Science (80-)*. 2011;334(6058):935–9.

[68] Duan C, Hook D, Chen Y, Tong J, and O'Hayre R. Zr and Y co-doped perovskite as a stable, high performance cathode for solid oxide fuel cells operating below 500 °C. *Energy Environ Sci*. 2017;10(1):176–82.

[69] Taniguchi N, Hatoh K, Niikura J, Gamo T, and Iwahara H. Proton conductive properties of gadolinium-doped barium cerates at high temperatures. *Solid State Ionics*. 1992;53–56(Part 2):998–1003.

[70] Dubal SU, Bhosale CH, and Jadhav LD. Performance of spray deposited Gd-doped barium cerate thin films for proton conducting SOFCs. *Ceram Int*. 2015;41(4):5607–13.

[71] Kühn S, Heinrich L, Turner H, *et al. Fuel Cell Stack, Fuel Cell Apparatus and Fuel Cell Vehicle* – International Patent No. WO2022013359A1 [Internet]. Germany; WO2022013359A1, 2022. Available from: https://patents.google.com/patent/WO2022013359A1/en. Accessed 17 December 2024

[72] Joghee P, Malik JN, Pylypenko S, and O'Hayre R. A review on direct methanol fuel cells – in the perspective of energy and sustainability. *MRS Energy Sustain*. 2015;2(1):1–31.

[73] Kamarudin MZF, Kamarudin SK, Masdar MS, and Daud WRW. Review: direct ethanol fuel cells. *Int J Hydrogen Energy*. 2013;38(22):9438–53.

[74] Kashyap D, Teller H, and Schechter A. Dimethyl ether oxidation on an active SnO_2/Pt/C catalyst for high-power fuel cells. *ChemElectroChem*. 2019;6(9):2407–14.

[75] Du L, Lou S, Chen G, *et al.* Direct dimethyl ether fuel cells with low platinum-group-metal loading at anode: investigations of operating temperatures and anode Pt/Ru ratios. *J Power Sources*. 2019;433:126690.

[76] Yoo JH, Choi HG, Chung CH, and Cho SM. Fuel cells using dimethyl ether. *J Power Sources*. 2006;163(1):103–6.

[77] Li Q, Wu G, Johnston CM, and Zelenay P. Direct dimethyl ether fuel cell with much improved performance. *Electrocatalysis*. 2014;5(3):310–7.

[78] Kishimoto M, Muroyama H, Suzuki S, *et al.* Development of 1 kW-class ammonia-fueled solid oxide fuel cell stack. *Fuel Cells*. 2020;20(1):80–8.

[79] Jeerh G, Zhang M, and Tao S. Recent progress in ammonia fuel cells and their potential applications. *J Mater Chem A*. 2021;9(2):727–52.

[80] Liu Y, Pan Z, Esan OC, Xu X, and An L. Performance characteristics of a direct ammonia fuel cell with an anion exchange membrane. *Energy Fuels*. 2022;36(21):13203–11.

[81] Lan R, Tao S, and Irvine JTS. A direct urea fuel cell – power from fertiliser and waste. *Energy Environ Sci*. 2010;3(4):438–41.

[82] Sayed ET, Eisa T, Mohamed HO, *et al.* Direct urea fuel cells: challenges and opportunities. *J Power Sources*. 2019;417(September 2018):159–75.

[83] Kim OH, Choi HJ, Kang SY, *et al.* Towards outstanding performance of direct urea fuel cells through optimization of anode catalyst layer and operating conditions. *J Electroanal Chem*. 2022;921:116661.

[84] Yu X, and Pickup PG. Recent advances in direct formic acid fuel cells (DFAFC). *J Power Sources*. 2008;182(1):124–32.

[85] El-Nagar GA, Hassan MA, Lauermann I, and Roth C. Efficient direct formic acid fuel cells (DFAFCs) anode derived from seafood waste: migration mechanism. *Sci Rep*. 2017;7(1):1–11.

[86] Fukunaga A, Kato A, Hara Y, and Matsumoto T. Dehydrogenation of methylcyclohexane using solid oxide fuel cell – a smart energy conversion. *Appl Energy*. 2023;348(May):121469.

[87] Nguyen VN, and Blum L. Reversible fuel cells. In: Barbir F, Basile A, and Veziroğlu TN (eds). *Compendium of Hydrogen Energy: Hydrogen Energy Conversion: Vol 3. Woodhead Publishing*; 2016. pp. 115–45.

[88] CORDIS European Commission. REFLEX project – Reversible solid oxide electrolyzer and fuel cell for optimized local energy miX – results [Internet]. 2018 [cited 2023 Nov 2]. Available from: https://cordis.europa.eu/project/id/779577/results.

[89] Spazzafumo G, and Gaggio G. Cogenerative plant with microbial indirect fuel cells fed by industrial wastes. In: *Proceedings of 7th World Hydrogen Energy Conference, September 1988, Moscow (USSR), vol 1*. Moscow, USSR; 1988. pp. 315–25.

[90] Katz E, and Bollella P. Fuel cells and biofuel cells: from past to perspectives. *Isr J Chem*. 2021;61(1–2):68–84.

[91] Götz M, Lefebvre J, Mörs F, *et al.* Renewable power-to-gas: a technological and economic review. *Renew Energy*. 2016;85:1371–90.

[92] Marcogaz. Overview of available test results and regulatory limits for hydrogen admission into existing natural gas infrastructure and end use [Internet]. 2019. Available from: https://www.marcogaz.org/publications/overview-of-test-results-regulatory-limits-for-hydrogen-admission-into-existing-natural-gas-infrastructure-end-use/. Accessed 17 December 2024

[93] Ullmann. *Ullmann's Encyclopedia of Industrial Chemistry*. 5th completely revised edn. Weinheim: Wiley-VCH; 1989.

[94] Stangeland K, Kalai D, Li H, and Yu Z. CO_2 methanation: the effect of catalysts and reaction conditions. *Energy Procedia*. 2017;105(1876):2022–7.

[95] Keith DW, Holmes G, St. Angelo D, and Heidel K. A process for capturing CO_2 from the atmosphere. *Joule*. 2018;2(8):1573–94.

[96] Köppel W, Götz M, and Graf F. Biogas upgrading for injection into the gas grid – quality aspects, technological and ecological consideration. *gwf-Gas Erdgas*. 2009;150(International Issue):26–35.

[97] Yentekakis IV, and Goula G. Biogas management: advanced utilization for production of renewable energy and added-value chemicals. *Front Environ Sci*. 2017;5(7).

[98] De Saint Jean M, Baurens P, Bouallou C, and Couturier K. Economic assessment of a power-to-substitute-natural-gas process including high-temperature steam electrolysis. *Int J Hydrogen Energy*. 2015;40(20):6487–500.

[99] Vandewalle J, Bruninx K, and D'Haeseleer W. Effects of large-scale power to gas conversion on the power, gas and carbon sectors and their interactions. *Energy Convers Manag*. 2015;94:28–39.

[100] Blanco H, Nijs W, Ruf J, and Faaij A. Potential of power-to-methane in the EU energy transition to a low carbon system using cost optimization. *Appl Energy*. 2018;232(October):323–40.

[101] Schaaf T, Grünig J, Schuster MR, Rothenfluh T, and Orth A. Methanation of CO_2 – storage of renewable energy in a gas distribution system. *Energy Sustain Soc*. 2014;4(1):1–14.

[102] Tschiggerl K, Sledz C, and Topic M. Considering environmental impacts of energy storage technologies: a life cycle assessment of power-to-gas business models. *Energy*. 2018;160:1091–100.

[103] van Leeuwen C, and Zauner A. D8.3— Report on the costs involved with PtG technologies and their potentials across the EU [Internet]. Brussels, Belgium: EU Horizon 2020 Project STORE&GO; 2018. Available from: https://www.storeandgo.info/fileadmin/downloads/deliverables_2019/20190801-STOREandGO-D8.3-RUG-Report_on_the_costs_involved_with_PtG_technologies_and_their_potentials_across_the_EU.pdf. Accessed 5 July 2020

[104] Gorre J, Ortloff F, and van Leeuwen C. Production costs for synthetic methane in 2030 and 2050 of an optimized Power-to-Gas plant with intermediate hydrogen storage. *Appl Energy*. 2019;253(July):113594.

[105] Gorre J, Ruoss F, Karjunen H, Schaffert J, and Tynjälä T. Cost benefits of optimizing hydrogen storage and methanation capacities for Power-to-Gas plants in dynamic operation. *Appl Energy*. 2020;257(October 2019):113967.

[106] Gao J, Liu Q, Gu F, Liu B, Zhong Z, and Su F. Recent advances in methanation catalysts for the production of synthetic natural gas. *RSC Adv*. 2015;5(29):22759–76.

[107] Méndez-mateos D, Laura Barrio V, Requies JM, and Cambra JF. Effect of the addition of alkaline earth and lanthanide metals for the modification of the alumina support in Ni and Ru catalysts in CO_2 methanation. *Catalysts*. 2021;11(3):353.

[108] Bartholomew CH, Agrawal PK, and Katzer JR. Sulfur poisoning of metals. *Adv Catal*. 1982;31:135–242.

[109] Bartholomew CH. Mechanisms of catalyst deactivation. *Appl Catal A Gen*. 2001;212:17–60.

[110] Foscolo PU, Rossi L, Gallucci K, and Micheli F. *Studio della valorizzazione dell'impianto di gassificazione di biomasse a letto fluidizzato internamente ricircolante per la produzione di metano. Parte B: Studio dei processi di purificazione finale del metano [Internet*]. Report RdS/2011/141. 2011. Available from: https://www.enea.it/it/Ricerca_sviluppo/documenti/ricerca-di-sistema-elettrico/celle-a-combustibile-biomasse/rds141-1.pdf. Accessed 6 November 2023

[111] Gao J, Wang Y, Ping Y, *et al.* A thermodynamic analysis of methanation reactions of carbon oxides for the production of synthetic natural gas. *RSC Adv*. 2012;2(6):2358–68.

[112] Assettati A, Bassano C, Barbarossa V, Battipaglia G, Manzi G, and Vanga G. *Sperimentazione di Sistemi di Trattamento e Conversione del Syngas*. Rome: ENEA; 2010.

[113] Bailera M, Lisbona P, Romeo LM, and Espatolero S. Power to gas projects review: lab, pilot and demo plants for storing renewable energy and CO_2. *Renew Sustain Energy Rev*. 2017;69(November 2016):292–312.

[114] Mazza A, Bompard E, and Chicco G. Applications of power to gas technologies in emerging electrical systems. *Renew Sustain Energy Rev*. 2018; 92(March):794–806.

[115] Jensen JH, Poulsen JM, and Andersen NU. From coal to clean energy. *Nitrogen+Syngas*. 2011;310(March-April):34–8.

[116] Fendt S, Buttler A, Gaderer M, and Spliethoff H. Comparison of synthetic natural gas production pathways for the storage of renewable energy. *Wiley Interdiscip Rev Energy Environ*. 2016;5(3):327–50.

[117] Haldor Topsøe. From solid fuels to substitute natural gas (SNG) using TREMP [Internet]. 2009. Available from: https://www.netl.doe.gov/sites/default/files/netl-file/tremp-2009.pdf. Accessed 17 December 2024

[118] Kopyscinski J, Schildhauer TJ, and Biollaz SMA. Production of synthetic natural gas (SNG) from coal and dry biomass – a technology review from 1950 to 2009. *Fuel [Internet]*. 2010 Aug 1 [cited 2023 Apr 5];89(8):1763–83. Available from: http://dx.doi.org/10.1016/j.fuel.2010.01.027.

[119] Rönsch S, Schneider J, Matthischke S, *et al.* Review on methanation – from fundamentals to current projects. *Fuel.* 2016;166:276–96.

[120] Ghaib K, and Ben-Fares FZ. Power-to-methane: a state-of-the-art review. *Renew Sustain Energy Rev.* 2018;81(August 2017):433–46.

[121] Porubova J, Bazbauers G, and Markova D. Modeling of the adiabatic and isothermal methanation process. *Environ Clim Technol.* 2011;6(1):79–84.

[122] Eigenberger G. Fixed bed reactors. In: *Ullmann's Encyclopedia of Industrial Chemistry, Vol B4 [Internet]*. New York: VCH; 1992. Available from: https://elib.uni-stuttgart.de/bitstream/11682/1848/1/eig16.pdf. Accessed 5 April 2023

[123] Bailera M, Lisbona P, Peña B, and Romeo LM. *Energy Storage – Hybridization of Power-to-Gas Technology and Carbon Capture.* In: Bailera M, Lisbona P, Peña B, and Romeo LM (eds). Cham: Springer; 2020.

[124] Sudiro M, and Bertucco A. Synthetic natural gas (SNG) from coal and biomass: a survey of existing process technologies, open issues and perspectives. In: Potocnik P (ed). *Natural Gas [Internet].* Intech Open; 2010. pp. 105–26. Available from: https://www.intechopen.com/chapters/11472. Accessed 17 December 2024

[125] Hervy M, Maistrello J, Brito L, *et al.* Power-to-gas: CO_2 methanation in a catalytic fluidized bed reactor at demonstration scale, experimental results and simulation. *J CO2 Util.* 2021;50:101610.

[126] Held M, Schollenberger D, Sauerschell S, Bajohr S, and Kolb T. Power-to-gas: CO_2 methanation concepts for SNG production at the Engler-Bunte-Institut. *Chemie-Ingenieur-Technik.* 2020;92(5):595–602.

[127] Liu Z, Chu B, Zhai X, Jin Y, and Cheng Y. Total methanation of syngas to synthetic natural gas over Ni catalyst in a micro-channel reactor. *Fuel.* 2012;95:599–605.

[128] Audi. Power-to-gas plant > g-tron > Technology > Audi Malta [Internet]. 2015 [cited 2023 Nov 3]. Available from: https://www.audi.com.mt/mt/web/en/models/layer/technology/g-tron/power-to-gas-plant.html.

[129] Power Technology. Audi e-gas project, Germany [Internet]. 2021 [cited 2023 Nov 3]. Available from: https://www.power-technology.com/marketdata/audi-e-gas-project-germany/.

[130] González MI, and Schaub G. Fischer-Tropsch synthesis with H_2/CO_2 – catalyst behavior under transient conditions. *Chemie-Ingenieur-Technik.* 2015;87(6):848–54.

[131] Lefebvre J, Götz M, Bajohr S, Reimert R, and Kolb T. Improvement of three-phase methanation reactor performance for steady-state and transient operation. *Fuel Process Technol.* 2015;132:83–90.

[132] Mutz B, Carvalho HWP, Mangold S, Kleist W, and Grunwaldt JD. Methanation of CO_2: structural response of a Ni-based catalyst under fluctuating reaction conditions unraveled by *operando* spectroscopy. *J Catal.* 2015;327:48–53.

[133] Pavlostathis SG, and Giraldo-Gomez E. Kinetics of anaerobic treatment: a critical review. *Crit Rev Environ Sci Technol.* 1991;21(5–6):411–90.

[134] Burkhardt M, and Busch G. Methanation of hydrogen and carbon dioxide. *Appl Energy*. 2013;111:74–9.

[135] Seifert AH, Rittmann S, Bernacchi S, and Herwig C. Method for assessing the impact of emission gasses on physiology and productivity in biological methanogenesis. *Bioresour Technol*. 2013;136:747–51.

[136] Peillex JP, Fardeau ML, Boussand R, Navarro JM, and Belaich JP. Growth of *Methanococcus thermolithotrophicus* in batch and continuous culture on H_2 and CO_2: influence of agitation. *Appl Microbiol Biotechnol*. 1988;29 (6):560–4.

[137] Stone KA, Hilliard MV, He QP, and Wang J. A mini review on bioreactor configurations and gas transfer enhancements for biochemical methane conversion. *Biochem Eng J*. 2017;128:83–92.

[138] Bargiacchi E, Candelaresi D, and Spazzafumo G. Power to methane. In: Spazzafumo G (ed). *Power to Fuel – How to Speed Up a Hydrogen Economy*. 1st edn.New York: Academic Press; 2021. pp. 75–101.

[139] Sauerschell S, Bajohr S, and Kolb T. Methanation pilot plant with a slurry bubble column reactor: setup and first experimental results. *Energy Fuels*. 2022;36(13):7166–76.

[140] Brunner C, and Thomas A. Arbeitspaket 6: Gasnetzanalysen und Wirtschaftlichkeitsbetrachtung. *Energ Wasser-Praxis [Internet]*. 2014;11 (November):56–9. Available from: http://www.dvgw-innovation.de/fileadmin/dvgw/angebote/forschung/ innovation/pdf/1411brunner.pdf. Accessed 17 December 2024

[141] Buchholz OS, Van Der Ham AGJ, Veneman R, Brilman DWF, and Kersten SRA. Power-to-gas: storing surplus electrical energy. A design study. *Energy Procedia*. 2014;63:7993–8009.

[142] E&E consultant, HESPUL, Solagro. Etude portant sur l'hydrogène et la méthanation comme procédé de valorisation de l'électricité excédentaire. 2014;238. https://librairie.ademe.fr/recherche-et-innovation/3043-etude-portant-sur-l-hydrogene-et-la-methanation-comme-procede-de-valorisation-de-l-electricite-excedentaire.html. Accessed 17 December 2024

[143] Müller-Syring G, Henel M, Köppel W, Mlaker H, Sterner M, and Höcher T. Entwicklung von modularen Konzepten zur Erzeugung, Speicherung und Einspeisung von Wasserstoff und Methan ins Erdgasnetz. *DVGW Forsch [Internet]*. 2013;350. Available from: https://www.dvgw.de/medien/dvgw/forschung/berichte/g1_07_10.pdf. Accessed 17 December 2024

[144] Schaaf T, Graf F, Götz M, Henel M, and Tichler R. Abschlussbericht: Technoökonomische Studie von Power-to-Gas-Konzepten. *Dtsch Verein des Gas- und Wasserfaches e V [Internet]*. 2014;287(November). Available from: http://www.dvgw-innovation.de/fileadmin/dvgw/angebote/forschung/innovation/pdf/g3_01_12_tp_b_d.pdf. Accessed 17 December 2024

[145] Bargiacchi E, Frigo S, and Spazzafumo G. Energetic and exergetic analysis of an innovative plant for the production of electricity and substitute natural gas. *Energy Procedia*. 2018;148:312–9.

[146] Bargiacchi E, Frigo S, and Spazzafumo G. From biomass and electrolytic hydrogen to substitute natural gas and power: the issue of intermediate gas storages. *Int J Hydrogen Energy*. 2019;44(38):21045–54.

[147] Frigo S, and Spazzafumo G. Comparison of different system layouts to generate a substitute of natural gas from biomass and electrolytic hydrogen. *Int J Hydrogen Energy*. 2020;45(49):26166–78.

[148] Bargiacchi E, Candelaresi D, Valente A, Frigo S, and Spazzafumo G. Life cycle assessment of substitute natural gas production from biomass and electrolytic hydrogen. *Int J Hydrogen Energy*. 2021;46(72):35974–84.

[149] Candelaresi D, and Spazzafumo G. Power to methanol. In: Spazzafumo G (ed). *Power to Fuel – How to Speed Up a Hydrogen Economy*. 1st edn.New York: Academic Press; 2021. pp. 103–22.

[150] Supp E. *How to Produce Methanol from Coal*. Berlin:Springer; 1990. 203 p.

[151] Studt F, Behrens M, Kunkes EL, *et al.* The mechanism of CO and CO_2 hydrogenation to methanol over Cu-based catalysts. *ChemCatChem*. 2015;7 (7):1105–11.

[152] Dieterich V, Buttler A, Hanel A, Spliethoff H, and Fendt S. Power-to-liquid *via* synthesis of methanol, DME or Fischer-Tropsch-fuels: a review. *Energy Environ Sci*. 2020;13(10):3207–52.

[153] Nielsen CS. Topsøe *methanol technology for coal based plants [Internet]*. 2009. Available from: http://topsoe.ru/sites/default/files/topsoe_methanol_coal_based_plants.ashx_.pdf. Accessed 5 June 2020

[154] Hansen JB. Methanol production technology: todays and future renewable solutions. *In: Methanol Workshop [Internet]*. Lund: Haldor Topsøe; 2015. Available from: http://www.lth.se/fileadmin/mot2030/filer/9._Bogild_Hansen_-_Methanol_Production_Technology_Todays_and_Future.pdf. Accessed 17 December 2024

[155] ASTM. ASTM D1152-06. *Standard Specification for Methanol (Methyl Alcohol)*. Conshohocken, PA: American Society for Testing and Materials; 2012.

[156] Pass G, Holzhauser C, Akgerman A, and Anthony RG. Methanol synthesis in a trickle-bed reactor. *AIChE J*. 1990;36(7):1054–60.

[157] Rahimpour MR, and Ghader S. Theoretical investigation of a Pd-membrane reactor for methanol synthesis. *Chem Eng Technol*. 2003;26(8):902–7.

[158] Gallucci F, and Basile A. A theoretical analysis of methanol synthesis from CO_2 and H_2 in a ceramic membrane reactor. *Int J Hydrogen Energy*. 2007; 32(18):5050–8.

[159] Arora A, Iyer SS, Bajaj I, and Faruque Hasan MM. Optimal methanol production via sorption-enhanced reaction process. *Ind Eng Chem Res*. 2018;57(42):14143–61.

[160] Terreni J, Trottmann M, Franken T, Heel A, and Borgschulte A. Sorption-enhanced methanol synthesis. *Energy Technol*. 2019;7(4):1801093.

[161] Doss B, Ramos C, and Atkins S. Optimization of methanol synthesis from carbon dioxide and hydrogen: demonstration of a pilot-scale carbon-neutral synthetic fuels process. *Energy Fuels*. 2009;23(9):4647–50.

[162] Pontzen F, Liebner W, Gronemann V, Rothaemel M, and Ahlers B. CO_2-based methanol and DME – efficient technologies for industrial scale production. *Catal Today*. 2011;171(1):242–50.

[163] Dang S, Yang H, Gao P, *et al.* A review of research progress on heterogeneous catalysts for methanol synthesis from carbon dioxide hydrogenation. *Catal Today*. 2019;330(June):61–75.

[164] Guil-López R, Mota N, Llorente J, *et al.* Methanol synthesis from CO_2: a review of the latest developments in heterogeneous catalysis. *Materials (Basel)*. 2019;12(23):3902.

[165] Lipscomb JD. Biochemistry of the soluble methane monooxygenase. *Annu Rev Microbiol*. 1994;48(October):371–99.

[166] Hanson RS, and Hanson TE. Methanotrophic bacteria. *Microbiol Rev*. 1996;60(2):439–71.

[167] Semelsberger TA, Borup RL, and Greene HL. Dimethyl ether (DME) as an alternative fuel. *J Power Sources*. 2006;156(2):497–511.

[168] Willems W, Pannwitz M, Zubel M, and Weber J. Sauerstoffhaltige Kraftstoffe im selbstzündenden Verbrennungsmotor. *MTZ – Mot Zeitschrift*. 2020;81(3):28–37.

[169] Ledesma C, López E, Trifonov T, Rodríguez Á, and Llorca J. Catalytic reforming of dimethyl ether in microchannels. *Catal Today*. 2019;323:209–15.

[170] Maus W (ed.). *Zukünftige Kraftstoffe – Energiewende des Transports als ein weltweites Klimaziel*. In: Maus W (ed.). Berlin: Springer; 2019.

[171] Schemme S, Breuer JL, Köller M, *et al.* H_2-based synthetic fuels: a techno-economic comparison of alcohol, ether and hydrocarbon production. *Int J Hydrogen Energy*. 2019;45(8):5395–414.

[172] Azizi Z, Rezaeimanesh M, Tohidian T, and Rahimpour MR. Dimethyl ether: a review of technologies and production challenges. *Chem Eng Process Process Intensif*. 2014;82:150–72.

[173] Arnold U, Haltenort P, Herrera Delgado K, Niethammer B, and Sauer J. Die Rolle von Dimethylether (DME) als Schlüsselbaustein synthetischer Kraftstoffe aus erneuerbaren Rohstoffen. In: Maus W (ed.). *Zukünftige Kraftstoffe*. Berlin: Springer; 2019. pp. 532–61.

[174] Bakhtyari A, and Rahimpour MR. Methanol to dimethyl ether. In: Basile A, and Dalena F (eds). *Methanol*. Amsterdam: Elsevier; 2018. pp. 281–311.

[175] Shoji K, and Terai S. *Process for Producing Dimethyl Ether – United States Patent No.* US2004/0034255A1 [Internet]. United States; US2004/0034255A1, 2004. Available from: https://patents.google.com/patent/US20040034255A1/en. Accessed 17 December 2024

[176] Ahlers B, Birke G, Koempel H, *et al. Method and Device for Producing Dimethyl Ether from Methanol* – United States Patent No. US8884074B2 [Internet]. United States; US 8,884,074 B2, 2014. Available from: https://patents.google.com/patent/US8884074B2/en. Accessed 17 December 2024

[177] Catizzone E, Bonura G, Migliori M, Frusteri F, and Giordano G. CO_2 recycling to dimethyl ether: state-of-the-art and perspectives. *Molecules*. 2017;23(1):31.

[178] Yaripour F, Baghaei F, Schmidt I, and Perregaarad J. Synthesis of dimethyl ether from methanol over aluminium phosphate and silica-titania catalysts. *Catal Commun*. 2005;6(8):542–9.

[179] Yaripour F, Shariatinia Z, Sahebdelfar S, and Irandoukht A. The effects of synthesis operation conditions on the properties of modified γ-alumina nanocatalysts in methanol dehydration to dimethyl ether using factorial experimental design. *Fuel*. 2015;139:40–50.

[180] Liu L, Huang W, Gao Z, and Yin L. The dehydration of methanol to dimethyl ether over a novel slurry catalyst. *Energy Sources A Recover Util Environ Eff*. 2010;32(15):1379–87.

[181] Peng XD, Toseland BA, Wang AW, and Parris GE. Progress in development of LPDME process: dinetics and catalysts. *In: Coal Liquefaction & Solid Fuels Contractors Review Conference, 3-4 September 1997, Pittsburgh, PA* [Internet]. Pittsburgh, PA; 1997. pp. 1–6. Available from: http://www.fischer-tropsch.org/DOE/_conf_proc/Coal Conferences/coal_liq_conf97/peng.pdf. Accessed 5 April 2023

[182] Gogate MR. The direct dimethyl ether (DME) synthesis process from syngas I. Process feasibility and chemical synergy in one-step LPDME^{tm} process. *Pet Sci Technol*. 2018;36(8):547–54.

[183] Mondal U, and Yadav GD. Perspective of dimethyl ether as fuel: part I. Catalysis. *J CO2 Util*. 2019;32:299–320.

[184] Goehna H, and Koenig P. Producing methanol from CO_2: the keys are a catalyst that can tolerate a high CO_2 concentration and an unusual process configuration. *Chemtech*. 1994;24(6):36–9.

[185] Marlin DS, Sarron E, and Sigurbjörnsson Ó. Process advantages of direct CO_2 to methanol synthesis. *Front Chem*. 2018;6:446.

[186] De Falco M, Capocelli M, and Centi G. Dimethyl ether production from CO_2 rich feedstocks in a one-step process: thermodynamic evaluation and reactor simulation. *Chem Eng J*. 2016;294:400–9.

[187] Harmsen GJ. Reactive distillation: the front-runner of industrial process intensification. A full review of commercial applications, research, scale-up, design and operation. *Chem Eng Process Process Intensif.* 2007;46(9): 774–80.

[188] Kiss AA. Novel catalytic reactive distillation processes for a sustainable chemical industry. *Top Catal*. 2018;89:479.

[189] Iliuta I, Larachi F, and Fongarland P. Dimethyl ether synthesis with in situ H_2O removal in fixed-bed membrane reactor: model and simulations. *Ind Eng Chem Res*. 2010;49(15):6870–7.

[190] Appl M. *Ammonia: Principles and Industrial Practice*. Weinheim: Wiley-VCH; 1999.

[191] Sakarika M, Spanoghe J, Sui Y, *et al.* Purple non-sulphur bacteria and plant production: benefits for fertilization, stress resistance and the environment. *Microb Biotechnol*. 2020;13(5):1336–65.

[192] Perl A. Power to ammonia and urea. In: *Power to Fuel - How to Speed Up a Hydrogen Economy*. Giuseppe Spazzafumo (ed.), Academic Press, London (UK) 2021. pp. 153–67.

[193] Amar IA, Lan R, Petit CTG, and Tao S. Solid-state electrochemical synthesis of ammonia: a review. *J Solid State Electrochem*. 2011;15 (9):1845–60.

[194] Chen A, and Xia BY. Ambient dinitrogen electrocatalytic reduction for ammonia synthesis. *J Mater Chem A*. 2019;7(41):23416–31.

[195] Yang J, Weng W, and Xiao W. Electrochemical synthesis of ammonia in molten salts. *J Energy Chem*. 2020;43:195–207.

[196] Imamura K, and Kubota J. Ammonia synthesis from nitrogen and water at intermediate temperatures and elevated pressures by using an electrochemical hydrogen-membrane reactor with supported Ru catalysts and phosphate electrolytes. *Sustain Energy Fuels*. 2019;3(6):1406–17.

[197] Nagaishi S, Hayashi R, Hirata A, Sagara R, and Kubota J. Ammonia synthesis from nitrogen and steam using electrochemical cells with a hydrogen-permeable membrane and $Ru/Cs^+/C$ catalysts. *Sustain Energy Fuels*. 2024;8(5):914–26.

[198] Sagara R, and Kubota J. NH_3 synthesis from H_2O and N_2 using electrochemical cells with molten NaOH–KOH electrolytes at 250 °C. *In: Book of Abstracts HYdrogen POwer THeoretical & Engineering Solutions International Symposium – HYPOTHESIS XIX*. Hiroshima, Japan; 2024. pp. 132–4.

[199] Sánchez A, and Martín M. Scale up and scale down issues of renewable ammonia plants: towards modular design. *Sustain Prod Consum*. 2018; 16:176–92.

[200] Ikäheimo J, Kiviluoma J, Weiss R, and Holttinen H. Power-to-ammonia in future North European 100% renewable power and heat system. *Int J Hydrogen Energy*. 2018;43(36):17295–308.

[201] Armijo J, and Philibert C. Flexible production of green hydrogen and ammonia from variable solar and wind energy: case study of Chile and Argentina. *Int J Hydrogen Energy*. 2020;45(3):1541–58.

[202] Bennani Y, Perl A, Patil A, Someren CV, Heijne L, and Steenis MV. *Power-to-ammonia: rethinking the role of ammonia – from a value product to a flexible energy carrier (FlexNH3) [Internet]. Project Systeemintegratie Studie – Final Report. Groningen, the Netherlands*; 2016. Available from: https://research.hanze.nl/en/publications/power-to-ammonia-rethinking-the-role-of-ammonia-from-a-value-prod. Accessed 17 December 2024

[203] Boggs BK, King RL, and Botte GG. Urea electrolysis: direct hydrogen production from urine. *Chem Commun*. 2009;(32):4859–61.

[204] Eppinger J, and Huang K-W. Formic acid as a hydrogen energy carrier. *ACS Energy Lett*. 2017;2(1):188–95.

[205] EPFL. The world's first formic acid-based fuel cell [Internet]. 2018 [cited 2020 Aug 15]. Available from: https://actu.epfl.ch/news/the-world-s-first-formic-acid-based-fuel-cell/.

[206] Zhang W, Hu Y, Ma L, *et al.* Progress and perspective of electrocatalytic CO_2 reduction for renewable carbonaceous fuels and chemicals. *Adv Sci*. 2018;5(1):1700275.

[207] Canfield D, and Frese KW. Reduction of carbon dioxide to methanol on n- and p-GaAs and p-InP. Effect of crystal face, electrolyte and current density. *J Electrochem Soc*. 1983;130(8):1772–3.

[208] Summers DP, Leach S, and Frese KW. The electrochemical reduction of aqueous carbon dioxide to methanol at molybdenum electrodes with low overpotentials. *J Electroanal Chem*. 1986;205(1–2):219–32.

[209] Frese KW. Electrochemical reduction of CO_2 at intentionally oxidized copper electrodes. *J Electrochem Soc*. 1991;138(11):3338–44.

[210] Popić JP, Avramov-Ivić ML, and Vuković NB. Reduction of carbon dioxide on ruthenium oxide and modified ruthenium oxide electrodes in 0.5 M $NaHCO_3$. *J Electroanal Chem*. 1997;421(1–2):105–10.

[211] Spataru N, Tokuhiro K, Terashima C, Rao TN, and Fujishima A. Electrochemical reduction of carbon dioxide at ruthenium dioxide deposited on boron-doped diamond. *J Appl Electrochem*. 2003;33(12):1205–10.

[212] Durst J, Rudnev A, Dutta A, *et al.* Electrochemical CO_2 reduction – a critical view on fundamentals, materials and applications. *Chimia (Aarau)*. 2015;69(12):769–76.

Chapter 6
Applications

Hydrogen is considered the fuel of the future. And this is probably correct. Many argue that hydrogen will not have a future because electricity is better. This is the wrong way to approach the energy sector. Both hydrogen and electricity are simple energy carriers and have different characteristics so they do not compete, but can support each other. Moreover, not only fuel and electricity are required, but also heat, and although heat can be obtained either from fuel or from electricity, it can be directly obtained, stored and distributed from solar, geothermal and nuclear energy. Finally, there could be alternative intermediate solutions before completely replacing fossil fuels.

Being the scenario so complex, it is necessary to analyse separately each sector using energy since the needs can be sometimes very different.

6.1 Transportation

Although currently it is not the main field of use of hydrogen, the transport sector, especially on the road, has always attracted the most attention. It is a very diversified sector: road, rail, sea and air transport, all with different specific characteristics.

6.1.1 Road vehicles

Road transport can be further divided into urban, suburban, public and private. For private urban vehicles, maybe hydrogen could not be the right solution, since electric vehicles have some advantages in terms of efficiency and cost. However, the comparison is not so obvious from the point of view of environmental impact.

But if you need a longer range, hydrogen should be preferable. A fuel cell electric vehicle (FCEV) can run up to 800 km before the next refuelling.

The ideal propulsion system should have the following characteristics:

- low fuel consumption;
- operation without harmful emissions;
- silence and absence of vibrations;
- lightness and compactness;
- flexibility of performance;
- ease of starting;

- simple, safe and fast refuelling and great range;
- safety and non-toxicity of the accumulation;
- reduced maintenance;
- low cost.

There are no combination of propulsion system and energy carrier capable to fully satisfy all these requirements. For example, the internal combustion engine, still the absolute dominator, should consume less, be safer and more silent. And the main fuel used, gasoline, allows for a much safer and faster refuelling than natural gas, but produces far more harmful emissions.

Not even fuel cells can meet all the requirements, but certainly there are considerable differences between the different types that make them more or less suitable for use in automotive. The automotive sector is the one that started the great development of fuel cells in the early 1990s.

Actually, the first fuel cell vehicle prototype was the General Motors Electrovan (Figure 6.1, [1]) presented in 1966, but was too heavy and with a limited space due to the large size of fuel cells and hydrogen and oxygen tanks. Few years later (1970), the Austrian chemist Karl Kordesch equipped his car with fuel cells, batteries and hydrogen tanks [2]. He drove this car for some years proving the technology. Unfortunately, the cost were too high and for such a reason the use of fuel cells was practically relegated for some decades to the aerospace.

A comparison between the characteristics of fuel cells is schematically reported in Table 6.1, and is carried out in relative terms among the various fuel cells.

Figure 6.1 General Motors Electrovan [1]

Table 6.1 Comparison among different kinds of fuel cells

	AFC	PEMFC	PAFC	MCFC	SOFC
Low fuel consumption	√	√		√	√
Operation without harmful emissions	√	√	√	√	√
Silence and absence of vibrations	√	√	√	√	√
Lightness and compactness		√			√
Air as an oxidant		√	√	√	√
Solid electrolyte		√			√
Flexibility of performance	√	√	√	√	√
Ease of starting	√	√			
Reduced maintenance	√	√	√	√	√
Low cost			√		

Figure 6.2 Toyota Mirai

The superiority of the proton exchange membrane fuel cells (PEMFCs) is until now evident as it is superior or comparable with the others in almost all the characteristics compared, and most of the car manufacturers focused the activity on this kind of fuel cells. Figure 6.2 shows an example of currently FCEV commercially available.

Solid oxide fuel cells (SOFCs) have good characteristics, but has a high operating temperature which prevents cold starting. However, Volkswagen although does not declare a strong involvement in hydrogen, is co-assignee with Kraftwerk Group of a patent on SOFC also for automotive application [3]. Probably they expect to overcome this problem in the coming years, which in any case does not appear insurmountable in current automotive technology.

Molten carbonate fuel cells (MCFCs) add to these disadvantages the need for CO_2 recirculation, which makes it the most complex among fuel cells.

Phosphoric acid fuel cells (PAFCs) are less expensive, but are characterised by the lowest efficiency and power density and no one is currently investing in its development.

Alkaline fuel cells (AFCs) are bulky and usually require pure oxygen: they can also be fed with air, but only after adequate purification.

It is commonly believed that as the market expands, the cost of PEMFCs will drop significantly. However, we must also consider that with the increase in demand, the availability of platinum will become less and less and the price of the catalysts will increase. It is therefore imperative to find a viable platinum substitute that is widely available and inexpensive. SOFC technology does not require expensive catalysts and could be a good alternative.

Some of the characteristics listed above specifically concern the fuel storage system.

The gas fed to the anode of the cell must be rich in hydrogen and therefore there are three alternatives:

- equipping the vehicle with a pure hydrogen storage;
- equipping the vehicle with a reformer for the process of the fuel stored on board;
- equipping the vehicle with an on-board production system based on chemical hydrides.

The choice of hydrogen offers the advantage of the total absence of polluting emissions, but suffers from the lack of distribution infrastructures. A relatively high density of refuelling stations, like that shown in Figure 6.3, is present only in Japan and South Korea, while numerically, the country with the most is China. In Europe, the Alternative Fuels Infrastructure Regulation (AFIR) forecasts a refuelling station every 200 km, but in the meanwhile Everfuel in Denmark closed or paused its refuelling stations [4], whereas Shell and Motive closed most of their refuelling stations in UK [5]. A constantly updated map is maintained by LBST (Ludwig-Bölkow-Systemtechnik) [6]. It also has a low volumetric energy density and therefore involves a bulky storage tank to allow distances comparable to those of traditional systems. The choice of a traditional fuel requires more components that can involve higher costs and dimensions, in addition to the emission of pollutants. The use of chemical hydrides is very interesting under several points of view, but it requires to regenerate the hydrides starting from the oxide or hydroxide produced during hydrogen evolution.

The choice of fuel therefore becomes very important. Although toxic, methanol seems to be the best among the various possible hydrocarbons. Its reforming process requires low temperatures and produces few harmful substances, Moreover, in some countries, there is a good distribution network and in any case, being

Figure 6.3 Hydrogen refuelling station

liquid, it can use the distribution network of traditional liquid fuels. Finally, its production is not linked to oil. There remains the problem of the production of carbon monoxide which could damage the cells and must be brought back to acceptable values. However, after some prototypes equipped with on-board reformers, vehicle manufacturers have returned to prefer the accumulation of hydrogen on board. Methanol could be of interest again in the event of further development of direct methanol fuel cells (DMFCs).

Thanks to the good response of the fuel cells, there is often no need for a battery for peak demand, which is however important for the recovery of braking energy, cold start, cost balancing. Instead, a DC/AC converter may be required if a synchronous motor rather than a DC motor is preferred or DC/DC converter to match different power output of the fuel cell with specific needs of the motors. And this choice seems to be the one followed by all manufacturers. A different concept is used on Riversimple Rasa: batteries are replaced by supercapacitors. It is a very light and aerodynamic car and a range of 450 km with 1.5 kg of hydrogen is claimed [7].

The auxiliary components include heat, water and reagents management systems. Hydrogen could be fed into a dead-end anode with periodic purging. Air must instead be supplied in excess and removed from the cathode outlet due to the accumulation of nitrogen and produced water. Therefore, the pressurisation of the cell if, on the one hand, provides a significant improvement in performance up to about 0.3 MPa, on the other it involves a not negligible consumption of energy by

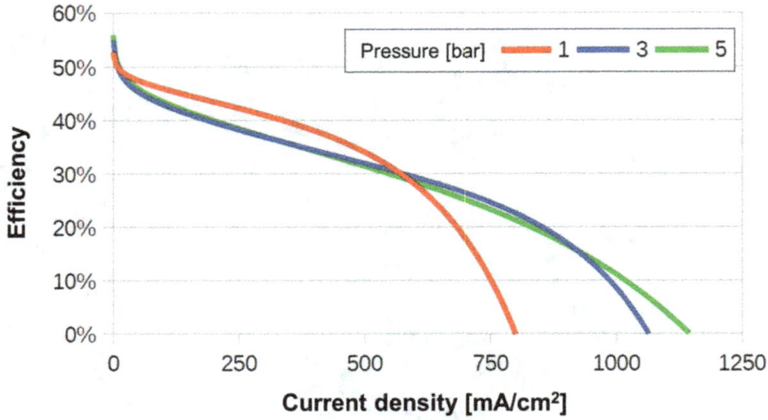

Figure 6.4 Comparison between the efficiencies of the stack/compressor system at different operating pressures

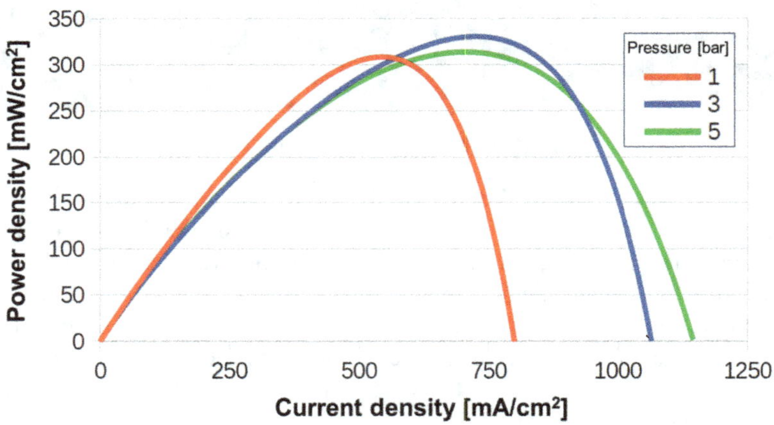

Figure 6.5 Power density vs. current density of the stack/compressor system changing operating pressure

the compressor. Taking into account the consumption of the compressor, having assumed an efficiency of 0.75, it can be seen in Figure 6.4 the comparison among the efficiencies of the stack/compressor system for different pressure values. The insertion of the compressor involves a lowering of the efficiency such that the systems at 3 and 5 bar become practically equal in the whole operating range and that the atmospheric stack is more efficient than the pressurised systems in all its operating range, i.e. for current densities not exceeding 0.5 A/cm^2.

Figure 6.5 shows that the presence of the compressor results in a maximum power density very close for the three different operating pressures, with a slight better performance for the intermediate pressure.

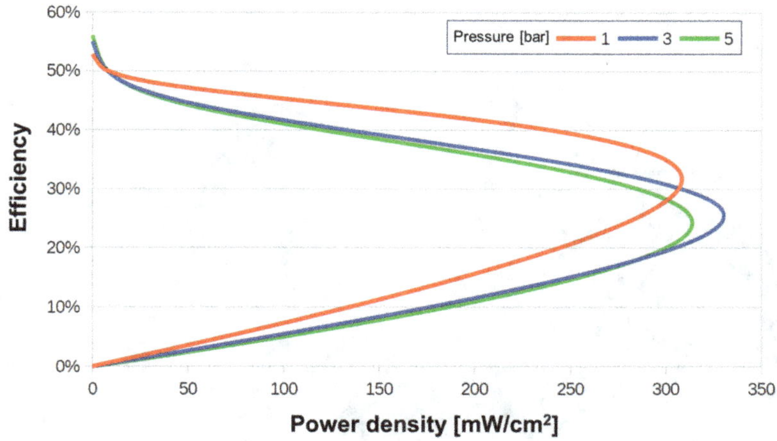

Figure 6.6 Efficiency of the stack/compressor system as a function of the power density changing operating pressure

Finally, in Figure 6.6, the efficiency comparison at different pressure levels is made in terms of power density: you can see how the additional advantage of pressurisation is considerably reduced when the pressure increases. Moreover, the presence of the compressor involves an increase in weight and cost of the vehicle.

The ability of a fuel cell vehicle to operate in any climate has been demonstrated since 2004 when the GM Opel Fuel Cell Marathon was organized and carried out: a 10,000 km journey through 14 countries and different climates, starting in May from Norway, and arriving in June in Portugal. Two OPEL-GM HydroGen3 cars were used for this experiment, one equipped with a compressed hydrogen tank and the other with a liquid hydrogen tank. As there were no hydrogen service stations along the route, the refuelling was carried out by mobile tankers.

Unfortunately, FCEVs are still very expensive and their market is currently rather negligible. For example, there are less than 500 FCEVs in the whole UK causing the closure of most refuelling stations [5].

Toyota, a leader with Hyundai in the sector of FCEVs, decided to invest money also in the development of a car equipped with a hydrogen-fuelled internal combustion engine (ICE) [8]. ICE is less expensive than FC, is able to use less pure hydrogen and could also run with mixtures of hydrogen and fossil derived fuels or with synthetic fuels produced from hydrogen and sequestered carbon dioxide. Therefore, it could have a larger market.

The behaviour of ICE with hydrogen and fuel mixtures containing hydrogen has been described in Section 5.1.2, whereas the use of synthetic fuel does not have substantial impact on traditional ICE.

Another option is the hybrid operation, that is with ICE and a battery. There are two different kinds of hybrid systems: parallel and series. The latter means that only the electric motor is connected to the transmission, whereas the ICE is

Figure 6.7 Hydrogen-fuelled bicycle [10] (copyright Pragma Industries)

connected to an electric generator which recharges the battery pack. In this way, the ICE does not work continuously, but always at its design point. Parallel means that propulsion is preferably electric when driving in the urban traffic and preferably thermal where speed is more constant. Moreover, electric motor can assist ICE during acceleration.

In addition to the technical aspects of propulsion, it is interesting to examine the environmental ones. This is feasible using, for example, the life cycle assessment methodology.

6.1.2 Bicycles and motorcycles

Despite the bulk of the fuel cell and hydrogen storage, prototypes of hydrogen bicycles have been developed for many years and in recent years some models have been offered on the market. An example is shown in Figure 6.7 and Table 6.2 reports its technical characteristics [9].

Kawasaki, Suzuki, Honda and Yamaha joined in the project HySE (hydrogen small mobility & engine technology) to develop small internal combustion engines especially devoted to motorcycles. And Kawasaki Ninja H2 SX is the first prototype equipped with a hydrogen-fuelled four-cylinder supercharged engine.

6.1.3 Buses, vans and trucks

Buses and trucks have larger dimensions and weights and, above all, a different management. These characteristics have a great influence on the possibility of using hydrogen as a fuel. On the one hand, the greater bulk and weight of the tanks has a lower impact, while on the other hand there is the possibility of organising refuelling in a limited number of hubs, at least for buses and trucks fleets.

Table 6.2 Alpha Neo technical characteristics

Maximum assisted speed	km/h	25
Working temperature	°C	−10/+45
Motor power	W	250
Rotating torque	Nm	36
Weight	kg	30
Dimensions	mm	1903 × 1052 × 647
Gearbox		Integrated, 5 speed
Brakes		Hydraulic disk
Maximum range	km	150
Electric supply	W	PEMFC 480 (supercapacitor hybrid)
Hydrogen purity	%	≥ 99.95
Tank capacity	l	3
Storage pressure	bar	300
Refill time	min	2

Table 6.3 Characteristics of Solaris Urbino 18 Hydrogen (data from [12])

Motor	Traction motor TSA TMF35-44-4 240 kW or electric portal axle ZF AVE130 2 × 125 kW
Hydrogen fuel cell	100 kW
Traction batteries	Lithium-ion
Hydrogen tanks	Composite tanks 5 × 312 l
Hydrogen tanks	Composite tanks 3 × 190 l
Hydrogen refilling port	Above and/or the 1st right wheel arch
Charging system	Plug-in

And indeed the industry of vehicles equipped with hydrogen-powered fuel cells seems more dynamic for these sectors.

The first fuel cell bus was built at Georgetown University and was an experimental prototype equipped with a methanol reformer and a 100 kW phosphoric acid fuel cell stack manufactured by UTC Fuel Cells LLC [11].

A recent example of commercial hydrogen fuelled bus is the Solaris Urbino 18 Hydrogen. In this Solaris bus, the PEMFC stack is powered by hydrogen and the electricity produced is transferred directly to the electric motor. The bus is also equipped with storage batteries which have a capacity of around 60 kWh and have the purpose of recovering energy during braking and then reusing it during acceleration transients. The vehicle's range is approximately 350 km with a full refuelling, that takes approximately 20 minutes. Depending on the configuration, this bus will be able to carry up to 138 passengers. Table 6.3 reports some of characteristics of this bus.

Another example is VanHool Exqui.City24 Fuel Cell whose main characteristics are reported in Table 6.4.

Table 6.4 Characteristics of VanHool Exqui.City24 Fuel Cell (data from [13])

Length (m)	23.820
Height (m)	3.300
Width (m)	2.550
Number of seats	60
Fuel cells	Ballard fuel cell
Type of drive train	Velocity HD100
Permanent magnetic drive motor	2 × Siemens PEM-1DB2016 160 kW

Figure 6.8 Fuel cell bus from Toyota operating in the city of Himeji, Hyogo prefecture, Japan

Figure 6.8 shows a hydrogen fuel cell bus currently in service in the city of Himeji, Japan. This bus was supplied by Toyota and has an on-board storage of 600 litres of hydrogen gas at 82 MPa. This storage consists of 10 cylinders mounted on the roof and feeds two PEMFC stacks located in the rear part of the roof. The refuelling time is 10 minutes and offers a range of up to 200 km. Figure 6.9 instead shows Fotonbus HC12 FCE with some details and a hydrogen refuelling column.

Figures 6.10 and 6.11 show a comparison in terms of range and refuelling time among buses with different kinds of propulsion system [14].

Figure 6.9 Fotonbus HC12 FCE

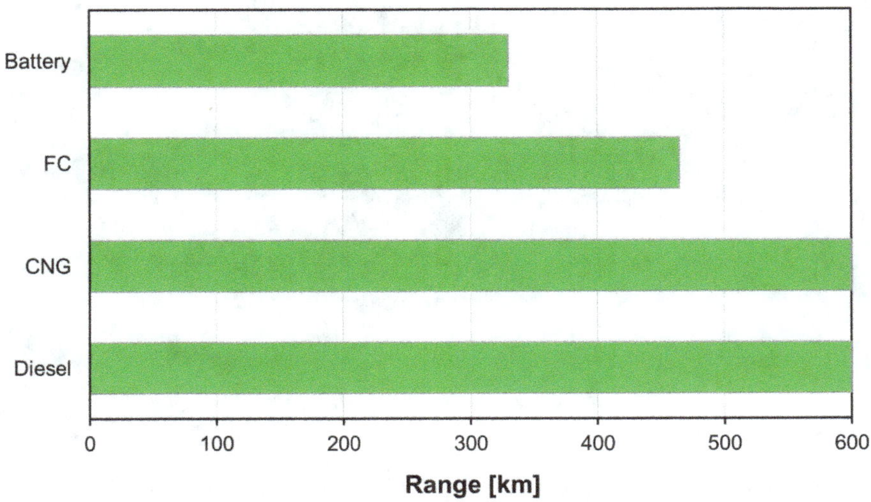

Figure 6.10 Range comparison among buses with different kinds of propulsion systems

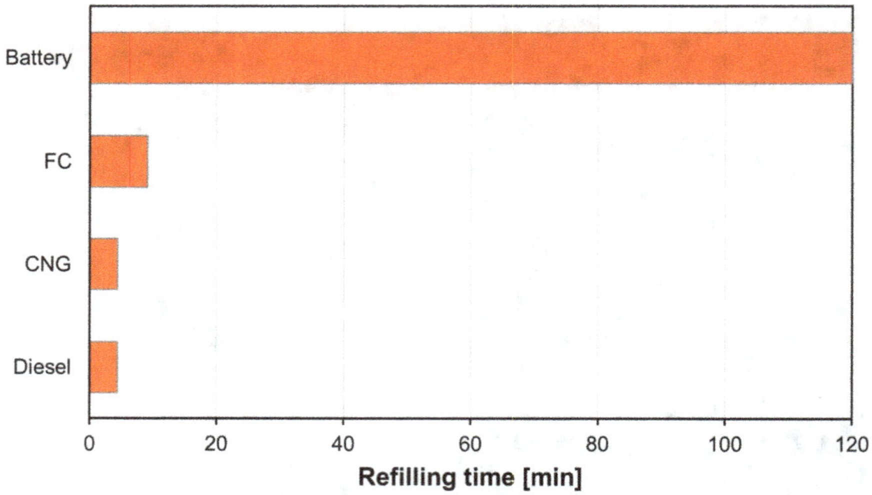

Figure 6.11 Refilling time comparison among buses with different kinds of propulsion systems

Figure 6.12 Iveco Daily [15]

Several car manufacturers are also very active on hydrogen-powered vans: they are preparing mass production around 2025. These include Stellantis with the Opel and Peugeot brands, in synergy with Symbio for fuel cell production. Iveco will present a hydrogen-powered Iveco Daily (Figure 6.12) with the support of Hyundai, which will supply the fuel cell system. Renault, in cooperation with Plug Power, under the Hyvia brand will present vans of various sizes with hydrogen power.

A study from Roland Berger GmbH and the Fuel Cells and Hydrogen Joint Undertaking [16] has analysed nine different case studies and concluded that by 2030 trucks equipped with hydrogen fuel cells can be economically competitive with diesel trucks while maintaining comparable performance in terms of payload capacity, daily range and refuelling time. More critical is the comparison in economic terms on short distances with battery-powered electric trucks, which are however characterized by less flexibility which in some cases can make the difference.

An example of commercially available truck is Nikola Tre FCEV.

As recent application, Daimler is developing hydrogen cryogenics fed truck (Gen H2) to reach the 1000 km maximum mileage obtainable with a single full refilling with two 40 kg cryogenic tanks. Considering that the 2023 consumer trials was successful, Daimler planned to start series production before 2030.

A sub-cooled prototype filling liquid hydrogen station was built at the development and testing Daimler centre in Wörth, to test a rapid refiling system.

As an alternative also buses and trucks could be equipped with ICE fed by hydrogen, hydrogen mixtures or synthetic fuels.

Almost all hydrogen fuel manufacturers have chosen the solution with gaseous storage at both 35 and 70 MPa (IVECO/Nikola, Hyzon, TEVVA, Hyliko, Volvo), Mercedes is the only manufacturer that aims to have cryogenic storage of liquid hydrogen for a range of up to 1600 km in order to exceed 1000 km of autonomy, completely comparable with the diesel trucks used for long distances.

Another area where hydrogen traction vehicles have been applied is in road transport within ports and logistics centres. A hydrogen yard truck (Figure 6.13) was

Figure 6.13 The hydrogen yard truck at Valencia Terminal Europa [17]

Figure 6.14 Volvo HX04 [18]

built in the framework of the European ‘H2Ports’ project and operates at Valencia Terminal Europa [17]. It is a hybrid (fuel cell/battery) vehicle for loading/unloading operations. The hydrogen is stored in the form of compressed gas at 35 MPa in four Type 3 cylinders for a total capacity of approximately 12 kg. A 70 kW Ballard fuel cell stack powers an electric motor in conjunction with a 25 kWh lithium iron phosphate battery that also provides energy recovery under braking or deceleration.

In the USA, a similar project was developed for port in California (ZECAP Project – Zero Emissions for California Ports).

There are also some early experiences with hydrogen-powered construction site vehicles equipped with fuel cells. These include excavators like JCB and Komatsu, and articulated hauler like Volvo HX04 shown in Figure 6.14.

The first application of fuel cells in industrial sector has been the forklift for use in close environment. As an example the Still RX20 is shown in Figure 6.15.

6.1.4 Trains

The situation is completely different when considering trains since the preferred option is electric power from the grid. However, the electrification of some railway sections is very expensive and diesel trains are generally used for these sections. In this case, fuel cell trains are gaining more and more interest notwithstanding the cost and the need for pure hydrogen. Some applications then cannot be electrified by overhead power lines such as locomotives in logistics centres or ports where goods are handled both by air and on land. In these situations, replacing diesel locomotives with hydrogen ones can be a viable alternative for decarbonisation.

The first FC passenger train to enter into commercial service was the Coradia iLint [20], in 2018. All devices are duplicated, one for each wagon. Hydrogen is stored as a compressed gas (35 MPa) in two tanks, each one with a maximum capacity of 94 kg distributed in 24 canisters. The tanks, as well as the two fuel cells stacks, are positioned on the roof, whereas all the propulsion and electric auxiliary

Figure 6.15 Still RX20 hydrogen fuelled forklift [19]

equipment is located under the floor, including two lithium-ion battery packs for electric power storing, pick request of energy and energy recovery during braking and two traction motors, each rated at 314 kW. A different position for hydrogen storage, that is in a specific carriage in the middle of passenger train, has been planned for Coradia Streams H a version designed for Italian railways [21].

Alstom itself and other manufacturers (e.g. CAF in Spain, Stadler in Switzerland, PESA in Poland, and Siemens in Germany) are proposing trains powered only by hydrogen (with supporting batteries) or mixed by catenary and hydrogen, for those trains operating on railway lines that are only partly electrified. Figure 6.16 [22–25] shows the hydrogen train by CAF.

As an alternative, train could also be equipped with ICE fed by hydrogen, hydrogen mixtures or synthetic fuels. Using a green synthetic diesel fuel for the already existing trains could be a more rational and less expensive option, although not zero emissions due to the presence of nitrogen oxides.

6.1.5 Boats

Moving to the sea the situation is rather inhomogeneous:

- tugs, which operate inside a harbour or along well-defined river routes and require just one or two refuelling hubs;
- ferries and large cruise ships which follow pre-established itineraries, requiring refuelling hubs in a limited number of harbours;
- private yachts and sailboats, which require a refuelling point in every harbour and probably synthetic fuels are more suitable for these kinds of boats.

Figure 6.16 CAF hydrogen train (copyright CAF)

Figure 6.17 Hydrotug [27] (courtesy of CMB.TECH)

An example of hydrogen fuelled tug is the Elektra designed to operate from Berlin to Hamburg [26].

At port of Antwerp will operate a hybrid hydrogen–diesel tug using an ICE (Figure 6.17). This 'Hydrotug' is driven by combustion engines that burn hydrogen

Figure 6.18 Hydra [30] (copyright Norled and Samferdselsfoto)

in combination with diesel. The engine reduces NO_x and PM (particulate matter) in comparison with current standard on the market. This tug is being built by Compagnie Maritime Belge (CMB).

Figure 6.18 shows MF Hydra [28], an 82.4 m ferry having a capability of up to 300 passengers and 80 vehicles. It is equipped with two 200 kW fuel cell stacks and a liquid hydrogen tank [29].

Emirates Team New Zealand uses a hydrogen powered foiling catamaran, Chase Zero (Figure 6.19). The catamaran is equipped with two Toyota 80 kW FC stacks, two 42 kWh batteries, two 220 kW engines and four tanks containing 8 kg of hydrogen at 350 bar, for a range of 180 km; the cruising speed is 30 knots and the maximum speed is 50 knots.

Another recent hydrogen-powered boat is a passenger catamaran about 49.9 m long and 10.4 m wide, equipped with a 500 kW hydrogen fuel cell stack coupled with a 1800 kWh lithium battery system (Figure 6.20).

Italian Fincantieri launched in 2022 the ZEUS (Figure 6.21): a 25.60 m long ship with a gross tonnage of about 160 tonnes. Its primary purpose is to be a laboratory ship designed to truly test the technologies and systems needed to use hydrogen as the primary energy vector for a naval vessel.

It is very interesting that some boats (e.g. sailboats) need ballast which could be replaced by metal hydrides as an ideal solution for on-board hydrogen storage.

Figure 6.19 Chase Zero [31] (courtesy of Team New Zealand)

Figure 6.20 Three Gorges Hydrogen Boat No.1 [32]

Usually lead is used, so that there is no need for a light hydride and this makes easier to find a not expensive solution.

Finally, in February 2024, Yamaha presented at Miami Boat Show a prototype of a Hydrogen Outboard Motor [34].

Figure 6.21 ZEUS [33]

6.1.6 Airplanes

In the world of air transport, the use of hydrogen is following two different paths. The first sees the use of hydrogen as a partial or total replacement for traditional carbon fuels in combustion engines used for aircraft propulsion (mostly turbines). This involves adapting a traditional well-known propulsion system to a new fuel without emission of carbon dioxide, but with not a complete zero emission system. A couple of past example were already reported in Section 4.6.5, whereas a current example is the collaboration among GE Aerospace, Airbus and CFM International [35].

The second path is to carry out a complete electrification of the propulsion system of an aircraft in which thrust is achieved with propeller or turbomachines driven by an electric motor. In this case, hydrogen would be used in a fuel cell to generate electricity, which in turn powers the aircraft's electric propulsion system. This second technology chain promises better overall efficiency and the complete absence of emissions, but it also represents a more demanding challenge for companies in the industry committed to introducing several new technologies in the sector, completely abandoning combustion for aircraft propulsion.

Recently, two different tests were conducted by ZeroAvia and Universal Hydrogen using hydrogen and fuel cells. ZeroAvia modified a 19-seat Dornier 228 (Figure 6.22), whereas Universal Hydrogen modified a 50-seat Dash-8. These first test flights lasted 10 and 15 minutes, respectively. In both cases, a prototype hydrogen–electric propulsion unit replaced the engine under one wing, maintaining the standard engine under the other wing.

The introduction of hydrogen in the world of aviation as alternative to carbon fuels is certainly at the centre of interest, not least because of the impetus coming

Figure 6.22 ZeroAvia Dornier 228 [36]

from the carbon taxation system (Emission Trading System), which both in Europe and on other continents has involved the world of aviation since 2012.

Whatever the final use of hydrogen on aircraft (fuel in engines or reagent in fuel cells), surely one of the most challenging issues remains that of on-board storage.

As a compressed gas (as in the main tests run) or as a cryogenic liquid, more activities on both research and regulatory side are need. At this time hydrogen storage system in form of cryogenic liquid uses a third of the energy content of the hydrogen in the liquefaction process. Moreover, air routes envelop the whole world, but it is unthinkable that all airports quickly equip themselves with hydrogen refuelling capacity. Perhaps the use of synthetic fuels would be more reasonable.

6.2 Portable systems

Only since the 1990s, thanks to the growing interest around PEMFCs for the automotive sector, the fuel cells power density made significant advancements. Therefore, the field of portable applications, which requires extreme compactness, has been the last, in historical terms, to have attracted the attention of fuel cell manufacturers.

However, while the idea of replacing cell phone and laptop batteries is very attractive and the market is vast, fuel storage density continues to be a major obstacle.

The first applications that have had some success are those relating to small auxiliary units for campers and small boats, units based on methanol fuel cells or, in some cases, powered by hydrogen stored in small metal hydride bottles. For these applications, the fuel cell can be used directly not only as an electricity generator but also as a battery charger and as a heat source.

Among portable fuel cell applications, several solutions are emerging in recent years to provide electricity in temporary situations such as construction sites where the electricity grid is not yet present, concerts or temporary gatherings of people, or disasters. In cases where there is more social attention (such as concerts and large events), the higher costs of installations with fuel cells and hydrogen can also be easily covered.

Typically, these are installations in containers or, in any case, in boxes, with baskets of cylinders inside them containing hydrogen under pressure (20–35 MPa) to supply a fuel cell, usually of the PEMFC type. The power involved varies from a few dozen to a few hundred kW of electrical power output, with autonomies generally not exceeding 10 hours. The French company Gaussin proposes such a solution (Figure 6.23).

Six gensets (3 × 25 kW and 3 × 100 kW) have been realised, tested and certified for real operating conditions in the EveryWh2ere project [38] financed by the Clean Hydrogen JU (Figure 6.24). For such a project, Power Cell Systems supplied its commercial stacks S2 and S3.

Hydrogen also seems to be an enabling technology to charge systems for electric battery cars in those situations where there is no electricity grid capable of supporting a fast-charging operation.

In fact, it may be more convenient than the increasing of electric power grid capacity, particularly in the early stages of market development, to install a fuel cell with hydrogen storage in a portable system that can deliver up to a few

Figure 6.23 H_2 power generator from Gaussin [37])

hundred kW of electric power for a few minutes to recharge cars. Examples are offered by the French company Gaussin, the US company Renewable Innovations (Figure 6.25, [39]), English company AFC Energy, Swiss company H2 Energy (Table 6.5).

Figure 6.24 A module from EveryWh2ere project [38]

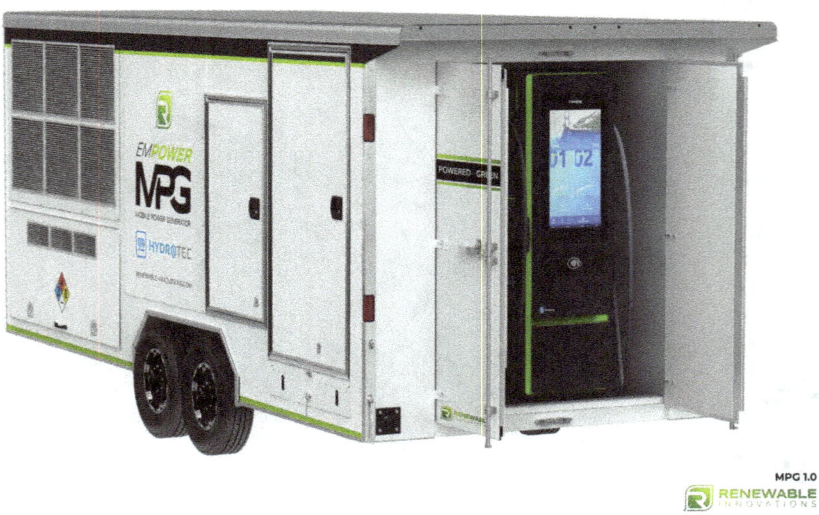

Figure 6.25 Renewable Innovations hydrogen-fuelled portable EV charger [39] (courtesy of Renewable Innovations)

Table 6.5 Main characteristics of kvyreen 80 and 160 (data from [40])

	kvyreen 80	**kvyreen 160**
Size ($W \times L \times H$) (m)	1.6 × 1.8 × 2.4	1.6 × 2.9 × 2.4
Weight (kg)	1800	2500
Hydrogen supply pressure (bar)	9–16	9–16
Charging power (kW)	80	160
Max. charge voltage (V)	800	800
Number of charging points	1	1

6.3 Residential systems

Pure hydrogen or hydrogen blend boilers are being studied and tested by several companies.

BDR Thermea group, with Baxi brand [41], as well as Worcester Bosch [42], produced and tested dozens of pure or blended hydrogen boilers (Figure 6.26) with useful dimensions for single-family flats (around 24 kW_t in heating and 28 kW_t in domestic hot water mode).

In the South Bank area of Teesside (North of England), Northern Gas Networks (NGN) started in 2022 to use a network of existing natural gas mains to carry out standard gas operational procedures under 100% hydrogen conditions for the first time. The hydrogen in the gas grid is used for domestic boiler.

In that area is currently fully commissioned the world's largest off-shore wind farm (around 2.5 GW in Hornsea One and Two projects) [44]. Work is underway to realise two more wind farm lots, which will lead to a total wind generation capacity of 6 GW. A huge non-programmable and non-dispatchable electricity flows will be handled also by the production of hydrogen from electrolysis to avoid curtailment of wind power generation. In that condition, a way to use the electricity in a relative cheap way, if there is no synchronous demand and production of electricity or if the electric grid could not manage all the power, is to create a demand for hydrogen produced by electrolyser.

Even if producing hydrogen by electrolysis to burn it in a boiler is not the thermodynamically best way to use electricity, it could be the cheapest one in a situation where renewable energy production and consumption are not simultaneous: a domestic hydrogen boiler is expected to cost the same as a methane boiler, certainly cheaper than a fuel cell system. If, moreover, it will be possible to distribute hydrogen with the current methane grid, the change of technology appears to be rapid at an acceptable cost in view of strict decarbonisation targets in 2030 and 2050. Hydrogen application concerns also co-generation of heat and power or, even better, tri-generation considering also air cooling.

The residential applications sector for fuel cells began to develop in the second half of the 1990s following the good results of co-generation with PAFC and the growing development of PEMFC technology.

The fuel cell has technological, energy and economic constraints, which makes its use possible only if placed within specific systems. Unless the type of fuel cell

Figure 6.26 Bosch hydrogen boilers installed in an Italian school [43]

used is suitable for direct conversion of the available fuel, at the very least, a fuel system is required that supplies a gaseous and hydrogen-rich fuel. Furthermore, unless you are dealing with small direct current users, you need an inverter for the electricity produced. Finally, depending on the operating temperature, a system that dissipates or recovers the heat generated is required.

Figure 6.27 ONSI corporation PC25 module

In the 1990s, ONSI Corporation introduced in the market the PC25 fuel cell module (Figure 6.27): a co-generation module powered by natural gas and capable of operating at external temperatures between −30 °C and 45 °C and at an altitude of up to 1500 m. The electrical power was 200 kW and the same amount was recovered as thermal power, half at 60 °C and half at 120 °C, with an electrical efficiency (referring to the LHV) of about 40% and a co-generation efficiency approximately double. During the 1990s, it was distributed in hundreds of units which operated for some tens of thousands of hours providing excellent performance with an availability exceeding 95%.

For single residential users, there is a tendency to use small 'boilers' based on PEMFC technology (3–5 kW_e) limiting the co-generation to hot water for sanitary purposes. Co-generating heat also for heating use would be more complex since the co-produced electric power would be higher than that which can be used for common residential use. However, there may be different management systems for the electrical and thermal components in relation to the possibility of connecting to the electrical grid and the capacity of the latter to absorb power.

However, the fuel currently available in houses is natural gas and this means that the system must include a reforming section, a shift section and a residual CO elimination section. For this reason, there are attempts to use different types of fuel cells, especially SOFC, which can use natural gas directly. An example in this direction was provided in 2005 by Sulzer Hexis which marketed a module, called Galileo (Figure 6.28), equipped with a 200 litres hot water accumulator for sanitary purposes.

A similar solution is proposed by SolydEra with their BlueGen [46] based on SOFC for use with methane that produce 24/7 up 1.3 kW of electric power.

Figure 6.28 Sulzer HEXIS Galileo [45]

Another scenario for the use of hydrogen in the residential sector is that hydrogen is used as a feedstock for methanation in a centralized chemical plant.

In this case, local CO_2 streams can be exploited with a capture and reuse system combined with hydrogen production from electrolysis or from imports from areas where technical, environmental and economic conditions allow hydrogen production and transportation at low cost.

The so-produced substitute of natural gas (also called e-methane if starting from electrolysis) could be a sustainable solution if distributed with the existing infrastructure and using with existing boilers.

6.4 Power plants

Hydrogen is globally considered an energy carrier and not an energy source. Usually, electricity is produced from energy sources, except in specific cases where it is more useful for various reasons to produce electricity from a possibly accumulated and stored carrier.

The production of electricity from hydrogen, in particular for feeding into an electricity transmission grid, is to be seen in this light: it is generally not recommended, unless there are specific situations that express a technical, economic or environmental advantage in this production. In fact, if hydrogen is obtained from fossil fuels, the intermediate conversion into hydrogen could generate more disadvantages than advantages. In the case where hydrogen was produced by non-programmable and not-dispatchable renewables electricity, there could be an opportunity to convert it back into electricity. Hydrogen in these cases is a storage and an intermediate carrier to better synchronise availability of renewable with electric demand: storing and transporting hydrogen (or other e-fuel) is a way at territorial or world scale to produce dispatchable renewable electricity. And among the different methods of energy storage, hydrogen can assert its advantages being a carrier capable to store even large amounts of energy for months, allowing seasonal storage of renewable energy. It is placed on the time scale after ultracapacitor, flywheel, batteries and hydro-pumped storage, which are excellent systems for seconds, minutes, daily and weekly storage and before possible storage with carbon-based gases or liquids (methane, methanol, ethanol and others) that can be considered as annual storage.

The described scenario is what usually called power-to-power system: hydrogen is one of the systems that are coming onto the market for the management of electricity generation with renewables.

The conversion of hydrogen into electricity can be easily achieved using both thermodynamic cycle or electrochemical pathway.

In the first case, this type of system is currently implemented using internal combustion engines, turbines or microturbines that use pure or methane blended hydrogen. All major producer of gas turbine for power plants are proposing hydrogen ready system that are able to use hydrogen in a different percentage of blending with methane.

The first gas turbine fuelled with hydrogen was installed in Fusina (Italy) in 2010. At Kobe in Japan, a co-generation hydrogen gas turbine was successful completed on 2018 by Kawasaki. Subsequently, a cargo of liquid hydrogen was embarked in Australia on the world's first ship, named Suiso Frontier, capable of such transport, to be unloaded in Kobe to create a first international supply chain of this kind. The vessel, manufactured by Kawasaki Heavy Industries (KHI), has a capacity of 1250 cubic meters, equating to 75 tons of liquid hydrogen [47]. The hydrogen for this first shipment was produced starting from brown coal and biomass as a pilot system: if it reaches commercial scale, the Australian consortium involved in the project would produce an estimated 225,000 tonnes of carbon-neutral liquefied hydrogen per year [48].

Another pathway for production of electricity by hydrogen is using fuel cells. This conversion is characterised by the higher efficiency in comparison with one step thermodynamic cycle, even at partial loads and also in distributed generation systems, i.e. those of limited size and located near the user, decreasing energy losses in electrical transmission.

There are some possible applications:

- isolated users;
- co-generation of electricity and heat;
- self-production, especially during peak hours;
- support to the network through positioning in strategic nodes of the grid poor of power generation
- offer of ancillary services to electricity grids (frequency regulation and support in daily demand transients)

In the case of use of hydrogen, the fuel cell also has other advantages in addition to the efficiency ones mentioned above: they are modular and have a linear cost with the installed power, so they allow to create systems that can be scaled exactly to the required size. They are also silent and have low emissions, so they can be easily positioned near any utility.

Acid cells can operate in dead-end anode mode releasing only small amounts of hydrogen during purges. If the mode is in open anode, the gas leaving the anodic compartment will in any case consist of only hydrogen plus traces of any impurities and, therefore, will be easily recyclable at the anodic inlet, also in this case with modest losses due to periodic purges. The case of cells with alkaline behaviour, for which water is produced at the anode, is different. In this case, however, the hydrogen will be easily separated by condensation of the vapour and recirculation of the gaseous phase at the inlet of the anode compartment. This obviously does not apply to MCFCs, whose anodic exhaust contains also carbon dioxide. However, this characteristic could be very interesting in the perspective of carbon dioxide concentration (see Section 5.2.6).

The fuel cell also generates heat. Therefore, it must be utilised or dissipated. The possibility of using it is closely connected to the operating temperature and the size of the system, as well as obviously to the presence of thermal users in the area.

However, when it comes to building power plants characterised by great powers and full load operation, the focus shifts to the cost of the plant. Since, in the case of fuel cell systems, the latter is almost linear with the installed power, it could be convenient to create systems based on turbomachinery, for which the cost of the plant grows less markedly with the size. Therefore, it is conceivable that fuel cell systems maintain an economic advantage up to sizes of the order of a few tens of MW, while for larger sizes we continue to resort to plants based on high-yield thermodynamic cycles. To date, the Shinincheon Bitdream Hydrogen Fuel Cell Power Plant (at the city of Incheon in South Korea), around 79 MW, is the largest based on fuel cells, fully commissioned in 2021 [49]. A little less powerful (77 MW) is the Western Incheon FC Power Plant.

A very interesting option during the transition to a full hydrogen-based energy system is that of attached cycles (see Section 5.1.4) which allows to obtain hydrogen conversion marginal efficiencies greater than those of fuel cells.

6.5 Industrial sector

To date, almost all of the hydrogen produced in the world (more than 95% of which comes from fossil sources) is used in industry, in particular as a feedstock in chemistry, petrochemicals, glass industry or as a fuel for niche applications such as the smelting of precious metals (such as gold). So hydrogen in many fields is a common element.

But, driven by the need to decarbonise industrial activities, hydrogen produced with low or no emission of CO_2 is considered one of the alternatives to replace the current fossil-based carbon fuels and feedstock in all sectors that nowadays does not use it.

Alternative solutions are direct use of electricity or use of fuels that are still carbon-based but with a circular carbon, i.e. preventing by reusing anthropogenic CO_2 emissions into the atmosphere. Those fuels are called often e-fuel or low-carbon fuel depending on production supply chain.

In many cases, the shift to hydrogen or to its alternative is often not technologically and commercially mature: to date very few applications are at TRL equal to nine, that is the level at what business Company is confident to invest.

The difficulty in abandoning fossil fuels is linked to a multiplicity of factors: technological and commercial (as mentioned introducing TRL), but also connected to hydrogen supply chain. In many cases, it is not clear what could be exactly the supply chain and related costs for the enormous amounts of green hydrogen necessary if you do not want to use fossil-derived energy.

Each type of industry has specific problems that can interfere differently with the use of hydrogen. For example, the Swedish pilot plant Hybrit has already demonstrated the possibility of using hydrogen for steel production. In that case, hydrogen is used as reductant for iron ores in alternative to coke. So the problems remain the cost for the new plant and the supply of a large amount of non-fossil hydrogen.

Recent technological and regulatory evolutions (Emission Trading System) are removing some barriers and the market is moving into building plants like that. In May 2023, Salzgitter Flachstahl, a company of the German steel group Zalzgitter, commissioned the Italian group Danieli together with Tenova, a plant for the production of steel with direct reduced iron technology which will use methane and hydrogen in a flexible way, up to 100% hydrogen depending on availability. The plant will have a capacity of 2.1 million tonnes per year and is expected to start operating in July 2026 in Lower Saxony (Germany) [50].

In June 2023, the Norwegian company Hydro produced a batch of aluminium in a plant in Navarre (Spain) using green hydrogen as an energy source to replace natural gas [51].

But for other kind of products, there could be also technological issues to be solved since hydrogen could affect the quality of the product or could reduce the lifetime of the machinery. For example, there are various experimental campaigns to supply furnaces with mixtures with natural gas/hydrogen to verify the maximum hydrogen content allowed without damaging, for example, the refractories of the furnaces itself. A solution could be the conversion of hydrogen into a synthetic fuel similar to that currently used.

Electrolytic hydrogen can be very useful to realise circular economy districts especially using waste materials as a feedstock. So, if the carbon is the same captured by biomass from atmosphere or captured directly from a chimney after combustion, starting from electrolytic hydrogen, it is possible to create a synthetic fuel, with no net carbon dioxide emission: the system is almost perfectly closed from the point of view of materials.

In May 2023 at Örnsköldsvik in Northern Sweden, the companies Ørsted, Siemens Energy, Carbon Clean, and Haldor Topsøe started the construction of a e-methanol plant with a capacity of 50,000 tonnes yearly, operational since 2025 [52]. Hydrogen will be produced by electrolysis from renewables electricity and carbon will be captured from the nearby Hörneborgsverket combined heat and power (CHP) plant, owned by Övik Energi. The CHP plant uses lignocellulosic biomass as fuel in main part from waste of the nearby Domsjö Fabriker plant that produce cellulose from local forest wood.

Many waste materials are organic, that is made of carbon, hydrogen, oxygen. They can be used as a carbon and hydrogen source to produce synthetic fuels. Nextchem, a subsidiary of Maire Tecnimont, is planning and designing many plants in Italy and Europe to produce ethanol, methanol and hydrogen from urban solid not recyclable waste. One at Rome to produce 1500 tonnes yearly of hydrogen and 55000 tonnes yearly of ethanol from 200,000 tonnes of urban local waste [53].

6.6 Production and use of hydrogen in a Renewable Energy Community in Europe

In December 2018, the European Parliament and the Council of the European Union have adopted a new version of the Directive on the promotion of the use of energy from renewable sources: the so-called RED II [54]. On 18 October 2023 the RED II was updated with the European directive number 2413, known as RED III, which did not substantially change the general framework, but only updated some objectives.

A comprehensive new common framework for all European Members was set, to lead the EU towards a more sustainable society using renewables. For the first time, Renewable Energy Community (REC) was defined as a legal entity, in article 2 comma 16 of the Directive.

'In accordance with the applicable national law, [REC] is based on open and voluntary participation, is autonomous, and is effectively controlled by shareholders or members that are located in the proximity of the renewable energy projects that are owned and developed by that legal entity'.

A first singularity is that the members of a REC have a geographical common feature that is the proximity to not only a renewable project, that is a renewable production plant, but also an energy vector production site, or an energy consumption site. Behind this, there is the willingness of EU to create legal entity, so a group of members, not only a single company or person, that manage a local energetic system. The underlying motive is to create local energy chains that produce and consume renewable energy harvested directly on the territories where members live, avoiding transmission or transport of energy across long distances and with inevitable energy losses, introduction of weaknesses, imbalances. This is also intended to promote the valorisation of renewable resources that have so far not been exploited, because they are low-energy-dense and economically unsustainable due to economic or social barriers (just think of the great heritage that the roofs of houses in small towns in marginal territories represent in potential photovoltaic production).

'The shareholders or members [of a REC] are natural persons, SMEs or local authorities, including municipalities'. So the legal entity could be composed by both private and public stakeholders, avoiding participation of big companies. A clear idea to involve new actors who normally only peripherally deal with energy-related issues, but whom the EU wants to make more aware of them, offering them an active role and benefits, as further specified in point (c) of the article defining the RECs.

'The primary purpose [of a REC] is to provide environmental, economic or social community benefits for its shareholders or members or for the local areas where it operates, rather than financial profits'. Thus, a group of members, public and private, who normally do not deal with energy, aggregate around energy projects in order to obtain not a profit, but a common advantage or the solution of a local problem.

In June 2019, the European Parliament and the Council of the European Union have adopted the Directive 2019/944 'on common rules for the internal market for electricity' [55]. A different legal entity was introduced: so called 'citizen energy communities' with few differences with REC, but with similarities that can equate to RECs.

In 'A European Strategy for Hydrogen', a European Parliament resolution of 19 May 2021, it is underlined that 'renewable energy communities can be involved in the production of hydrogen' [56]. The European Parliament itself therefore highlights how hydrogen in RECs can be a tool to achieve the aims of RECs themselves: to create a local benefit and a solution to the problems of communities living in an area, using local renewable energy resources.

It was realized that hydrogen, as an energy carrier, can also be produced locally from various marginal renewable resources, and can help create value for local communities on various issues, such as security of energy supply (e.g. on islands), local transport, dispatchable renewable electricity production.

Being the RED II a directive, it is not immediately applicable in EU Member States. It has to be adopted by the individual national parliaments of the 27 Member States, which have the power and need to coordinate it with national laws and their own specific environmental, economic, technological and social characteristics.

Currently, the directive is not yet fully operational in all Member States, and with important implementation differences. For example, considering the production and consumption of electricity from renewable sources, the distance between producers and consumers who can be part of the same REC is defined differently. In Spain, they can currently be at most 500 m apart; in Italy, they must be powered by the same electrical substation (secondary in the transitional phase or primary with the definitive construction of the Italian regulatory framework). In Italy, therefore, there is no fixed distance for defining the nearness or proximity of REC members.

In the RECs, it is possible to produce and use different forms of energy: the most interesting one for hydrogen is certainly electric energy which is preparatory to generation through electrolysis.

In all the States of RED II adoption, however, economic incentives are provided for the production and instantaneous local consumption of renewable electricity in a REC. There are different interpretations of the concept of instantaneousness. In Italy, for example, we talk about production and consumption at the same hour, expanding the possibility of receiving incentives with respect to the exact simultaneity. This is to facilitate less complex technological solutions especially from the point of view of control system.

In some states, the incentive is a cash flow to the REC and not to its members, while in others it is a direct savings in the electricity bill of the REC members. In all situations, however, electricity must be produced from new renewable plants (all sources included) and consumed locally among REC members in a synchronous manner. The simultaneity of production and consumption is not taken for granted and technically not easily feasible, given that renewables are often not programmable and consumption follows logic dictated by SME's needs or personal lifestyles, not dependent on the availability of renewables. But the European Commission, precisely to make people aware of this difficulty and to stimulate changes in consumption styles among citizens and SMEs, has decided to encourage with incentives a new way of consuming electricity.

Hydrogen, as a storable energy carrier, can enable various technological solutions to improve the local exploitation of energy renewable resources. In fact in a REC, in cases where there is not sufficient simultaneous electricity consumption in the face of a renewable production from non-programmable sources (solar, wind) or difficult to modulate one (biomass), an electrolyser can follow the electricity excess and convert it into hydrogen, to be stored and used for other purposes. The same can also be done with batteries: self-discharge phenomena will have to be considered, which can limit their performance, in particular thinking of seasonal accumulations to exploit the generation surplus that occurs in some cases (photovoltaic in summer compared to winter).

By means of a power-to-gas system, a REC can produce hydrogen from excess renewable electricity by electrolysis, and then decide to sell the hydrogen for other local purposes. For example, a second energy community can be powered by a fuel cell with the hydrogen produced by the first [57], or hydrogen can be sold to refuel hydrogen vehicles.

In the first scenario, renewable energy can be made dispatchable in the REC which is not fed directly by renewables, maximizing the incentive to instantaneous consumption of electricity by the introduction of a modular electric load such as the electrolyser. In the second scenario, there is a local source of decarbonization of transport: REC members can locally produce energy for hydrogen vehicles, gaining the local benefits that a fleet of hydrogen vehicles can bring (e.g. air quality).

The limit in the application of these scenarios is given by various factors. In some national implementations of RED II, hydrogen produced from renewable sources is not considered to feed a REC. Furthermore, the costs of hydrogen generation, storage and consumption systems are still currently too high to guarantee sufficient economic sustainability, despite the incentives. But in some scenario, currently it is possible to produce hydrogen from renewables using REC incentives, at a price that is competitive with hydrogen from fossil source [58]. In the coming years, it is expected that mass production of components for the hydrogen supply chain will enable these scenarios.

References

[1] Autoevolution, Panait M. *GM Electrovan FCV concept hits the big five-oh [Internet]*. 2016 [cited 2023 Nov 5]. Available from: https://www.autoevolution.com/news/gm-electrovan-fcv-concept-hits-the-big-five-oh-111983.html#agal_5. Accessed 5 November 2023

[2] Kordesch K, Hacker V, Reichmann K, Cifrain M, Hejze T, and Aronsson RR. The safe and economic revival of alkaline hydrogen/air fuel cells with circulating electrolytes, recommended for vehicles using battery hybrid systems and H_2 from ammonia crackers. *ECS Trans*. 2008;11(32):167–85.

[3] Kühn S, Heinrich L, Turner H, *et al. Fuel Cell Stack, Fuel Cell Apparatus and Fuel Cell Vehicle* – International Patent No. WO2022013359A1 [Internet]. Germany; WO2022013359A1, 2022. Available from: https://patents.google.com/patent/WO2022013359A1/en. Accessed 17 December 2024

[4] Hydrogen Insight. Hydrogen vehicles in Denmark left without fuel as all commercial refuelling stations shuttered [Internet]. 2023 [cited 2023 Nov 5]. Available from: https://www.hydrogeninsight.com/transport/hydrogen-vehicles-in-denmark-left-without-fuel-as-all-commercial-refuelling-stations-shuttered/2-1-1519914.

[5] Innovation Origins. First Shell, now Motive, hydrogen fuel station clusures continue in the UK [Internet]. 2023 [cited 2023 Nov 5]. Available from: https://innovationorigins.com/en/first-shell-now-motive-hydrogen-fuel-station-closures-continue-in-the-uk/.

[6] H2Stations.org. H_2-Stations [Internet]. 2023 [cited 2023 Nov 5]. Available from: https://www.h2stations.org/.

[7] Riversimple. Sustainability [Internet]. [cited 2023 Nov 5]. Available from: https://www.riversimple.com/sustainability.

[8] Toyota Europe Newsroom. Toyota showcases experimental hydrogen-powered GR Yaris [Internet]. 2021 [cited 2023 Nov 5]. Available from:

https://newsroom.toyota.eu/toyota-showcases-experimental-hydrogen-powered-gr-yaris/.
[9] Pragma Mobility. Alpha Neo bike [Internet]. 2024 [cited 2024 Sep 2]. Available from: https://pragma-mobility.com/en/vehicule/velo-alpha-neo/.
[10] Pragma Mobility. Our hydrogen-powered vehicles [Internet]. [cited 2024 Sep 2]. Available from: https://pragma-mobility.com/en/our-hydrogen-powered-vehicles/.
[11] Romano S, and Larkins JT. Georgetown University fuel cell transit bus program. *Fuel Cells*. 2003;3(3):128–32.
[12] Solaris Bus. Hydrogen [Internet]. 2022 [cited 2023 Nov 5]. Available from: https://www.solarisbus.com/en/vehicles/zero-emissions/hydrogen.
[13] Van Hool. Exqui.City24 Fuel Cell [Internet]. 2023 [cited 2023 Nov 5]. Available from: https://www.vanhool.com/en/vehicles/public-transport/hydrogen/exquicity24-fuel-cell.
[14] Zeroemissionbus.org, Ballard. Zero emission bus – interactive exhibit – powered by Ballard [Internet]. [cited 2023 Feb 10]. Available from: http://web.archive.org/web/20230329085636/http://zeroemissionbus.org/.
[15] Iveco Group. IVECO and Hyundai present the first fuel cell large van at IAA in Hanover as their partnership develops [Internet]. 2022 [cited 2024 Sep 2]. Available from: https://www.ivecogroup.com/media/corporate_press_releases/2022/september/iveco_and_hyundai_present_the_first_fuel_cell_large_van_at_iaa_in-hanover_as_their_partnership_develops.
[16] Ruf Y, Baum M, Zorn T, Menzel A, and Rehberger J. Fuel cell hydrogen trucks – heavy-duty's high performance green solution – study summary [Internet]. Brussels, Belgium; 2020. Available from: https://www.roland-berger.com/publications/publication_pdf/roland_berger_fuel_cells_hydrogen_trucks.pdf. Accessed 17 December 2024
[17] H2PORTS.eu. The Hydrogen Terminal Tractor of the H2PORTS project disembarks at the Port of Valencia [Internet]. 2023 [cited 2023 Nov 5]. Available from: https://h2ports.eu/the-hydrogen-terminal-tractor-of-the-h2ports-project-disembarks-at-the-port-of-valencia/.
[18] Volvo Construction Equipment. Volvo CE starts testing of the world's first prototype hydrogen articulated hauler [Internet]. 2022 [cited 2023 Nov 5]. Available from: https://www.volvoce.com/global/en/news-and-events/news-and-stories/2022/volvo-ce-starts-testing-of-the-worlds-first-prototype-hydrogen-articulated-hauler/.
[19] STILL UK. Hydrogen forklift truck: Corstyrène relies on STILL [Internet]. 2021 [cited 2024 Sep 2]. Available from: https://www.still.co.uk/solution-competence/references/use-cases/detail/hydrogen-forklift-truck-corstyrene-relies-on-still.html.
[20] Alstom. Alstom's hydrogen train Coradia iLint completes successful tests in the Netherlands [Internet]. 2020 [cited 2023 Nov 5]. Available from: https://www.alstom.com/press-releases-news/2020/3/alstoms-hydrogen-train-coradia-ilint-completes-successful-tests.

[21] Alstom. FNM and Alstom present Italy's first hydrogen-powered train [Internet]. 2023 [cited 2023 Nov 5]. Available from: https://www.alstom.com/press-releases-news/2023/10/fnm-and-alstom-present-italys-first-hydrogen-powered-train.

[22] Rollingstockworld.com. Talgo and CAF continue testing of hydrogen rolling stock [Internet]. 2023 [cited 2023 Nov 5]. Available from: https://rollingstockworld.com/passenger-cars/talgo-and-caf-continue-testing-of-hydrogen-rolling-stock/.

[23] Stadler Rail. FLIRT Wasserstoff [Internet]. [cited 2023 Nov 5]. Available from: https://www.stadlerrail.com/it/flirt-h2/details/.

[24] PESA.pl. SM42-6Dn Hydrogen [Internet]. [cited 2023 Oct 10]. Available from: https://pesa.pl/en/produkty/hydrogen-ready/sm42-6dn-hydrogen/.

[25] Siemens. Deutsche Bahn and Siemens enter the hydrogen age [Internet]. 2020 [cited 2023 Nov 5]. Available from: https://press.siemens.com/global/en/feature/deutsche-bahn-and-siemens-enter-hydrogen-age.

[26] EST-Floattech. The world's first emission free hydrogen push boat "Elektra" [Internet]. [cited 2023 Nov 5]. Available from: https://www.est-floattech.com/projects/elektra-first-hydrogen-canal-tug/.

[27] Port of Antwerp Bruges. Hydrogen-powered tug is world first for Port of Antwerp [Internet]. [cited 2023 Nov 5]. Available from: https://newsroom.portofantwerpbruges.com/hydrogen-powered-tug-is-world-first-for-port-of-antwerp#.

[28] FuelCellsWorks. Norway: MF "Hydra", The world's first hydrogen operated ferry wins ship of the year 2021 [Internet]. 2021 [cited 2023 Nov 5]. Available from: https://fuelcellsworks.com/news/norway-mf-hydra-the-worlds-first-hydrogen-operated-ferry-wins-ship-of-the-year-2021/.

[29] Adda247 Current Affairs. "MF Hydra": world's first liquid hydrogen-powered ferry gets operational [Internet]. 2023 [cited 2024 Sep 2]. Available from: https://currentaffairs.adda247.com/mf-hydra-worlds-first-liquid-hydrogen-powered-ferry-gets-operational/.

[30] Norled. MF Hydra sails on zero-emission liquid hydrogen [Internet]. 2023 [cited 2024 Sep 2]. Available from: https://www.norled.no/en/mf-hydra-sails-on-zero-emission-liquid-hydrogen/.

[31] Emirates Team New Zealand. Emirates Team New Zealand take flight in hydrogen powered foiling chase boat [Internet]. 2022 [cited 2023 Nov 5]. Available from: https://emirates-team-new-zealand.americascup.com/en/news/549_EMIRATES-TEAM-NEW-ZEALAND-TAKE-FLIGHT-IN-HYDROGEN-POWERED-FOILING-CHASE-BOAT.html.

[32] China Classification Society – CCS.org.cn. "Here Comes the Hydrogen Boat"! "Three Gorges Hydrogen Boat No. 1" with technical support from China Classification Society (CCS) for is launched [Internet]. 2023 [cited 2023 Nov 5]. Available from: https://www.ccs.org.cn/ccswzen/articleDetail?id=202303280227441152.

[33] Fincantieri SI. Fincantieri SI | Innovation [Internet]. [cited 2023 Nov 5]. Available from: https://www.fincantierisi.it/innovation.

[34] Yamaha Motor. Yamaha Motor to exhibit prototype hydrogen outboard motor at 2024 Miami International Boat Show – accelerating development to achieve carbon neutrality on the water – News releases [Internet]. 2023 [cited 2024 Sep 2]. Available from: https://global.yamaha-motor.com/news/2023/1207/proto-obm.html.

[35] GE Aerospace. Hydrogen takes flight: Airbus and CFM International to partner on hydrogen-fueled demonstration – the GE Aerospace Blog | Aviation & Flight News [Internet]. 2023 [cited 2023 Nov 5]. Available from: https://blog.geaerospace.com/sustainability/hydrogen-takes-flight-airbus-and-cfm-international-to-partner-on-hydrogen-fueled-demonstration/.

[36] ZeroAvia. ZeroAvia makes aviation history, flying world's largest aircraft powered with a hydrogen-electric engine [Internet]. 2023 [cited 2023 Nov 5]. Available from: https://zeroavia.com/do228-first-flight/.

[37] Gaussin. H2 powered generator [Internet]. [cited 2023 Nov 5]. Available from: https://www.gaussin.com/h2-powered-generator.

[38] Everywh2ere.eu. Gensets [Internet]. [cited 2023 Nov 5]. Available from: https://www.everywh2ere.eu/the-gensets.

[39] Renewable Innovations. Hydrogen mobile power generator (MPG) [Internet]. [cited 2023 Nov 5]. Available from: https://www.renewable-innovations.com/mpg.

[40] H_2 Energy AG. kvyreen – H_2 energy – Wasserstoff Verbraucher [Internet]. [cited 2023 Nov 5]. Available from: https://h2energy.ch/en/kvyreen/.

[41] BDR Thermea Group. Hydrogen boilers [Internet]. [cited 2023 Nov 5]. Available from: https://www.bdrthermeagroup.com/en/products-and-services/products/hydrogen-boilers.

[42] Worcester Bosch. Hydrogen-fired boiler [Internet]. [cited 2023 Nov 5]. Available from: https://www.worcester-bosch.co.uk/hydrogen.

[43] PHAM news – Plumbing Heating & Air Movement. Hydrogen boilers ready to go on field trials [Internet]. 2020 [cited 2023 Nov 5]. Available from: https://www.phamnews.co.uk/hydrogen-boilers-ready-to-go-on-field-trials/#!prettyphoto/0/.

[44] Ørsted, Hornsea projects. Hornsea offshore wind farms – where are they? [Internet]. [cited 2023 Nov 5]. Available from: https://hornseaprojects.co.uk/.

[45] BHKW-Prinz.de. HEXIS AG: Galileo 1000 N Brennstoffzellen-Heizgerät | BHKW-Prinz.de [Internet]. [cited 2023 Nov 5]. Available from: http://www.bhkw-prinz.de/hexis-ag-galileo-1000-n-brennstoffzellen-heizgerat/693.

[46] Energie-experten.org. SOFC-Brennstoffzelle Bluegen von Solidpower – Technische Charakteristika der Bluegen-Technologie. 2018 https://www.energie-experten.org/hersteller/solidpower/bluegen. Accessed 17 December 2024

[47] S&P Global Commodity Insights. Japan's Suiso Frontier sets sail for Kobe with liquid hydrogen cargo [Internet]. 2022 [cited 2023 Nov 5]. Available from: https://www.spglobal.com/commodityinsights/en/market-insights/latest-news/energy-transition/012822-japans-suiso-frontier-sets-sail-for-kobe-with-liquid-hydrogen-cargo.

[48] Global Australia.gov, Australian Government – Australian Trade and Investment Commission. Australia exports world's first shipment of liquified hydrogen to Japan [Internet]. [cited 2023 Nov 5]. Available from: https://www.globalaustralia.gov.au/news-and-resources/news-items/australia-exports-worlds-first-shipment-liquified-hydrogen-japan#:~:text=Australia is sending the world's,sea to an international market.

[49] FuelCellsWorks. World's largest hydrogen fuel cell power plant jointly built by Doosan fuel cell put into service [Internet]. 2021 [cited 2023 Nov 5]. Available from: https://fuelcellsworks.com/news/worlds-largest-hydrogen-fuel-cell-power-plant-jointly-built-by-doosan-fuel-cell-put-into-service/.

[50] HydroNews.it. Le aziende italiane Danieli e Tenova forniranno a Salzgitter un impianto per la produzione di DRI completamente 'hydrogen ready' [Internet]. 2023 [cited 2023 Nov 5]. Available from: https://hydronews.it/le-aziende-italiane-danieli-e-tenova-forniranno-a-salzgitter-un-impianto-per-la-produzione-di-dri-completamente-hydrogen-ready/.

[51] Hydro.com. World's first batch of recycled aluminium using hydrogen fueled production [Internet]. 2023 [cited 2023 Nov 5]. Available from: https://www.hydro.com/en/media/news/2023/worlds-first-batch-of-recycled-aluminium-using-hydrogen-fueled-production/.

[52] Ørsted. Ørsted opens a new era in green shipping by breaking ground on Europe's largest e-methanol project [Internet]. 2023 [cited 2023 Nov 5]. Available from: https://orsted.com/en/media/newsroom/news/2023/05/13682622.

[53] Greenreport.it. Idrogeno ed etanolo da rifiuti, a Roma al via la fase d'ingegneria per l'hydrogen valley – Greenreport: economia ecologica e sviluppo sostenibile [Internet]. 2022 [cited 2023 Nov 5]. Available from: https://greenreport.it/news/economia-ecologica/idrogeno-ed-etanolo-da-rifiuti-a-roma-al-via-la-fase-dingegneria-per-lhydrogen-valley/.

[54] European Union. *Directive (EU) 2018/2001 of the European Parliament and of the Council on the Promotion of the Use of Energy from Renewable Sources [Internet]*. Official Journal of the European Union. Strasbourg: European Parliament and Council; 2018. 82–208 p. Available from: https://eur-lex.europa.eu/legal-content/EN/TXT/PDF/?uri=CELEX:32018L2001&from=EN. Accessed 17 December 2024

[55] European Union. *Directive (EU) 2019/944 of the European Parliament and of the Council of 5 June 2019 on Common Rules for the Internal Market for Electricity and Amending Directive 2012/27/EU [Internet]. Official Journal of the European Union*. Brussels, Belgium; 2019. Available from: https://eur-lex.europa.eu/legal-content/EN/TXT/PDF/?uri=CELEX:32019L0944. Accessed 17 December 2024

[56] European Commission. A hydrogen strategy for a climate-neutral Europe. *Communication from the Commission to the European Parliament, the Council, the European Economic and Social Commitee and the Committee of the regions – COM(2020) 301 Final* [Internet]. Brussels; 2020. Available from: https://ec.europa.eu/energy/sites/ener/files/hydrogen_strategy.pdf. Accessed 17 December 2024

[57] Raimondi G, and Spazzafumo G. Exploring renewable energy communities integration through a hydrogen Power-to-Power system in Italy. *Renew Energy*. 2023;206:710–21.

[58] Spazzafumo G, and Raimondi G. Economic assessment of hydrogen production in a renewable energy community in Italy. *e-Prime – Adv Electr Eng Electron Energy [Internet*]. 2023;4(November 2022):100131. Available from: https://doi.org/10.1016/j.prime.2023.100131.

Index

www.ingramcontent.com/pod-product-compliance
Lightning Source LLC
LaVergne TN
LVHW011845200926
848484LV00002B/400
* 9 7 8 1 8 3 9 5 3 4 4 5 4 *